Appraising building defects

Appraising building defects:

Perspectives on stability and hygrothermal performance

Geoffrey K. Cook

Dr. A. John Hinks

Longman
Scientific &
Technical

Longman Scientific & Technical.
Longman Group UK Limited,
Longman House, Burnt Mill, Harlow,
Essex CM20 2JE, England
and Associated Companies throughout the world.

First published 1992

British Library Cataloguing in Publication Data
Cook, Geoffrey K.
  Appraising building defects: Perspectives on
  stability and hygrothermal performance.
  I. Title II. Hinks, A. John
  690

  ISBN 0–582–05108–8

Set in 9/13 Palatino

Produced by Longman Group (FE) Limited
Printed in Hong Kong

This book is for:

Ruth, Sally and Jasmine.
Rosie, Ruth and Rebecca.

Contents

Contents

Differential movement – Long-term durability – Flat-roof problems –
Condensation – General defects with the industrialised and system-built
forms – Rain penetration and condensation.

Contents

Differential movement – Thermal movement – Workmanship – Tolerance
– Thermal breaks – The design wrangle – Materials and life expectancy –
Planning maintenance – Moisture control

Contents

Contents

Contents

Preface

The range of potential defects that can occur with buildings is very broad. Apparently similar symptoms can be evidence of different causes. Conversely, similar underlying causes can produce different symptoms in different buildings. The task of the defect expert in assessing these problems is enormous and fraught with red herrings or misinformation.

This book is not intended to offer definitive solutions to defects, nor is it expected that it shall be used in this manner. Individual situations dictate unique solutions. Nor does it purport to hold all the answers for defect analysis. Individual solutions defy categorisation, and it is this very facet that complicates the defect expert's task, and makes it essential that he/she approaches the problems from first principles. It is not intended in any way to substitute for a sound knowledge of construction principles. Rather, it is designed to offer and debate a strategy for defect analysis more closely aligned to the process followed by the expert than previous books on the subject, and to give a starting source of the information required to allow the defects expert to view the situation from a logical standpoint, and so to make more reasoned decisions.

Therefore, whilst the principles discussed and solutions suggested are the product of careful study and referred research, the authors and publisher cannot in any way guarantee the suitability of solutions or principles for individual problems, and they shall not be under any legal liability of any kind in respect of or arising out of the form or contents of these articles or any error therein, or any reliance of any person thereon.

Acknowledgements

We are indebted to the following for permission to use copyright material:

Dr W A Allen of Bickerdike Allen Partners for our Fig. 13.3; Dr W A Allen, and Building (Publishers) Ltd. for our Fig. 6.10 from Building Failure Sheet 7, *House built in a wood* by Bickerdike, Allen, Rich and Partners, T O'Brien published by George Godwin; The Architects' Journal for our Fig. 6.9 from *Soft Building and Trees* by John Benson; Professor F G Bell for our Fig. 4.6 and our Tables 4.1 and 4.2 from *Site Investigations in Areas of Mining Subsidence* published by Newnes-Butterworths; Butterworth Architecture for our Fig. 6.5 from the *AJ Handbook of Building Structures* by Allan Hodgkinson; Ian Chandler for our Fig. 1.4 from *Low Rise PRC System* published by the Birmingham TERN Project, and for our Fig. 8.6; R M E Diamant for our Fig. 1.7 from *Industrialised Building 2* and *Industrialised Building 3* published by Iliffe Books Ltd.; W F Hill for our Fig. 6.8 from his paper *Tree Root Damage to Low Rise Buildings* published in Structural Survey Vol. 1, No. 3; Table 2.1 is reproduced from Building Research Establishment Digest 251, by permission of the Controller, HMSO: Crown Copyright; Our Figs 4.1 and 4.4 are Crown Copyright and are reproduced from Building Research Establishment (UK) Digests 322 and 346; Dr N E Hickin for our Table 16.1 from *The Woodworm Problem 3rd Edition* published by Rentokil Ltd.; Macmillan, London and Basingstoke for our Table 15.2 from *Building Maintenance* by Professor I H Seeley; Dr Eric Marchant for our Fig. 4.2 from his paper *Fire Engineering and Smoke Control*; Ian A Melville and Ian A Gordon for our Fig. 6.11 from *The Repair and Maintenance of Houses* published by Estates Gazette; Philip Rainger for our Table 3.1 from *Movement Control in the Fabric of Buildings* published by Batsford Academic and Educational Ltd.; Rentokil Ltd. for our Fig. 16.1; R Michael Rostron for our Fig. 3.5 from *Light Cladding of Buildings* published by Butterworths Architecture; Henry Stewart Publications for our Tables 1.1–1.5 from *Weaponry in Structural Surveys 2. Common Building Defects and their Diagnosis* by Christopher Urbanowicz; Worcester Evening News for our Fig. 1.1.

AUTHORS' ACKNOWLEDGEMENTS

The authors would like to express their thanks for the use of the facilities at Birmingham Polytechnic, Wolverhampton Polytechnic and Reading University. Thanks especially to Ian Horton BA for photographic consultancy and image production, Gordon Whittaker for advice and assistance with the Macintosh Artwork, and Jack Hinks for his painstaking proof reading.

Introduction

Need for systemic approach

The analysis of defects in buildings must be based on a sound and extensive knowledge of building construction. This knowledge has traditionally been gathered from a combination of textbooks, teaching and experience. Unfortunately the general approach has followed an elemental path, where defects have been categorised into either materials, components or elements. A more reasoned approach is to consider the behaviour of the total building, since it should perform as an integrated system. The experiential part of defects analysis will use the earlier elemental knowledge to give an overall appraisal of the buildings defects. This is, therefore, recognised as an important part of the analysis process. This book attempts to provide information in a suitably systemic way so that the considerations of total building performance can be used in defects analysis.

Historic basis for the book

The historical baseline of this text has not been drawn for any particular year: the evolutionary aspects of construction would make this difficult or impossible. However, the majority of the text is concerned with buildings from the late 19th century up to the present. This general period has been adopted for three reasons. Firstly, because of the current relevance of defects in buildings of this period: secondly, because a wide range of different constructional forms are present; and finally, because the appraisal of older historic buildings demands a different approach.

Classification of building defects as structural technology, hygrothermal, comfort and environmental performance

Developments in construction technology are seen to lead the developments of building defects. These defects can be classified under broad generic groupings, so as to allow for a more integrated analysis of them to occur. This, in turn, will intrinsically demand an overall view of defects to occur. These aspects have been summarised in Table 1. Time has been taken as the vertical axis, and three broad headings have been given to the sources of defects.

Structural technology includes the construction solutions to building forms. Defects in this area can be caused by deficiency in the structure, its materials, and a failure to appreciate the effects of the external environment

Table 1 Phrases in technology development and their domain

Phases	Domains		
	Structural technology	**Hygrothermal**	**Comfort and environmental performance**
Initial	Empirical knowledge	Weather resisting	Limited user expectations
	Natural materials	Poor seals	Many buildings unhealthy
	Traditional structural forms	Reliance on mass/thickness	Basic heating and ventilation
	Wide design margins		
	Evolutionary trade construction tradition	Seasonal changes tolerated	
Development	Growth in scientific knowledge	Weather-proofing	Increasing user Expectations
	New structural forms and materials	Partially sealed	Some buildings unhealthy
	Reduction in design tolerances	Emphasis on lightweight construction	Improved heating and ventilation
	Innovation of new construction	Emergence of insulation as response to energy crisis	Safety issues emerge for users
	Traditional and innovatory mix in new construction forms		
Current	Continuing growth in scientific knowledge	Totally sealed buildings	High standards of comfort demanded by users

Table 1 *continued*

Phases	Domains		
	Structural technology	**Hygrothermal**	**Comfort and environmental performance**
	Refinement of existing constructional forms	External envelope modified following earlier failures	Safety issues assume increased importance
	Consequential demands on technology	Insulation factors begin to dominate the external envelope	Growing concern over unhealthy buildings
	High-technology buildings with minimal design tolerance emerge		

upon them. Many of the elemental texts on building defects are exclusively based upon these aspects of cause and effect.

A deficiency in the basic functional requirement of keeping out the weather, and other sources of water entry, should not be analysed in isolation from the thermal behaviour of the building. When the hygro-thermal deficiency of the building is considered, the influence of both factors can be more effectively appraised.

The demands made by building users of their buildings is increasing with time. This can be seen by the increasing emphasis on the provision of comfort for the user. Buildings which perform badly in a structural and hygrothermal sense are unlikely to be comfortable, although many are tolerated. In this sense they can be assessed as defective. Unfortunately this is not the complete answer. This can only be achieved when the causes of deficiency in comfort and environmental performance are considered. The role of health and safety in this assessment is obvious, since it is concerned with the total performance of the building. The environmental performance of the building is also linked to the provision of the building services systems within it. They are, of course, subject to a wide range of deficiencies. Although important, this text has not attempted to deal with them, since there are so many topics associated with defective comfort. The authors recognise that a need for such a text exists.

Initial phase

The early developments in construction technology were based around simple structural forms produced from natural materials. These materials, timber, stone and clay, are still performing acceptably in many buildings, although suffering from a range of defects. The early craftsman may have been responsible for the selection and manufacture of the materials used, and this carried with it a responsibility for its suitability. This evolution of trades practices has produced the framework of the present industry. It also partially explains why defects are traditionally considered in an elemental manner.

Although designers may have been the exception, the construction process was based on empirical knowledge. Buildings were erected, and those that did not immediately, or subsequently, fail, were (and still are) considered structurally sound. Foundation failures were common and may have been due to poor understanding of the soils and/or the way the building would perform on them. The margins for structural error were variable, because no detailed methods of design were available, and a full structural analysis could not be undertaken. Experience was essential and the knowledge of methods and practices was carefully guarded. What development did occur, occurred through a gradual evolutionary process which lasted many years.

In time new materials were developed, and these led to new constructional techniques. The provision of a hygrothermal performance requirement was recognised, since shelter from the weather was essential. Groundwater entry could be minimised by raising floors or massive construction, but this option was not universally available. The general standard of comfort was in tune with the aspirations of the building users, and in comparison to the present this was minimal.

Development phase

During the 19th and early 20th centuries, the need for constructionally efficient buildings was demanded by clients who recognised a need for economic efficiency. This could only be achieved by a refinement of the design process, which would achieve structural soundness, for minimal use of material. In this way the wide safety/structural margins, common with earlier constructional forms, were reduced. This in turn reduced the threshold of tolerance for deficiency. Scientific knowledge of materials and structural forms was taking over from traditional empirical knowledge, as the basis for design and construction decisions. Some of this knowledge was gained from research, which could be considered as part of the industry's response to meet the new demands of efficiency in construction. Unfortunately, where the development phase of construction research was omitted, the risk of failure and incidence of defects increased. The defects associated with industrialised construction systems are examples of this.

Enhancement of the hygrothermal performance of buildings during this development phase was in response to user demands, as well as an attempt to remedy some of the inherent defects of earlier construction. Damp-proofing and cavity walls are empirical solutions to some of these problems. The environmental barrier provided by the building was becoming more complex, and allowed a greater degree of environmental difference to be achieved between inside and outside. This in turn placed increasing, and different, demands upon the hygrothermal performance of the barrier. The provision of increasing standards of comfort for building occupiers continued throughout this development phase and was linked to the general social health programme and an increasing standard of living.

Hybrid forms of construction, which mixed traditional and innovative technologies, could suffer from the deficiencies of both. They were designed so that the traditional craft, and unskilled processes, could be efficiently integrated to reduce production times. This was part of a drive for economic efficiency, and, in certain cases, was the industry's response to political pressures, particularly in the substantial provision of low-cost housing. At the time of construction these buildings were praised; their current status is very low.

Current phase   There are increasing consequential demands upon construction technology. This is redrawing the scale which measures the amount of deficiency in modern buildings, and demands that defects analysis is kept up to date with new developments. The application of the empirical knowledge concerning defects found in earlier buildings has influenced some aspects of modern construction, although the feeding back of knowledge about defects to the designer may not be totally effective, due to a range of problems (Hinks and Cook 1986). Economic pressures continue to dictate the width of the design margins, and thereby the threshold of tolerance to defects of the building. The adoption of new materials is encouraged by the move towards performance based specifications, rather than the older prescriptive approach. This brings with it a need to research and develop them effectively, prior to use.

The need to provide increasing standards of comfort and environmental performance, to satisfy users' requirements, must be seen in a context which recognises the importance of energy conservation. This has produced high-technology buildings, where the integration of services and structure is considerable. The traditional methods of defects analysis may not be totally effective in these instances, since the environmental defects may produce buildings which are 'unhealthy'. Unhealthy buildings are not a new phenomena, although they appear to be so as a result of the extensive application of design processes which are deemed to provide the required level of comfort.

Summary   The avoidance of defects is impossible, and this text is not going to describe the ideal way of reducing them, although it will be of use to the designer as a source of feedback information. It does attempt to explain the cause and effect of a range of defects contained within the structural and hygrothermal domain of building performance. These have been chosen because they cover topics where a considerable number of defects exist, and also because they are more relevant for older building types. The elemental approach, which deals with defects in packages relating to materials, components, or elements, has not been followed. It is important that the interrelationship between all of these factors is considered when defects are analysed, and their compartmentation would not assist in this process.

The range of defects associated with the relatively new areas of comfort and environmental performance are outside the scope of this text. Although it will provide a framework for the appraisal of defects found in a range of building types to be undertaken.

Part I Stability of the structure

1 Structure and stability

Stability and structural form

Primary and secondary structural defects

Primary defects may produce or contribute to the ocurrence of further defects, termed secondary defects. It is not possible to generalise about the scale and importance of both, since minor primary defects may lead to major secondary defects. It is essential to recognise firstly the rarity of primary structural defects occurring completely in isolation. The basic structural principles of building are aimed towards the production of a series of accessible and connected working platforms, frequently located at different levels. These must be structurally interlinked, or in some cases separated, in such a manner that the internal resolution of the static and dynamic forces retains satisfactory stability in the building. This can be ensured using a wide range of materials and techniques.

The traditional loadbearing brickwork structures were reliant on the external envelope and usually the internal cross-wall constructions for stiffness and loadbearing capacity. The self and imposed loads of floors would be transferred out to the external loadbearing envelope, which would also bear the dynamic external forces such as wind, and where necessary any vibration or seismic activity. The stiffness of the external structure was significant in determining the reaction of the building to the dynamic external forces, and also its long-term dimensional stability.

Stiffness and structural stability

Stiffness is also achieved in relatively modern structures by the use of a skeletal frame, conventionally produced using concrete or steel, around which an enclosing envelope is attached. This form of construction has been made practical only with the advent of consistently high-quality materials and sophisticated structural design concepts. Generally, modern building design is more refined and has less of the wasteful redundancy that contributes to the tolerance of deficiency in older structures.

Loadbearing and non-loadbearing envelopes

The modern envelope may be loadbearing in the sense that it carries loads other than its self-weight and wind pressures, or a non-loadbearing cladding. Even today there is disagreement over the implications of this term. In rare instances the envelope has been designed to contribute to the overall stiffness of the structure by acting as a stressed skin. Nevertheless, non-loadbearing cladding carries and transmits live loads in the form of wind loads. Distortion of the framework may also impose loads on the cladding.

Stiffness, flexibility and structural form

Another technique for constructing stable buildings is to designate particular zones as being either stiff or flexible, with the flexible zones transmitting loads through to the stiff zones. This is the essence of the shear core or shear wall construction forms, which provide central stiff sections around which the flexible parts are attached and restrained. Wind loads on the side of the buildings are transferred through the external envelope and via the relatively flexible floor and wall constructs to the stiff central core or cross-walls. The rigidity of the whole built form depends on its connectivity.

The influence of the external environment

Since the wind pressures on the walls may be positive or negative and rapidly cycling, a common area of failure is where the stiff zone and either the external envelope or the internal cross-walls are not adequately connected to dynamic loading. Lozenging or racking can occur with redirected loads. Where this is a consequence of inadequate construction, the level of safety may be indeterminate, and the nature of response of the building unpredictable.

Stiffness, orthogonality and connectivity

In the tall UK systems buildings of the 1960s a multitude of combinations were used, and for the notable exceptions there was neither a skeletal frame nor discrete stiffening zones. Rather, the stiffness of the whole structure was the product of orthogonality and connectivity. These constructions became known as the house-of-cards forms after Ronan Point, a system-built high-rise block, fell down very publicly in 1968 only shortly after its completion (Byrol 1981). The reliance for stiffness and transferral of load is totally based for some forms on the interconnectivity between the site-assembled precast panels.

There were many contributory causes to deficiencies in these buildings: the quality of the workmanship, often because of inadequate supervision, also novelty construction unsuitable for our environment or climate.

Internal and external forces

In all modes of construction the stiffness and stability of the structure is dependent on the reaction of the building to internal and external forces. The safe resolution of these forces to the ground must involve those parts of the structure designed to carry load, and exclude or bypass elements or zones not designed for loading.

Thus the principal areas of failure in the structural stability of buildings arise where the design or construction is inadequate. This can result from oversight, inconsistency or misuse, or following alterations. Furthermore, the external forces may exceed those expected. Consider for example, the high winds of recent years.

Structural instability

There may be failure of the ground under excessive loading or internal seismic instabilities. This may be the consequence of a failure to carry out a proper soil survey, which in turn leads to identifiable imperfections in the soil conditions falling outwith the tolerance of the foundation design. There may of course be a deficit within the superstructure itself, or failure of the foundations.

The following list summarises the factors leading to structural instability:

Improper soil investigation
Inadequate design
Unforeseeable loading
Unforeseeable environmental conditions
Use of substandard materials
Poor workmanship
Poor supervision
Inadequate maintenance

Defects in the stability of the structure may emerge primarily as relative settlement in the building, relative movement between elements or components, or dislocation of elements or sections of the structure. Disintegration may occur through overloading or long-term fatigue from flexing or vibration.

Progressive instability

The complexity of the interrelationships between components and the dynamic nature of loadings means that primary defects may lead to secondary deficiency as forces that are inadequately supported or transferred by the now-deficient structure are redistributed, possibly causing greater instability. Cracking or distortion of the structure due to primary defects such as movement may lead to failings of other functions.

Serviceability and stability

It is quite possible for a loss of serviceability to lead to a reduction in structural adequacy. Moisture penetration, for example, is an unserviceability in buildings which may follow initial structural inadequacy. The collection of leaking water in the structure can produce overloading, and is a common catalyst in the failure of timber flat roofs. It could also lead to serious corrosion or degradation of materials, and precipitate further and more serious structural instability. Persistent leakage is a very common factor in fungal attack of timber roof eaves, for example.

Distortion and implications for stability

Usually less serious, but nevertheless aggravating, are serviceability faults arising from structural inadequacy, include the jamming of doors and windows in distorted frames. Indeed, the cracking of brittle components such as glazing can be a useful indication of early distortion. Service pipes spanning across the boundaries of differential movement may fracture. This is a common consequence of oversights during the insertion of movement joints in buildings.

Progressive collapse may occur as the structure overstresses itself through distortion. The changing loading circumstances may cause deterioration in the structural materials. The residual strength of old materials is another area of uncertainty. This whole process may be slow or very rapid.

Tolerance of instability

The tolerance of the structure to deficiency and distortions in the structural frame (taken in the broadest sense) will vary. In some cases it is indeterminate, and the stage at which the building becomes unsafe is unpredictable (BBC 1984). There is the distinct possibility that apparent distortions or deficiencies will lead to a loss of confidence in the structure. Structurally insignificant cracks have rendered stable and serviceable buildings unsaleable.

Cracking, stability and constructional modes

The soft masonry constructions of the 1800s cope well with minor movement and distortion. Indeed, the constructions are designed to accommodate movement. Generally, the soft brickwork tolerates movement reasonably well. Following the initial settlement of the structure, minor seasonal changes in the ground conditions will transfer through to the structures with few problems. Distortions taking the wall close to the one-third-thickness criterion for instability (see later) need not produce the degree of structural unsoundness or the alarming cracking characteristic of the harder forms of construction (Benson 1984).

Tolerance and the traditional structures

The tolerance of movement in the older structures is likely also to produce long-term and staged deformations. The response of the structure to further movement is not likely to be easily predictable.

With simple house construction there is usually overprovision, which produces a useful safety margin against structural deficiency. The materials are somewhat variable, however, and this margin can differ with masonry forms and with the deterioration of age. With brickwork there is commonly a period of prior indication before structural instability becomes critical. The dislocation of the masonry bonding produces clearly interpretable patterns, and it is reasonably practicable to distinguish the symptoms of twisting, shear or other movements produced by tensile or compressive forces.

Intolerance and concrete frames

Concrete structures generally respond poorly to unanticipated deflection and eccentricity compared with the masonry forms. This is partially because they have fewer joints to accommodate the movement stresses, so cracking is concentrated and frequently occurs at the more critical points. Similarly to masonry failures, cracking is indicative of excessive tensile stresses. The quality and accuracy of placing of reinforcement is a key factor to establish. Checks for the omission of shear reinforcement at beam and column connections may be a useful lead towards identifying the potential causes of cracking at beam and column connections in *in-situ* work.

Failure and warnings

Failures in concrete tend to be sudden and unannounced. To compound this, the composite reinforced concretes are relatively intolerant of cracking. In contrast, timber and steel usually give some warning in the form of marked deflection. This may be accompanied by cracking of rigid or brittle components attached to or enclosed by the frame. Both materials tend to produce distress noises as they are over-stressed and this can provide further and easily verifiable evidence.

The implications of deflection

Steel frames are designed to make use of a greater proportion of their strength in modern buildings. As the elastic limit is approached, so the flexibility of the structure and deformation under quite ordinary and safe loadings becomes significant. This is a problem if the external skin and the internal partitioning are relatively rigid and hence intolerant of the movement. Secondary defects may arise without any fundamental instability in the frame.

Physical movement in structures

As large-panel forms of construction and cladding have become popular and economically advantageous, so the forms of construction have followed this path. Potential physical movement is greater, the combinations of materials with markedly different properties is broad, and structural combinations follow any of a number of concepts. The symptomatic emergence of deficiency is entwined with structural form and is hence highly variable.

Stability of the structure

Because the loadbearing capacity of the structure as a whole is dependent on a variety of connections between components, it is at these points where movement and rigidity are most variable, and where the majority of symptoms emerge. This does not mean that the deficiency originates there, however, and the approach to analysis must start with the assessment of the structural character of the building, followed by judgement on what defects the apparent symptoms could reasonably correlate with.

Structural interrelationships

A partially deficient structure may cause a redistribution of the internal forces which makes it difficult to evaluate. The complexity of structural interrelationships within most buildings will often produce secondary defects as a consequence of a primary fault. There may of course be other faults too, which need not be causally related. On the one hand they may be red herrings in the assessment of fundamental structural instability. At the other extreme they may of course represent a synergistic combination of deficiency which threatens the overall stability of the building in a totally unpredictable manner.

Having established that the symptoms of failure can be complex and the analytical process tortuous, we have to make some form of appraisal other than in hindsight. Above all else, it is critical that a reasoned and systematic approach is adopted for the identification and analysis of the defects. The only hope of making repeatable and realistic judgements is to make use of all the available evidence.

This should all go without saying, but why not be systematic about stating the obvious also? It is comfortable to focus on the easily identifiable symptoms to the neglect of the possibly more important, if less obvious, faults. It is difficult to avoid preconceived ideas about causes of symptoms adversely influencing the thoroughness of investigation, or clouding and colouring the interpretation of evidence. Both are critical, however, and must be overcome.

Instability and appraisal

Progression and constancy

Firstly, a critical distinction needs to be made between those defects which are complete and those which are progressive, since their immediate and long-term significance for the building will differ, as must the response to their discovery. With either mode of failure, the assessment of the reliable residual stability is crucial.

Obviously, the distinction between an invariable or progressive fault is often impractical to assess immediately. The wise and responsible surveyor takes second advice when it is needed, and does not play with uncertainty.

Another general rule worth adopting is to be wary of giving an instant assessment merely to satisfy the impatient concerns of a client.

Symptoms and movement

Secondly, the movement of the structure is usually only symptomatic of the underlying cause(s). The symptoms of structural instability, in particular, will be uniquely representative of the response of each building, its materials and structural form, and general condition.

Restraint and absence of restraint

The movement of restrained elements may be more critical to the stability of the building than movement in free elements, since there will be a direct impact upon the surrounding structure. For instance, movement in the primary rigid elements such as shear cores or shear walls, which perform some restraint or other stabilising function, may be symptomatic of fundamental instability in the structure.

Movement of an unrestrained or a relatively free element may be indicative of a relief zone for stresses induced by thermal or moisture movement. These tend to occur at the boundaries between stiff and flexible parts of structures. Secondary defects are likely.

Evidently it will be essential to visualise and understand the structural character(s) of the building. The character is influenced also by connections with adjacent structures, and in such instances the building cannot be viewed in isolation. For example, dominant and recessive structures frequently exhibit movement deficiencies differently.

In essence, interpretation of the structure is the key to understanding structural behaviour and instability. Whilst the functional causes of defects are somewhat limited, their symptomatic emergence should be assumed virtually unique.

Resolution of forces

Finally, the emergence of symptoms may also represent an incomplete resolution of forces. Indeed, the discovery of a structurally critical defect may not be the result of any directly related symptoms.

Be prepared to alter your opinions in the light of new evidence. The first and last steps in analysing the building defect should be considered thought.

The role of monitoring

In instances where there is no immediate structural urgency to demolish or rebuild, it may be appropriate to establish medium- or long-term monitoring of the building. This facilitates the measurement of seasonal or diurnal changes in cracks, and time to distinguish between progressive and stable defects.

Stability and deficiency

Defects in the context of expectations

Structural defects are the main reason for identifiable failures in buildings. Such failures can vary from those which affect the short-term utility of the building to those which threaten total collapse.

An extensive range of constructional defects may develop without buildings collapsing, and few buildings if any exist without some damage (BRE 1981a). Meanwhile, public perception of defects rarely differentiates between the serious and non-serious structural fault. Hill's comment (1982) that in the current atmosphere it is often ignored that minor seasonal cracking is to be expected in most buildings, is probably even more apt today. Although many buildings do crack, whether this constitutes a deficiency obviously depends on user expectations as well as the standard criteria used for designing and assessing structural stability.

Differential movement

Cracking is but one apparent symptom of structural distress or, in many instances, only superficial deficiency. Very often the appearance of a minor crack will represent the resolution of stresses within a structure, and a re-establishment of some stability. Progressive failure is obviously more significant, and the distinction between minor and major cracking or movement depends largely on the context of their occurrence and expectations of the owner.

The appearance of the building may be marred by cracking, also the durability of materials adjacent to the fault. The weathertightness of the envelope may require separate consideration.

These are the obvious secondary effects, and their importance to the building owner may differ. The image of a modern prestige office block is indeed very important, and the owner will not readily tolerate imperfections. Less extrovert structures may tolerate greater imperfection quite graciously. Instability of the frontage will be significant for either, however.

Deficiency and economics

Attitudes and urgency of concern may be more related to foreseeable consequences and liability to third parties than devotion to the building. Budgetary provision can influence the acceptability of deficiency. Regardless of this, inattention to the needs of the building will not halt the advance of deterioration.

Significantly, Davison and Davison (1986) point out the prohibitive economics of producing buildings that do not crack. And whilst deformations or wholesale movements of the structure can obviously be problematic, many of the real difficulties appear to arise with differential movement, or rather its accommodation. Advice on such assessments may be sought from the defects expert and structural engineer. It is necessary to be able to distinguish between the various causes of a range of symptomatic defects in the structure.

Structural defects The causes of inherent structural building faults may be inadequate attention or adherence to design, materials or workmanship specification, or ignorance of the conflicting needs of the elements and sub-components. There may be, and there frequently is, a combination of causes. Commonly though, a design-stage reluctance to consider the building and its surroundings as a harmonic system can be identified. Assumptions about the external forces affecting the building may be incorrect or incomplete. In time, they may also become outdated.

Faults may also arise from abuse of the building rather than inherent deficiency. In instances of change of use, overloading is a frequent cause of problems. Other changes made to buildings without adequate consideration of the structural implications are commonplace. Figure 1.1 illustrates the consequences of an insensitive conversion of two terraced houses into one.

Symptoms and causes Defects will produce symptoms characteristic of the building structure and its relationship with its surroundings. It is imperative to appreciate that similar causes in combination with different circumstances may produce contrasting or even conflicting symptoms. Further problems arise from the unresolved eccentric forces existing in some buildings as a result of poor design, movement and distortion during their life. These can produce an escalating instability.

Deficiency is contextually dependent Superficially similar cracking symptoms may have different underlying causes. Deficiency is contextually dependent, and since many of the basic structural functions are common to all buildings, the performance criteria may also be contextually dependent on building use. A building that fails to fulfil one or more of its functions is probably defective.

Instability and deficiency

Resistance to instability To avoid the occurrence of instability in or between the component parts of a building, the loads imposed by any foreseeable combination of forces must be resisted. Whilst the sub-components and the system as a whole must be of adequate strength, it is important not to conjoin automatically structural stability and rigidity. Many building features demand movement accommodation in order to perform satisfactorily. Transference of loads within the

Fig. 1.1 Sagging roof and bulging front elevation, following removal of loadbearing party wall and roof timbers in an attempt to convert two Victorian terraced houses into one. (Photo: *Worcester Evening News*.)

structure and to the ground may involve reasonable elastic deflections and other movements of the building or its sub-components. This requires a degree of dynamic tolerance that must not descend into distortion or produce other instabilities.

Regulations and construction textbooks have traditionally emphasised these aspects lightly, but are modern buildings really more susceptible to stability-related defects than their traditional counterparts?

The variability of form

Firstly, the techniques for constructing modern buildings have certainly changed rapidly from the comparative stagnation of the 18th and 19th centuries. This presents the defects analyst with a range of highly variable circumstances. It is obviously important to understand the structure of the building. The concepts for this have changed in fundamental ways, of which modern materials and constructional forms are merely symptomatic. Dating the building and establishing the constructional form are closely related. This is an art in itself.

Trends in constructional technique and type of material emerge with buildings from different eras and geographical location. The variability of constructional form is reinforced by the individuality of buildings and their scope for alteration after construction. There is of course no guarantee even that new buildings are constructed in accordance with the design plans, and this is an obvious source of possible defects. It is this essence of unique form that makes the building defects analyst's job interesting and necessary.

Changing expectations

Secondly, the changing expectations of modern (non-traditional) materials have probably been both the cause and effect of technological and design evolution. The range of materials available for selection is now vast, as are the properties and modes of combination. As chemically or physically incompatible materials are integrated, so dimensional and other coordination problems attend the benefits. It seems unlikely though that the building industry, heavily dependent on the large-scale supply of basic resources, will ever be able completely to benefit from the opportunities scientific advancement offers: namely, the widespread design of materials to meet specific functional needs. With the exception of the state-of-the-art buildings, it will probably be necessary to continue to assign functions to those materials that are cheap and readily available. In this sense at least, such buildings bear little relationship to the architecture of the ordinary person.

This reversal of the logical approach to design is the legacy of an *ad hoc* construction technology. Accepting this as a necessity, it must at least be recognised that incompatibilities should be integrated into design philosophy rather than excluded from it. It is the ignorance or reluctance to do this that is responsible for many of the faults with buildings. The emergence of a greater range of materials merely complicates the issue.

Thirdly, increased user expectations leading to the identification and quantification of performance criteria are 20th-century developments. The integration of servicing into buildings influences both design and ease of construction. As changing lifestyles and expectations of greater comfort produce additional new problems, the existing building technologies adapt uneasily.

Empirical and scientific developments

Fourthly, building knowledge appears historically to have trailed behind speculative experimentation. Defects and wholesale failure were inevitable as the traditional methods emerged and became established. Science superseded empirical technology through a process of trial and error. Reid (1973), for example, reports the emergence of applied science for building structures in the 18th and 19th centuries. Looking at the further rapid changes in 20th-century construction generally, Reid (1973) marks the realisation of the uncertainty in building:

> Because there was no clear understanding of why established methods were successful, it was quite impossible to predict what would be the effect of substituting a new material for an old one in order to speed up the process; apparently small changes were found to make the difference between success and failure.

In essence, the 'robust', tried and tested technologies had formed a reliable technical precedent (Groak and Krimgold 1988). The partial emergence of scientific and technological solutions, and their implementation, brought new and greater problems. With little understanding of the exact nature of the performance of the robust technologies, the departure and adoption of non-traditional construction methods has often been wanting in caution. Particular difficulties arise when empirical and scientific methods are intermixed. In many instances of rapid development the basic technological knowledge appears to have been be lost or forgotten.

Innovation and the emergence of non-traditional forms

With the benefit of hindsight, it is obvious that rapid technological changes, such as have occurred with building in this century, are accompanied by a higher incidence of defects. Frequently, this is the cost of 'innovation without development' (BBC 1984). Meanwhile, technological innovations quickly become the basis for further experimentation before they themselves have passed the test of time. Uncertainty is endemic.

The prolonged use of non-traditional and systems methods to minimise the skill and quantity of labour resources produces an industry ill-prepared to cope with boom periods. Materials production response may become limited (BBC 1989), but the real significance for the occurrence of defects is the de-skilling caused by the prolonged departure from traditional methods. During a boom as seen in the late 1980s the emergence of another generation of well understood and avoidable constructional faults may well be in the making to join those arising from earlier innovation.

Perhaps most significantly though, in the last 50 years we have seriously started to study the occurrence of defects; to catalogue their incidence, symptoms and causes. The variability in design and construction and materials used complicates the analytical process.

Defects and constructional trends
Comparison of defects in buildings from different eras shows that there are a range of distinctive symptoms that occur with the various construction technologies (Richardson 1985b). These are not totally exclusive, and some faults are common to many building types.

Overview of domestic construction before the First World War

The surveyor or building analyst will find much work associated with the domestic forms of construction of the last two to three centuries, since this forms the largest proportion of the building stock. The occurrence of defects in this vast sector of the market is subject to both commonalities and characterisations. The types of defect occurring in any particular building will be characteristically symptomatic of the structure, and correct assessment will require a thorough understanding of the structural form. With unusual or non-traditional buildings this may involve considerable research, perhaps consulting experts in local building history. Any simple review of the constructional eras can only be indicative, and the local and individual variability in buildings obviously renders any statements completely general. This discussion concentrates on the buildings of this century and reviews some general building forms back to the Georgian and Queen Anne periods commencing 1700.

Pre-First World War
The vast range of property styles and methods of construction prior to the First World War is done a severe injustice by the single Pre-War categorisation. It incorrectly implies an irrelevance or simplistic similarity in constructional detail in the design of buildings before the 1900s. Indeed, the chronological definition of building eras is in at least one sense only an indirect description of the different eras of constructional form and/or materials being used.

In reality the reverse is evident. The implied uniformity of construction methods has only become truly attainable with the development of effective communication, prefabrication, and a practical transportation network for materials and other resources. The creation of national (and now international) codes complemented and encouraged the emergence of local commonality in the built form. Similarity in constructional detailing in the North, South, Midlands or on the coast has produced its own consequences, however, for both the effectiveness of design and the nature of defects. Not least is the erosion of the vernacular character of buildings.

Empirical developments

The empirical development of construction technology did not of course guarantee defect-free forms, nor did it protect materials against the degradation occurring with age. The Elizabethan and Jacobean buildings, for example, are particularly prone to insect attack where roof timbers have been strutted using unstripped branches (Staveley and Glover 1983), also to dampness in floors placed directly in contact with the earth. In this latter instance the lifestyle and requirements of the construction itself for dampness were in harmony with the detailing. Specifically, cob construction requires dampness to prevent it cracking and disintegrating.

Recall the observation of Reid (1973), that the development of construction technology historically had been an empirical process until the advancement of science, and industrial pursuit influenced the materials and methods of building in the last one and a half centuries.

It is to be expected that buildings spanning a number of political and social eras would differ in their priorities and the functions expected of them. Some characteristic design and constructional details emerge, also the materials used in different eras. Construction methods will tend to differ slowly with time and maintain the traditional form. There will be vernacular peculiarities which the analyst will probably only gradually appreciate.

Constructional trends and eras

During this Pre-War era there are obvious and definite constructional trends, each producing its own characteristic defects.*

* The reader is advised to study the following references for more specific details on the historic development of construction methods and the characteristics of buildings produced during the various social periods:
  (1) Melville I A, Gordon I A 1979 *The Repair and Maintenance of Houses*, Chapter 1: Historical Background, Estates Gazette: 10–80.
    An excellent and authoritative review of the constructional forms and design influences. Valuable background perspectives on the social climate across different periods.
  (2) Staveley H S, Glover P V 1983 *Surveying Buildings* Chapter 15: Reports on Older Buildings, Butterworths: 148–63. Review of landmark constructions from social periods and an outline discussion of the characteristics faults.
  (3) Richardson C 1985 *Structural Surveys* Architectural Press. A valuable review of some of the common and less common defects in traditional structures (post–1750).

The move from timber frames

The traditional construction of buildings with timber-framed structures was being overtaken by loadbearing brickwork in the early to mid-1700s (Richardson 1985b). In London, the response to the Great Fire had generated statutory requirements for stone or brick structures to limit contribution to fire growth and spread (Melville and Gordon 1979).

At the turn of the 18th century the loadbearing brick houses of suburbia were characterised by a subdued symmetry. The sash windows are usually shallow-set and the frontage reflects the geometric uniformity employed throughout the design. The buildings were generally well built and sturdy loadbearing box forms. There may be half basements. The external brick walling may be now only two bricks thick at the ground storey stepping down to one and a half bricks thick in the upper storey as the loading decreases above the floor level. This relatively slender walling was made possible by improvements in the production accuracy of the bricks, allowing better bonding. The stepped thickness usually provides the location for the floor joists. The 'M Roof' appears now, to allow for deeper houses with more rooms. The centre of the building is likely to confirm this, with loadbearing walling running parallel to the roof valley. This is an area of the roof that is particularly difficult to inspect and also prone to waterproofing defects.

Queen Anne and Georgian periods

The Georgian styles of construction were becoming established, and the trend of brick construction that started during the reign of Queen Anne continued. These types of buildings frequently have bonding timbers concealed within the solid walls and may be subject to mid-storey bulging as the timbers compress on rotting. Stone was still popular where it was available and its use has continued in some constructional form or other into the present century.

Some of the Queen Anne period buildings were rather marginally designed, and the slenderness of the solid walls produced problems with stability, and the side-effect of high rain penetration (Staveley and Glover 1983).

Georgian domestic building

There was much domestic building work during these periods as the population expanded very rapidly. The high-density Georgian terraces emerged in the new towns, although the methods and designs remained fairly traditional because of the craft-based structure of the construction industry at the time (Richardson 1985b). These towns contained arrangements of terraced housing set out to create regular plans forming circuses and crescents, landscaped with tree planting. Their appearance was very ordered and the relationship with nature was definitely one of dominance. Inside, the designs were uniform also. The living quarters were raised above the smelly and unhygienic conditions of the street. The somewhat ironic development of having kitchens at semi-basement level with deep areas appeared. Significantly, the main difference between the town housing for the poor and rich was in scale rather than fundamental characteristics of appearance.

Throughout the 1700s several changes in the party-wall regulations occurred and the minimum thickness was increased. This was further advanced by the Building Act of 1774, and external timber ornamentation was banned for fear of contribution to fire spread.

Window details   Window positioning and external details were strictly controlled to give uniformity to the speculatively built dwellings. Obvious but not exclusive external details of the era include the sash windows introduced sometime in the 1600s, now with a deeper reveal to protect the building and the window frame against water penetration (Potter and Potter 1973, Hollis 1982a). Around the mid-18th century the size of glazing bars was also reduced. Particularly in Scotland it was common practice to recess the window frame behind a stepped reveal and so give extra protection to the frame against water penetration. The crown glazing commonly used in good-quality windows superseded the early opaque and bubbly glazing and was cut from a blown and spun disc. That which remains is generally credited to the Victorian era, however.

Openings and cracks   Stepped reveals will give cracking indications of any movement around the opening. Cracks may appear at the vertical joint between the window frame and the reveal, also through the head and sill to the opening. See Fig. 1.2. The openings and their surrounding construction are usually the weakest point in the walling, and because of the transferral of loading they provide little restraint to movement. They are consequently a likely location for failure through overstressing, and this can be a useful indication of the stability of the elevation.

Clues about the age of a building can also be gained from the detailing around the openings. It is general to find stone or timber lintels across openings up until the Second World War, after which metal started to take over (Hollis 1982a), spurred by the timber shortages and the redesigned non-traditional systems.

Sometimes timber lintels were incorporated into solid walls to relieve apparently conventional brick arches of the loading across the opening, and are concealed. Rotting failure of the concealed lintel is likely if the pointing to the brickwork is weathered and porous. The archway may sag and produce the classic arched cracking. See Fig. 1.3.

Walls and foundations The walls are likely to be built directly onto the ground or from large padstones about 250 mm thick and at a depth of about 225–450 mm below the lowest floor level (Davey *et al* 1981). These padstones may be found laid to a levelling fill. Initial settlement of the infill may have produced a leaning of the building or cracking in the facade, perhaps without involving any residual instability. In addition, there were generally no damp-proof courses (DPCs) in the walls of this era, so rising damp can be problematic.

Roofs At the roof the incidence of leakage and other defects is high and the effects serious because of the difficulty of inspection. Valley gutters to 'M-roofs' are renowned because of the concealed concentrations of rainwater and leakage problems and their consequences for the structure may be chronic and severe. One of the more serious consequences is rotting to the supporting beams, usually indicated first by their deflection.

The roofs were rarely underfelted and loosening or loss of pargetting will allow further rain penetration into the roof space. The roof structure is usually sturdy, however, based on a kingpost or queenpost truss. The vulnerability to insect attack characteristic of the Elizabethan and Jacobean eras remains (Staveley and Glover 1983), depending indirectly on geographical location. Costs of repair can obviously be high because of access difficulties.

Construction and finishing materials The type of stone or brick used for the construction of the buildings will have been determined by local availability and the economics of construction of individual buildings. The good-quality stones will have survived with little obvious erosion or blunting of details; the softer and cheaper stones will be considerably less durable. The need for the internal panelling to the damp and draughty stone walls was waning by the mid-1700s, and for a while the ornamental panelling remained, to be superseded by the adoption of wallpaper (Potter and Potter 1973). This itself can provide useful background information for dating a property, if original wallpaper is found. For instance, Potter and Potter (1973) report the fashion appeal of Chinese wallpaper around 1760 following successful trade developments with the Far East. Faced with this, perhaps it would be worth contacting your building historian again.

The influence of transportation At home, the opening up of the transport network made the use of cheaper stone marginally less geographically dependent. By the mid-19th century it was to become indisputably more economic to import some of the major building materials across the established rail and canal routes rather than use the local stock. The vernacular architectures, materials and styles were definitely convergent.

Fig. 1.2 (a) Cracking passing through the head and sill of the opening is indicative of the redistribution of stresses around the opening. Stressing exceeds the capacity of the stonework, and it fractures. Problems will arise if there is a direct passage for water into the building or if the distorted slope of the sill fails to discharge water.

Iron and steel   Meanwhile, in the mills and warehouses of the late 1700s the use of cast iron was emerging, helped by the success of the Colebrookdale Iron Bridge in 1779. The material was used for the 'fireproof' framed structures and survived as a constructional material into the 1900s. This was rare, however, and the availability of wrought iron brought about by the advancement in ironworking techniques of the 1800s soon made way for the much better mild steel with its improved tensile qualities. This was the turning point

Fig. 1.2 (b)

also for the progression to the large steel-framed structures of the 1900s. The general appearance of the multi-storey building with non-loadbearing brick cladding was imminent.

The Regency period emerges

The Regency style of the early 1800s had been the forerunner to the Industrial Revolution with its new needs and trends. Stucco had been in

Fig. 1.3 Slippage of the brick arch is a frequent occurrence. Accompanying movement in the surrounding wall may make it impractical to realign the arch. Check the construction to ensure that it was intended that the arch supports the wall loading, rather than a concealed timber lintel that may have rotted.

ascendance in the town-house developments of this era, and the neoclassical styles appeared in London and on the coast. The houses are readily distinguished by their Greek and Roman quotations and recessed frontal elevations (Potter and Potter 1973) and Gothic arches. The traditional cornice and concealed roof was evolving into the overhanging eaves detail or plain parapet to keep the roof hidden.

Meanwhile, the terraces were losing their appeal as the earlier demand for property was incorrectly interpreted as having been satisfied, and the Georgian new towns were re-interpreted as being too structured and uninspiring. There was significant discontent as the new social structure and economy evolved and by the middle of the 19th century the Industrial Revolution was starting to influence building design. The advent of the proto-functionalist era was heralded by the regularity and mechanistic appeal of Paxton's Crystal Palace (Ferebee 1970).

This complemented the dominant Victorian architecture, which in the mid-19th century remained firmly Gothic revivalist. Still in this era, inspection of the building layout and the style of construction will reveal design for the servanted society. The predominant style now varies less throughout the country. There was to be neither the buoyant proto-functionalism nor Gothic revivalism for the working class, however.

Victorian developments: the mining and steel towns

In the mid-1800s the move occurred from the country to the villages and towns, soon to become cities. There was the creation of the early transport network for the movement of materials and products to support the industrial impetus. The focus was on the development of high-density housing to satisfy the housing needs of the industrial labour. The era was one of political *laissez faire*, however, and under limited controls the new housing predictably emerged as small and low-quality units to minimise capital outlay (Melville and Gordon 1979). They were of course grouped in high-density collections, with separating spaces limited to that needed for access or transport, and to satisfy the relevant minimum regulations.

Back-to-back slums

The back-to-back houses have been cleared now to a large extent, either by bomb-damage during the Second World War, or under slum-clearance policies. They were characterised by a lack of sanitary provisions and the severe overcrowding in unpleasant conditions. For the first time there was the realisation that such conditions could have been wilfully created.

Birmingham was a particularly desperate area and the experience of dealing with this influenced much of the legislation. In 1944 the Town and Country Planning Act identified areas for redevelopment. The shortage of property meant that the buildings had to be maintained for some time and redeveloped gradually.

Constructional form was now usually brick with slated roofs, made practical by cheap labour, easier transport of materials and the repeal of slate duty in 1831.

Problems with slates

The most common problem with roofs to these properties is the delamination of the slates and the rotting of fixing nails, the latter of which can be virtually entire. Tired nails will allow the slate to slip slightly and show the lead clips. Slates will eventually become dislodged and allow entry of rainwater.

It is also worth checking how the last row of repaired slates was fixed. The final slates cannot practicably be nailed since they require to be pushed up under the two upper layers of slating. The practice of fixing with a lead tie nailed to the lath below gives a repair that may fail under snow loading or high winds. Also be wary of narrow replacement slates substituted for the correct size which tend to leak at the edges where there is inadequate or no sidelap.

Controls over building height and street width

In complete contrast in London, the Building Act of 1844 was setting standards and mechanisms for the control of street widths and building heights, necessary because of previous policies to build houses of several floors in streets narrow and unpleasant as a means of meeting the housing shortage.

The result was a dull, close-packed uniformity in the lower- and middle-class properties, contrasting with the emergence of the architectural influence on flamboyant building design for the upper classes. Construction was more solid than in previous eras, and the use of tied cavity walls was strictly regional and infrequent. Damp penetration remained a problem; consequently the slow adoption of slate DPCs was beginning to solve some of the headaches over rising damp.

Sanitation and the need for the Public Health Act

The problems with sanitation in the back-to-back housing of the early Industrial Revolution became more and more unacceptable with each cholera epidemic. In 1875 the Public Health Act together with building by-laws was produced with an emphasis on minimum standards, little considered until then. These developments influenced the types of domestic building found in the inner suburbs of many cities. They are generally identifiable by their restricted frontal width and tend to be deeper and asymmetrical at the back. Other dimensional controls continued to relate to minimum street width and the dull and monotonous layout of towns. Basement construction was unsurprisingly popular, and is a prime target for flooding and dry rot.

The principal effects of the Public Health Act, in terms of constructional form and likely deficiencies, include the requirement for damp-proof courses, the emergence of sanitation regulations covering drainage and sewage disposal provision, and alterations to the minimum wall thicknesses (Potter and Potter 1973). The party wall penetrating the roof line in the terraced properties was another distinguishing feature.

The tenements

This is also the period around which the tenements started to appear. The first tenement was built in London in 1868. These were a revolutionary approach to accommodation. High-density populations in multi-storey buildings were at the time a comparative luxury compared with the back-to-back slums. There were attempts to introduce sanitation and washing facilities to compensate for lost use of minimal ground area. The early structures are loadbearing brick, the widespread introduction of the steel frame did not take place until considerably later.

The turn of the century

Around the turn of the century the designs were simplified somewhat in the towns, but the estates constructions continued virtually unaltered until 1909, and in many cases until the First World War. At the turn of the century the controls were still in terms of street widths and heights of buildings. Town-planning reform had to wait for the Town Planning Act of 1909 to introduce the concept of 'density of housing population' and the Tudor Walters Committee report to the Government during the First World War (Smith 1971, Melville and Gordon 1979). There were other milestone recommendations relating to the area of internal space minima* and internal sanitation. These recommendations were countered to some extent by the 1923 Housing Act[†], and consequently the average size of houses fell in the inter-war period. In reviewing environmental deficiency, it is important to appreciate that the majority of properties built between the wars had below 800 sq. ft. of floor space: this itself creates a limitation on the provision and upgrading of facilities.

The garden cities

As the first garden cities emerged after the First World War, the planners could now practise the planned coordination of industry and housing. The open areas contrasted markedly with the uniformity and density of the Victorian slums. The monotony of similarity in housing plan form and appearance was being shed, moreover the layout of the domestic zones was becoming relatively free-flowing. The buildings are varied and favour earlier styles. Traditional materials were the norm. Potter and Potter (1973) report that the emergence of the non-traditional materials was for cheapness in ornamentation of the facades rather than as any fundamental change in constructional method.

It is not intended to discuss the earlier building forms further. To do so uncomprehensively would be of little value. Detailed knowledge is required of the traditional forms of construction if accurate assessments are to be made of residual strength. Obviously the buildings themselves may be indicative of the dates of original and subsequent construction work, also the building methods. But this can form very specific areas of expertise,

* 900 sq. ft. floor area for three-bedroomed houses without parlours, implemented in the 1919 Housing Manual on the recommendations of the Tudor Walters Committee (AMA 1985).

† The qualifying criterion for construction subsidy was set at a minimum of 620 sq. ft. of floor space and a maximum of 950 sq. ft. for three-bedroomed houses. In practice it was reported that average floor areas fell to between 750 and 850 sq. ft. (AMA 1985).

outside the principal intention of this text and of only general interest to the majority of readers. Remember it is easy to be hoodwinked by constructional quotations from other eras or by alterations to the structure.

Established deficiency

It is of course quite possible to find buildings with an established history of deficiency and repair. It will be vital to assess what is original structure and what has been altered, appended, repaired or removed. In stone buildings for example, damaged cornices can be skilfully repaired with mortar to give a reasonable match, but the appearance will differ from the stone with ageing as the erosion rate varies (Davey *et al.* 1981).

The influence of modernisation

Much damage has also been done to existing structures by the unsympathetic 'modernisation' of structures and interiors. There was frequently little regard reserved for the main structure. Established structures are disturbed, producing disequilibrium. The Victorians in particular are renowned for their bastardised buildings.

Overview of domestic construction after the First World War

The inter-war period

The First World War produced a considerable housing shortage as the preoccupation with the war produced a period characterised by little building work. The existing properties were neither numerous enough nor suitable for early 20th-century society.

New systems

Immediately after the war the costs of materials and labour increased markedly because of shortages and the wartime inflation. Consequently house building revived only slowly. The speed and economy of house construction became paramount. To produce the 'homes for heroes', experimentation with new methods was widespread. The garden city developments referred to earlier now emerged strongly.

The birth of non-traditionalism

The common forms of non-traditional construction used during this period were the Boots pier and panel, some Boswell houses which were limited to Birmingham, Wolverhampton and Liverpool, and an early version of the Laing's Easiform house, of which about 2000 were built. For those who could afford it, the new architect-designed houses of the era were more flexible and generally more attractive than the local authorities and speculative builders could offer (Penn 1954). There was still some degree of variability.

Structural steel and concrete frames

The structural-steel-framed buildings erected during the 1920s and 1930s exploited the concepts of loadbearing framing and non-loadbearing infill, similar to those of the timber-framed styles of the early periods. It was common during this period to assume that the outer skin of brickwork would protect the steel frame from corrosion, and frequently the frame and brickwork were built up tight to give maximum support. If the brickwork or the pointing is sufficiently porous, and of course it usually is, the steel will be wetted frequently. The expansive corrosion may be as much as 30 mm and will produce vertical cracking in the brickwork immediately in front of the column.

In the reinforced concrete structures erected from this period onwards the problems with inadequate cover to the reinforcing steel are virtually timeless (Bond 1983).

The 1930s characterised

The 1930s were characterised by the re-adoption of the traditional construction methods. About four million houses in total were constructed between the wars, of these about one and a half million represented public housing and were mostly two-storey semi-detached or terraced properties with three bedrooms (Smith 1971).

Return to traditionalism

This return to traditional construction methods was partly a reaction to the problems of the 1920s experimentation, but was principally because full employment was a social and political necessity. In addition many of the by-laws were still too traditionally orientated to allow the easy adoption of some of the experimental approaches (Whittick 1957). The concept of new housing completed using rapid, efficient building systems was simply not politically acceptable. The cost of building fell together with interest rates, and probably as a consequence of this, domestic buildings of this era were poorly built under the speculative profit motive. The first age of the 'jerrybuilt' properties dawned.

Supply, demand and overcrowding

Although the costs of housing may have fallen at the start of the decade, as the economic depression hit they were still beyond the reach of much of the population. Many properties modest by today's standards remained for sale for long periods (Potter and Potter 1973). Standards were dropped in the 1930s to make rental feasible. The shortage of affordable housing made the 1936 Overcrowding Act virtually unenforceable, and the intentions to ensure less overcrowding and minimum room sizes were optimistic in the event. It was not until after the Second World War that the improvements in the low floor areas were to come about.

Stability of the structure

No-fines

Those forms of experimental construction used conventional or no-fines concrete, either *in situ* for single or cavity construction or precast in the form of slabs and piers interconnected by *in-situ* concrete. Other methods tried included solid timber or timber-framed construction. There was some experimentation with metal framing and metal cladding of dwellings.

A general feature of this era was the use of the bituminous DPC in preference to slate. The principal advantages were cost and flexibility, both in the sense of construction and performance in use. This has now exceeded its effective life and will often be found to have perished and cracked.

The emergence of the cavity wall

The cavity wall was an expensive form of construction compared with the solid wall and for smaller houses was economically unfavourable. Strength was not in contention, since the 9″ solid wall was usually stronger than necessary. Consequently, up to the start of the Second World War, houses were generally built of solid brick walling (Penn 1954).

Exceptions and rules

It is difficult to generalise and there are notable exceptions in the appearance of cavity walling. For instance, houses have been discovered dating from the early 1800s with cavities, and in London it is quite possible to find solid-walled constructions dating around 1950. There is considerable regional variation, however, and the severe climates of the north and coastal regions appear to have encouraged earlier use of the cavity. The cavity wall was not universally popular but its eventual adoption was inevitable because of chronic damp problems with the 9″ solid wall form (Whittick 1957). Early examples have been found with the outer leaf in mock Flemish bond to conceal the fact that the wall has a cavity.

In the large houses, cavity walling may be found considerably earlier (Whittick 1957). It was both more economically justifiable, and usually less of a consideration at the design stage. The 9″ external leaf endured, with the addition of 4.5″ inner leaf. The now common use of the 11″ cavity wall for housing did not occur for a while. In the 1920s and 1930s the use of poorly galvanised or ungalvanised wall ties in the early cavity walls has been recorded, and the problems of concealed corrosion in the cavity is arising (Staveley and Glover 1983). This is discussed further later.

Generic faults with traditional construction

In a sense the sheer variability of form and constructional method of the pre-war building stock is more variable than that after the Second World War. The designs were by no means universal and the pressures on design were largely from subsidy in the public sector and speculative profit in the private market. Such limitations or motivating forces emphasise the social influence on design and construction, but alone are not a very good method of interpreting structures.

Very loosely speaking, then, the traditional form of construction was brick walling with a tiled or slated covering of a timber roof and timber upper floors. Ground floors were commonly timber prior to the Second World War. The need for economy of materials had changed the basic design concept however, in the sense that more lightweight structures were built to economise on material needs.

Table 1.1 cites common defects and their causes in traditional housing.

Roof styles in the inter-war period

Although there was experimentation in constructional methods there was little widespread deviation from the traditional forms of roofing. Throughout the 1930s it was usual for the roof to be pitched and of tiled or slated covering. The materials used then for roof coverings are coming to the end of their useful life now. It is therefore important to assess the foreseeable as well as immediate costs involved in repair or re-roofing.

Another more fundamental problem arises with the roof structure. In this period roof timbers were infrequently treated with preservative. Indeed, the widespread treatment of timber was not commonplace for another twenty years or so. In addition, unseasoned timber was used (Staveley and Glover 1983). The consequence was to produce roofs susceptible to warping early in their life, and in the long term to insect attack.

The post-Second World War period

Immediately after the war there was a dire need to replace the existing obsolescent dwellings. In addition, there was pressure from overcrowding brought about by large numbers of people who had married during the war and now required separate accommodation (Whittick 1957). This housing shortage was more extreme than in the inter-war period, and post-war expectations were greater.

Knowledge and aspiration

The new housing was required to satisfy the aspirations of the new era whilst meeting very tight restrictions on cost and size (Penn 1954). These problems had been foreseen during the war, so there had been a period of advanced thinking ready for the challenge. There was also the advantage of more knowledge about design and construction, analysed and collated by the Building Research Station.

Design criteria

An estimated house lifetime of 60 years was set as a reasonable renewal period for permanent housing. The widely accepted estimate of housing requirements was 4,000,000, which suggested the adoption of new non-traditional systems to augment the scarce availability of skilled labour and traditional materials. Recommendations on the minimum floor area of

Table 1.1 Common defects in traditional housing

Components	Symptom	Typical causes	Aids to diagnosis
External masonry walls	Cracking, bending bulging or bowing	Subsoil movement, foundation movement	Crack gauge glued over crack
		Expansion of brickwork Chemical attack on brickwork	Plumbing Sample and test
		Failure of wall ties in cavity Spread of roof structure Mortar too strong or too weak	Endoscope survey Visual examination Sample and test
	Dampness on inside leaf of cavity wall	Water bridging of cavity due to insufficient width of cavity, mortar droppings on ties and other obstructions in cavity (eg insulation) Omission of DPC trays	Endoscope survey
	Crumbling mortar	Incorrect mix; frost attack, chemical action	Sample and test
	Crumbling bricks	Frost action; chemical attack	Sample and test
	Cracks in render	Differential shrinkage of render coats Expansion of mortar due to chemical attack Movement or cracking of masonry underneath Lack of bond to masonry	Location excavation Sample and test Local excavation Hammer testing
Brick parapet walls	Spalling brickwork	No DPC under coping. Use of incorrect bricks	Visual examination
	Crumbling mortar	Poor design of coping details	
Flat roofing	Blistering and rippling	Water trapped under mastic asphalt or felt roofing, vaporising	Visual inspection and local excavation
	Splitting and cracking	Asphalt surface oxidised and embrittled Thermal movement of roof structure Moisture penetration of insulation	Sample and test Visual inspection and local excavation

Table 1.1 *continued*

Components	Symptom	Typical causes	Aids to diagnosis
	Sagging of roof	Excessive span/depth ratio of roof joists	Visual inspection
		Inadequacy of main supporting beams	Visual inspection
		Lack of bracing; ponding of water	Check falls and levels
		Defects in structural timbers (see below)	Visual inspection Sample and test
Pitched roofing	Sagging and deformation	Overloaded timbers, horizontal movement of feet of rafters, removal of internal support – purlins and structs. Insect attack. Dry rot	Visual inspection Sample and test
	Slipping roof tiles and slates	Frost action on roof tiles; loss of fixings	Visual inspection
		Atmospheric pollution attack on slates; damage and corrosion of fixings	
Flooring	Warping and shrinking of timber boards	Use of undried boards; moisture absorption	Check moisture with meter
	Collapsed timber floors	Fungal attack. Damp rot Woodworm Excessive floor loading Ground movement	Visual inspection
	Lifting and curling of concrete screeds	Incorrect mix used	Sample and test
		Inadequate curing time, poor conditions	Visual examination
		Laid in too large bays – shrinkage	Visual examination
		Underlying concrete distortion	Visual examination
	Cracking of screeds or crumbling	Incorrect mix used; mix too weak	
	Concrete floors cracking	Chemical attack due to unstable aggregate	Sample and test
		Frost heave of underlying soil	Sample and test

Stability of the structure

Table 1.1 *continued*

Components	Symptom	Typical causes	Aids to diagnosis
		Chemical attack due to contaminated fill	Sample and test
	Lifting of Clay tiles	Expansion of clay tiles	Visual examination
		Thermal movements, too strong grout bedding mortar Shrinkage of underlying screed	Sample and test
	Lifting and curling of plastic tiles	Lack of DPM, passage of moisture into concrete screed	Moisture test
		Lack of bond	Visual inspection
External joinery	Surface deterioration	Insufficient protection to timbers	Visual inspection
		Use of inappropriate grades of timber	Sample and test for type
		Lack of maintenance	Visual inspection
		Moisture ingress into timber	Moisture meter test
	Distortion	Varying moisture content causing expansion and contraction	Moisture meter test
		Opening up of poorly formed joints	Visual inspection
	Delamination	Moisture ingress to poor grade of plywood	Moisture meter test
Windows	Cracking of glass	Corrosion of steel window frames	Visual inspection
		Excess deflection of lintel – distortion of frames	
Foundations	Varying degrees of cracking of various types seen on walls	Shrinkages and heave of clay soils, building on open ground, or near existing trees, or on sites newly cleared of trees	Visual inspection Use of tell-tale crack gauges
		Floor slab movement due to poorly compacted fill Chemical attack by aggressive fill materials Instability of sloping ground Poor consolidation of made up ground	Excavation and measuring Sampling and testing of soil

Table 1.1 *continued*

Components	Symptom	Typical causes	Aids to diagnosis
		Mining and other geological subsidence Poor and inadequate foundation design	Geological records
Lintels	Cracked concrete	Corrosion of reinforcement due to lack of cover and/or use of chlorides	Covermeter survey Sample and test
	Cracked brickwork	Insufficient bearing area; eccentric lintel bearings; unsuitable bearing materials	Visual inspection
	Rust staining	Corrosion of galvanised steel lintel due to abrasion of protective coating	Visual inspection

houses matched or exceeded those of the Tudor Walters Committee*.

Systematic testing was carried out to assess proposed systems, and from about 1400 proposals 83 non-traditional systems were granted full licences (under The House Construction Reports 1944, 1946 and 1948). This allowed the local authorities to finance the system dwellings as permanent houses under a 60-year life expectancy assumption (Smith 1971). Some had been designed in conjunction with the representatives of the local authorities.

At the time it was optimistically expected that the building industry could provide the required number of houses using the new systems in a period of about ten or twelve years (Whittick 1957). Because of the extreme shortage, the production of housing was proposed as a national emergency for 1946 and 1947 (Smith 1971). There was evidently sufficient incentive to make the experimental, non-traditional methods work (Whittick 1957), and this is undoubtably a key factor in their persistence. In the meantime 'prefabs' were constructed as temporary housing along with repaired war-damaged buildings and the release of government property to ease the immediate shortages.

Prefabrication revisited The mass production of prefabricated housing would be more sensitive than traditional housing to economies of scale. Indeed, the political and economic

---

* The Dudley Report recommended increasing the minimum floor area to 900 sq. ft. exclusive of stores. Implementation of the 1944 Housing Manual also covered cooking facilities and the need for sculleries, and for families of five or more two W.C.'s were required. There was a requirement for the cavity wall (AMA 1985).

success or failure of the systems hinged on large-volume production to make repayment of the capital investment in the manufacturing equipment possible. In the event, under 2,500,000 were actually produced in this period of twelve years (Whittick 1957), partially due to the optimistic evaluation of actual capacity of the industry, but also to a slow start to production. Consequently, the systems appeared at first not to produce the amount of housing expected of them, nor at the cost savings anticipated.

The new materials The technology of the solution was new housing based on rapidly erected frames clad with thin walls acting as insulating screens. The philosophy of the new systems relied on a selection of large prefabricated panels designed as standard components. This was both their strength and weakness. The emergent systems were based on the use of concrete, steel or timber. There was also some use of prefabricated brickwork and asbestos or plastic materials (Whittick 1957). Felt DPCs were still being used and there is the question of assessing their residual useful life now. In the cheapest buildings the cavity tray was bitumen, otherwise lead, zinc or copper (Penn 1954). The ventilation of the cavity was considered a poor detail by 1954, although airbricks in the top and base of the wall may be found in the earlier cavity constructions. The concrete floors commonly constructed because of the timber shortages were also frequently inadequately damp-proofed.

The magnesite floors A common material for 1950s floorcoverings was a magnesite oxychloride compound (also known simply as magnesite). This material contains organic and inorganic fillers bound with a composition of magnesium chloride and mixed with calcined magnesite (Ransom 1981). The magnesium chloride component is deliquescent and causes the floor surface to sweat in apparently dry conditions. It also provides a source of damp for the rusting of metal. The floorcovering disintegrates in the wet conditions produced either by the sweating or poorly damp-proofed and uninsulated floors. Frequent cleaning or condensation on the cold floor further accelerates the deterioration.

Hardwood timber flooring was another common choice of covering to concrete floors. If the hardwood flooring was installed unseasoned, which was not unknown, its dimensional instability combined with any irreversible shrinkage in the flooring easily produced buckling and lifting of the surface. In the long term there is a risk of rot or cracking (Staveley and Glover 1983).

The systems were designed to eliminate the variability of the traditional materials and methods that caused delays in production. The emphasis was on rapid assembly on site using large-scale components instead of the traditional small-scale units.

The dry forms Off-site manufacture and minimal on-site work would ensure continuity of production throughout the year. The avoidance of moisture in the production would (theoretically) overcome the traditional problems of fit and distortion, and largely eliminate drying delays in components or elements critical to the progression of the building. It also minimised the traditional disruptions due to cold weather. These possibilities had been seen in the mid-1930s, but the time had not been right.

Prefabrication of components There was some experimentation with the prefabrication of timber doors and windows, staircases and roof trusses. The skills developed of necessity during the war were exploited commercially to produce standardised steel or aluminium structural components. Sheet materials were introduced on a large scale for the first time, and the prefabrication concept was soon extended to whole wall panels. Eventually complete floors or entire buildings were prefabricated and transported to site for minor finishing.

Thermal insulation and soundproofing Significant changes were made in the provision of thermal insulation and soundproofing. These were necessary as the concentration on building materials shifted to concrete and glass for the building fabric, both of which are excellent conductors of heat and sound.

U values* for 1945 were 0.15 W/m^2K for external walls to the living room, the external walls in the remainder of the house were set at 0.20 W/m^2K. The ground floor U value was 0.15 W/m^2K and the roof (incorporating the ceiling) was set not to exceed 0.20 W/m^2K (Penn 1954). The performance of the systems buildings differed highly, however.

Material shortages Most of the materials used had not been tried in any large scale for building (Penn 1954). Timber was in acute shortage after the war because of restrictions on importation of products for economic reasons. Recurring shortages in the supply of bricks, steel and cement made design and its implementation difficult. Frequently houses were completed with whatever material was available at the time, and this variability makes the job of the defects analyst more difficult. It is possible to find houses of this era in a single street with significant differences in minor and even major detailing, as well as in the use of materials.

* The U value or thermal transmittance can be defined as the quantity of heat that will flow through a material unit surface area, in unit time, due to 1 degree C temperature difference between the inner and outer surfaces of the material.

Quality and architecture for the common man

The profound implications for design and coordination evidently also extended to quality. Many of the systems were to produce structural problems on an unprecedented scale. Such was the volume of production that if a design or repetitive constructional fault arose, the consequences were of course vast. The optimism about maintaining flair in the new architecture for the common man was hollow (Whittick 1957), and around 1950 there was a reduction in the standards of public housing (Penn 1954). During the 1950s there was a return to reductions in the minimum allowable floor areas*, implemented to try to concentrate scarce resources into the production of more housing units (Smith 1971).

The speculative private market was relaxed and materials were no longer licensed, but remained drastically more expensive than before the war. This made control of the economy difficult and the volume of public housing was decreased in the second half of the decade. Meanwhile, concerns over slum conditions in the older properties produced the slum-clearance programmes discussed earlier.

More new systems

The mid-1960s were characterised by a frenetic subscription to the systems approach. Some of this appears to have been at the ignorance of established knowledge about building construction. Limitations on building land and acute shortages of modern public housing in the cities led to even more optimistic construction targets than earlier. These were endorsed by electoral mandate and high-rise construction was in ascendance. Mostly European, the systems were ill-matched to our climate. They were constructed under incentives to heights and a scale for domestic construction unseen before in this country (Diamant 1990). In 1968 Ronan Point collapsed and the bubble burst.

Low-rise forms of non-traditional construction

Concrete houses with solid wall forms

The solid concrete constructional forms were difficult to weatherproof and in this sense presented little improvement over solid brick walling. The no-fines concrete construction was somewhat better than conventional solid concrete (Whittick 1957), since the voids in the walling, because of the lack of fines, limited moisture transport across the section. This of course only holds for good-quality no-fines. Typically the walls are 12″ thick produced *in situ* using a modular formwork. A weakness of the design could be the compaction and quality of concrete below openings formed in the shuttering, such as at windows. There were about 70,000 of the Wimpey no-fines houses built (Whittick 1957).

---

* Standards for minimum floor areas fell from 1050 sq. ft. for three-bedroomed public housing in 1951 to 897 sq. ft. by 1960 (AMA 1985).

Concrete cavity-wall construction

Cavity-wall construction in concrete was designed to have the same advantages as brick cavity. The walls were reinforced in some systems and this may be a source of corrosive failure now. The reinforcement in the Easiform house, for example, was through the walling, but was increased in thickness to act as an integral lintel at openings. Easiform houses may have been built using ballast concrete and may be subject to sulphate attack.

The early inter-war structures were no-fines clinker concrete. The latter forms of construction from the mid-1920s onwards included a cavity of 2″ with 3″ leafs (increased to 3.5″ post-1945), although there was an overlap. Here the cavity form was based on an external skin of dense aggregate reinforced concrete and an inner skin of unreinforced clinker, Lytag or other waste aggregate concrete. Instances of variability in the cavity width and wall width have been recorded. Reinforcing was placed horizontally at 2-foot centres and there were also some vertical connection bars and corner bars.

Concrete blocks

The use of concrete blocks also emerged. The system houses produced using these appeared to give little advantage either in terms of their construction or thermal insulation. They were lighter though when used as cavity or hollowed-out blocks. In larger constructional forms, there should be *in-situ* piers at about 8-foot centres.

Loadbearing panel construction

The loadbearing panel systems were seen as the most potentially advantageous methods of construction. They were precast and assembled on site using a variety of systems of design and connection. Systems were usually based on either small manageable components, such as the Tarran, or large panels designed for crane handling such as the Wates systems.

Problems with fixings and movement accommodation

The larger panel systems have implications in terms of fixing location and are critically similar to the small systems. However, the accommodation of physical movement is of greater import. The inner linings of the houses may be wood wool slabs for ease of secondary fixings and thermal performance (since generally there are fewer problems with acidity and rot with the wood wool placed in this position). The joints were reliant on the quality of application and durability of flexible compounds.

Overview of a selection of other systems

The Stent system was also based on storey height panels (as was the Wates) used as an external skin. The connections between the 12″ wide panels were sealed with mortar and hessian and the internal lining was variable, but commonly clinker block.

Stability of the structure

A Bryant system appeared based on a combination of precast and *in situ* concrete. An inner twin-leaf construction of precast clinker or foamed-slag concrete (for its insulating properties and low cost) was then clad with an outer leaf cast *in situ* from high-grade concrete incorporating steel reinforcement.

In 1948 a quicker system was developed, the Reema. This used precast storey-height panels with *in-situ* columns cast at their connection and the corners. A trough mechanism was used as formwork for the *in-situ* floors. The walls were fair faced with a 0.5″ insulation boarding to produce an 'excellent level of insulation' (Whittick 1957).

Reinforced concrete frames

The reinforced concrete frame systems that were developed used a post-and-panel form of construction. Examples include the Duo-Slab, Boots, Wright and Underdown as systems devised in the 1920s (Whittick 1957). It was anticipated that the Boots system would be good because it had a continuous cavity unlike the other systems. Orlit and Airey forms of construction were also considered to be good forms of construction at the time. The Orlit house relied on bolted connections between precast panels, and the Airey systems used horizontally orientated slabs placed between posts at 18″ centres. There is a double air space and connections were copper. Roof systems could be either flat or pitched.

Woolaway and Unity systems are both based around the use of external and internal slabs inserted into posts to form a cavity wall, externally rendered. There were 600 Woolaway houses by 1957. The Unity house was similar although the constructional dimensions obviously differ and the inner leaf was clinker concrete. There were about 17,500 of these houses by 1957.

Further forms included the Cornish Unit, characterised by its mansard roof. Some 30,000 units were built.

Boots pier and panel houses

The structure of the pre-Second World War Boots pier and panel houses is based on twin piers spaced at about 3-feet centres and intermediate cavity panels of clinker or breeze concrete connected using a tongue-and-grooved joint. The cavity wall was rendered externally and plastered internally. The floor construction is traditional timber. At first-floor and eaves level there are concrete ring beams; at the first floor the beam is precast. The eaves were cast *in situ* to stabilise the structure. The stability of the structure relies on ring beams and their connectivity with the rest of the structure, and the connections between the piers and panels to form a composite wall. The condition of the mild-steel wall ties connecting the twin piers is structurally important, as is the internal construction. Openings were not provided with cavity trays, although the cavity was continuous (BRE 1983c). Approximately 8000 Boots houses were built in the UK by 1930.

Recorded faults: lateral instability

Doubts have been expressed about the lateral stability of the walling if the two leaves become disconnected, following wall tie failure or failure of the outer columns through carbonation or spalling, for example. There have also been some problems of differential expansion between the components of the Boots houses, and there have been instances of columns being totally carbonated. Examples of carbonation damage can be seen in Fig. 8.6. Any disintegration of the ring beam in conjunction with localised column failure may produce instability. The performance of the building and its structural integrity would be dependent at least in part on the response of the internal partition construction on the loading it could carry.

There are important variations between the English and Scottish versions because of the aggregate used for the concrete. The English Boots houses (by far the majority) were generally constructed using clinker concrete. Some of these structures have shown evidence of deterioration of the reinforcing steel following carbonation of the cover concrete. See Fig. 8.6. The amount of damage is related to the exposure, hence the corner columns will commonly be the worst affected. A common indication may be vertical cracking following the line of reinforcement. Any relative movement at the joint between the ring beams and the normally storey-height columns will show as a surface cracking. These may be used to give an advanced warning about the state of the remaining construction.

Problems with clinker

The conventional dense aggregate forms of concrete (Scottish) of the Boots construction do not appear to be affected in the same way. The sulphates in the clinker exacerbate the carbonation effects. There have also been instances recorded of omitted wall ties.

Failures at connections

Any deterioration of the structure is most likely to cause problems at locations of high stress, such as around openings, high loadings or where there is minimal restraint from the wall connections. The structural interconnection within the structure produces problems where related components that rely on each other for support fail. This is akin to a progressive failure mode. Examples include ring beam construction and column connections. Consequently the critical elements to examine in addition to normal inspection are the ring beams and the column structures, the condition of the internal structure, and the connectivity between interdependent elements.

Rendering and cracking symptoms

The external rendering can give valuable indications of movement between components. This may be seen commonly as vertical cracking mirroring the connections between the columns and infill panels where differential

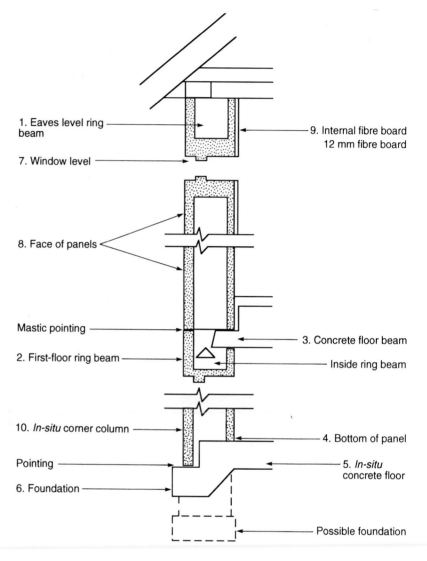

Fig. 1.4 (a) Examples of inspection points for Reema houses. (Originally published as figure 4 in Chandler I 1988 *Low Rise PRC System* The Birmingham TERN project. Reproduced by permission of Ian Chandler).

movement has occurred. Note that these types of cracking symptom are also produced by expansive corrosion of the buried steelwork. Horizontal cracking can occur at the connections between the ring beams and the remaining walling. In advanced deterioration there may also be evidence of spalling of the surface rendering.

Easiform houses

The Easiform system of house construction was used to produce houses both between the wars and particularly after the Second World War into the early 1970s. In total 90,000 houses were built, the majority after the Second World War. There is a wide variability in layout because of the flexibility in

Fig. 1.4 (b) Examples of intersection points for Unity houses. (Reproduced by permission of Ian Chandler 1988).

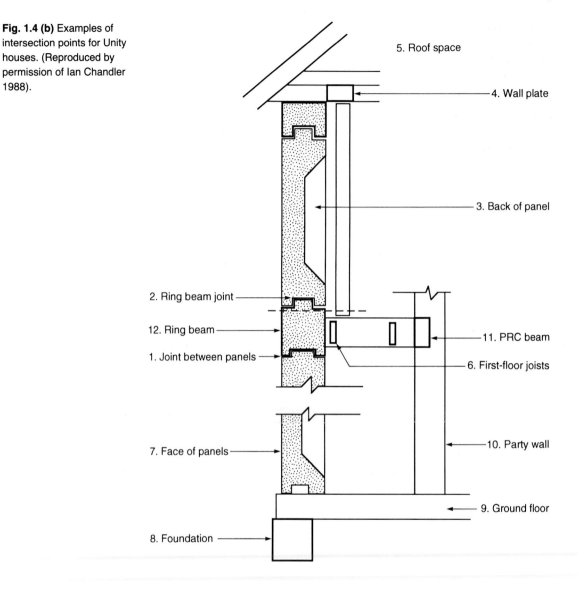

5. Roof space

4. Wall plate

3. Back of panel

2. Ring beam joint

12. Ring beam

1. Joint between panels

11. PRC beam

6. First-floor joists

7. Face of panels

10. Party wall

9. Ground floor

8. Foundation

design, and the system has been used for buildings up to four storeys and with variable appearances including partial brick cladding.

Steel-framed systems A few steel houses were experimented with in the 1920s, and after 1945 some were built. More were based on a light steel frame with a variety of claddings or infilling. Particularly distinctive is the Coventry house built using a tubular frame. Recent studies by the BRE into several types of steel-framed houses system including the Atholl, Crane, Dorlonco, Falkiner-Nutall, Trusteel MkII, and Trusteel 3M have broadly indicated that there is little evidence of serious structural corrosion even in buildings over 60 years

Fig. 1.4 (c) Examples of inspection points for Wates houses. (*Source*: Ian Chandler 1988).

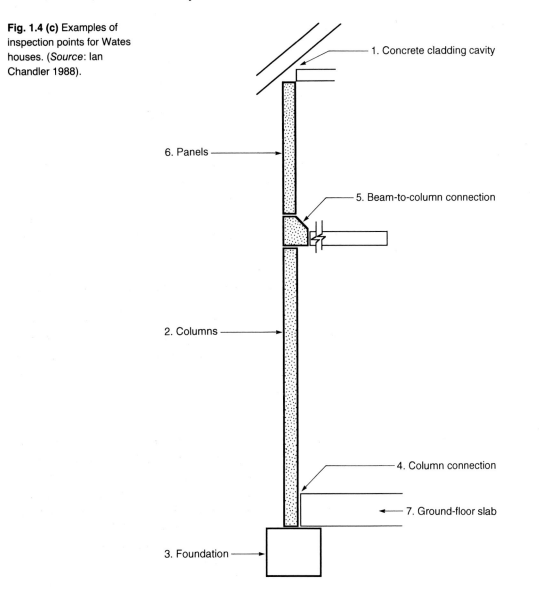

1. Concrete cladding cavity

6. Panels

5. Beam-to-column connection

2. Columns

4. Column connection

7. Ground-floor slab

3. Foundation

old. There is variable rusting of the steel-clad forms however, and this can be an extensive defect.

Timber-framed systems   The timber-framed housing boom of the 1970s and early 1980s was relatively short-lived. One of the commonly cited reasons for its premature demise was a *World in Action* programme which gave timber-framed construction considerable negative publicity. This film was criticised within the industry as being unfair and biased, but for the speculative housing market the damage had been done.

Trends in defects   As with any type of building there are faults, and characteristic trends of common defects have emerged (Urbanowicz 1987). One of the obvious and distinguishing features is the relationship between the internal loadbearing frame and the external cladding, traditionally assumed to be non-loadbearing. Differential movement occurring between the internal frame and the cladding can give rise to cracking in the structure.

Timber and rot   The potential degradation of wood has been a consistent threat in buildings. The employment of timber for structural elements merely makes the consequences of failure more severe. The movement with time to the use of thinner sections of timber more reliant on preservative makes the issue of protection through detailing and natural durability critical. Adequate preservative treatment is essential for all structural timber in the frame. These treatments were frequently of the odour-free, colourless type. They are difficult to identify by observation alone, and require chemical analysis.

There were also potential problems with mismatching of the standards of the country of origin of the timber and British Standards. For instance, in the mid-1970s certain types of American plywood were imported containing small percentages of white-rot fungus. These were used in some areas of the United States without the rot emerging, but were unsuitable in certain areas of the UK for timber-framed construction.

Timber frames and the problems of damp penetration   In modern timber-framed buildings it is conventional for the structural frame to rest on and transmit its loads through a sole plate. Both the frame and sole plate should be isolated from dampness coming from the ground or penetrating the external cladding. Problems of rotting in the sole plate may occur due to the DPC being punctured by the fixings. Where terraced or semi-detached houses are at different levels, the lower section of the party wall can be exposed to penetrating damp from the ground if adequate vertical damp-proofing is omitted or breached.

It is of course also important that any dampness produced internally is kept away from the structural frame, particularly since moisture vapour is very pervasive. The symptoms of rot in the timber frame are usually obvious only when it has reached a sufficiently advanced stage to allow unscheduled movements in the frame or between the frame and its attached components. For instance, the rotting of the frame will cause problems as gaps form around door and window openings.

Settlement and shrinkage of the frame   There is also the need to appreciate that the timber frame may settle and shrink marginally as it bears long-term loading. The creeping effects in timber are obvious and can be extreme without necessarily involving

Stability of the structure

Table 1.2 Common defects in non-traditional timber frame

Components	Symptom	Typical causes	Aids to diagnosis
External walls	Cracking of structure	Differential movement between cladding and frame	Visual inspection
		Inadequate tying-in of internal and external leaves	Endoscope survey
	Rotting timber frame	Movement causes gaps around window and door frames	Visual inspection
	Rotting timber cladding	Incorrect protection to interior timber planks	Moisture meter
	Water penetration	Joints between tiled cills and timber cills broken due to downwards shrinkage of timber frame creating back falls	Visual inspection Check levels
	Loss of heat	Compaction of thermal insulation material within cavity and generally poor insulation	Cavity endoscope survey
	Moisture ingress to inner leaf	Vapour barriers punctured or not continuous	Cavity endoscope survey
	Spread of fire	Omission of fire barriers in the cavity of external walls, and in party walls. Lack of internal plasterboard lining	Cavity endoscope survey
Roofs	Leakage of flat roofs	Differential movements between wall and roof	Leak detector
	Condensation	Insufficient insulation; poor design	Visual & infra-red meter inspection
	Ponding on roofs	Differential movement between roof and rainwater outlets, affecting falls	Visual inspection Use levels and gauges
		Settlement of foundations Deflection of supporting beams	Visual inspection Visual inspetion

structural instability in the period properties. But in the modern timber-framed buildings with rigid external brick claddings the accommodation of the perhaps structurally insignificant adjustment of the frame requires careful design and construction. Relative movement between the timber frame and the external cladding can affect the discharge of water from window sills, and the new falls may direct water into the cavity or onto the frame directly.

Table 1.2 summarises defects and their causes in timber-framed buildings.

Sound insulation   Sound insulation between dwellings is a commonly cited problem with the early timber-framed structures. The relatively low mass and a lack of appreciation of the alternative techniques of sound reduction led to excessive noise levels.

Fire spread   The spread of fire through the cavities of timber-framed buildings with cavity barriers omitted has led to much misinterpretation over where cavity barriers are required, and there are still misunderstandings over whether cavity barriers are required in masonry cavity walls.

Industrialised and system-built forms of construction

Development of forms   The realisation that production advantages could be obtained from changes in construction methodology was one of the factors which influenced the development of industrialised forms of construction. 'Factory' methods of construction were considered superior to those occurring on site because advanced manufacturing methods could be used. Unfortunately this was not fully realised, due to deficiencies in quality and supervision. To compound this, the maintenance aspects of these systems appear to have been given a lower priority than their methods of construction at the time.

In the 1960s it was seen as being in harmony with the technology of the time to develop 'systems' of construction (Diamant 1965). The number of different forms of systems available endorses this view, indeed in the early 1960s the UK was at the forefront of industrialised building development. This can be contrasted with the view that by 1973 industrialised building no longer existed (McKee 1973).

Political influences   The governmental view at this time was that the demand for buildings could not and would not be met without an increased degree of industrialisation. The areas where this was most critical in terms of high-rise

Fig. 1.5 These precast columns should be connected at storey intervals. The head of the column shows an inverted T-profile. The *in-situ* concreting is critical to give continuity of bonding to the reinforcement and to allow the frame to accommodate the vertical and lateral loads without loss of stability. Evidently the *in-situ* work was never completed correctly. Core sampling has been carried out on the column.

buildings was housing, hospitals and offices. In order for the investment in industrialised buildings to make sense, a continuing programme of construction was needed, and this was provided by government and the behaviour of the economy.

Construction density   One factor in the debate concerning the decision to build tall blocks of flats and offices was simply that the technology was available. It could be seen in

Fig. 1.6 Schematic diagram of a faulty *in-situ* concrete and precast reinforced concrete column connection in a high-rise building. The precast external columns are storey-height. The T-shaped detail at their head should be filled with *in-situ* concrete packed around the connecting bars. The design also incorporates a small horizontal gap between the head and tail of adjoining columns. This should have been grouted to provide structural continuity. In this instance the *in-situ* concrete is missing completely.

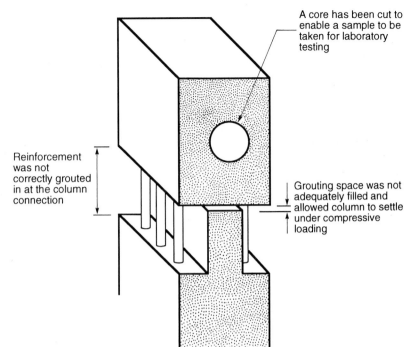

A core has been cut to enable a sample to be taken for laboratory testing

Reinforcement was not correctly grouted in at the column connection

Grouting space was not adequately filled and allowed column to settle under compressive loading

many European countries, and appeared to be both an economic and effective solution. The population was growing and the land mass was not, therefore high construction density was the logical outcome (Diamant 1968).

Construction systems

The proliferation of systems did not share a common dimensional coordination framework, although in the UK the 4-inch module was partially adopted. This did not harmonise with the 10-centimetre module of Europe, and increased the risk of dimensional mismatching between components. Much debate during the evolution of these systems centred around the adoption of open or closed systems. In practice the production of a truly open system never occurred, although the moves towards the dimensional coordination of the present started at around this time. An opportunity for an industry-wide code on dimensions, tolerance and fit was not taken.

In many cases the actual construction method or system was almost exclusively in the domain of the constructor. As this has emerged, so has the realisation that no-one knows how they really were built. This will complicate the analysis of their defects, and the construction may only be fully understood if a similar building is carefully taken down (Webb 1984).

Framed systems   These systems reduced the vertical loadbearing elements into columns, allowing the floor area to be divided into spaces to suit the dimensional base of the system. External panels were designed to fit the range of storey heights. In certain cases the structural integrity of the system could not be determined by reference to the code of the day (CP 116:1965), so specialised testing was carried out. The National Building Frame is an example of this.

Loadbearing concrete   A large number of loadbearing systems were produced, although broad
panel construction   differences existed. These included those where concrete was cast in an on-site or off-site factory; those where precast panels were used exclusively; and those where some *in-situ* construction was used. Totally *in-situ* systems were also available. The relative differences in terms of concrete quality are difficult to quantify, although many of the systems are currently regarded as being suspect. A sandwich of external facing concrete, inner core of polystyrene insulation, and internal quality concrete was common. This could achieve a U-value of less than 0.2 (Diamant 1968). A cross-section through a Balency system wall panel is shown in Fig. 1.7(a).

Unlike some of the low-rise examples, all of the construction processes required the use of cranes. The load-carrying capacity was linked to the construction system. Fram used a maximum of around 6 tons, Spacemaker and Wates 4 tons (Diamant 1968).

Total precast systems   The fixing and jointing together of the precast (PC) panels is the key feature in the construction of these systems. This has been carried out in a variety of different ways. Sections through wall/floor junctions of the Tersons system are shown in Fig. 1.7(b) and (c). The Terson system used lacing bars to link the hooped reinforcement projecting from the PC floor panels. Floor panels were bedded on wall panels, with their joint region occurring above the lower wall. The wall panels were fixed by means of storey-height rods being threaded onto their counterparts in the panel below. This was a special feature of this system, although in many others it was common for all of the jointing regions to be filled with concrete. A representation of the floor/wall joint in the Jespersen system is shown in Fig. 1.7(d) and (e). This illustrates the lack of reinforcement linkage between the precast floor units and the *in-situ* concrete common in some systems.

Partially precast   External facing and spandrel panels were normally precast, with the *in-situ*
systems   elements being walls and/or floors. A section through a wall panel of the Bossert system is shown in Fig. 1.7(f). The MSC system used a steel structural frame which incorporated a metal decking as permanent formwork for the floors. External panels being positioned by means of a

Fig. 1.7 (a)–(g)
Diagrammatic
representation of a range of
system specific
constructional details as
used in industrialised
buildings. (*Sources*: R M E
Diamant 1965 and R M E
Diamant 1968).

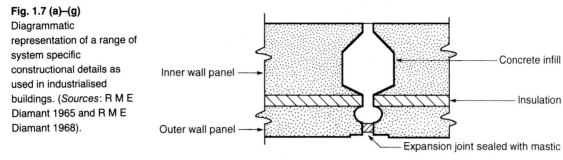

Inner wall panel

Concrete infill

Insulation

Outer wall panel

Expansion joint sealed with mastic

(a) Detail of joint between sandwich wall panels (Balency)

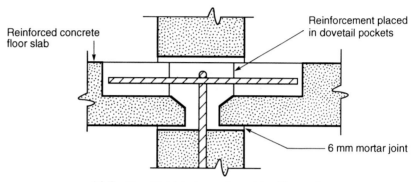

Reinforced concrete
floor slab

Reinforcement placed
in dovetail pockets

6 mm mortar joint

(b) Section of joint between floor slabs (Tersons)

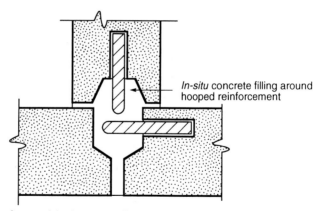

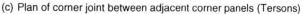

In-situ concrete filling around
hooped reinforcement

(c) Plan of corner joint between adjacent corner panels (Tersons)

Fig. 1.7 *continued*

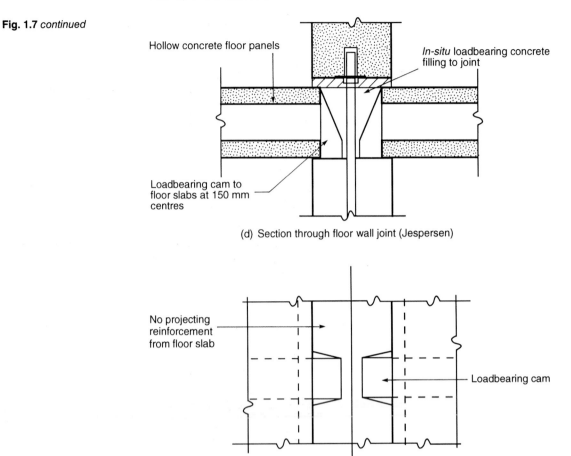

Hollow concrete floor panels

In-situ loadbearing concrete filling to joint

Loadbearing cam to floor slabs at 150 mm centres

(d) Section through floor wall joint (Jespersen)

No projecting reinforcement from floor slab

Loadbearing cam

(e) Plan of floor wall joint (Jespersen)

12 g galvanised string course welded to the structural framework. The Spacemaker system (Fig. 1.7(g)) used *in-situ* concrete for some of the internal loadbearing walls. The use of *in-situ* concrete for the lift shafts and stairs walls was a feature of the Willet system. External cavity wall construction could be used as an alternative to the PC panels.

In-situ systems   These systems were designed to perform as a single monolithic structure. An example is the Sectra system. The formwork was designed for multi-use and commonly fabricated from heavy-duty steel. Dimensional accuracy was maintained by the positioning of the shuttering on successive floor slabs. The system could provide for floors and walls to be cast in one operation, and curing times reduced by the application of steam (Diamant 1968).

Fig. 1.7 *continued*

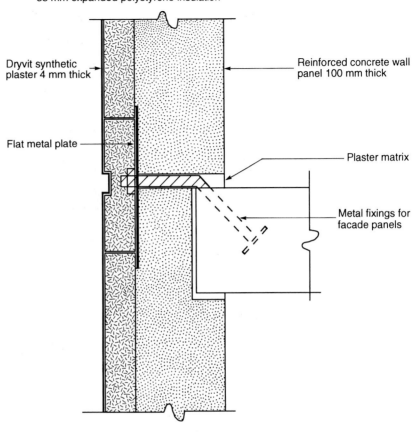

(f) Vertical section through facade panel joint (Bossert)

Decoration could be applied directly to the concrete surfaces, since they were fair faced.

Problems of positioning and fixing

Bearing surfaces of the precast panels were relatively small, and panels were relatively large (approximately 5740 × 2590 mm for the Fram system) (Diamant 1965). This has implications for the criticality of their positioning and fixing. External walls varied from approximately 300 mm (Wates) to 180 mm (Jespersen) thickness, and were positioned by means of cranes and raking props. The panel-bearing dimensions were related to the accuracy of construction of the panels, and in certain cases these were measured according to engineering construction (Diamant 1965). Dry joints between walls and floor panels demand a high degree of accuracy of construction, although the use of a mortar bedding permitted tolerances to be extended.

Fig. 1.7 *continued*

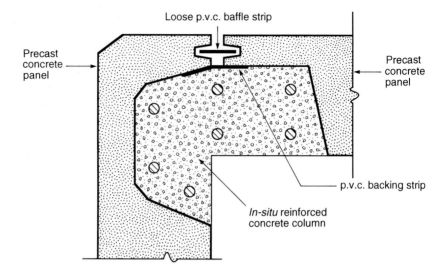

(g) Horizontal section through external corner (Spacemaker)

In the light of current evidence there appears to have been a significant mismatch between the dimensional tolerances acceptable for the production of panels, and their positioning. Notably, the Balency system used an *in-situ* floor in order to overcome problems that the accumulation of dimensional errors could produce.

Poor concrete The provision of poor-quality concrete in external panels has led to a variety of defects. Carbonation rates are increased, which can lead to corrosion of the reinforcement, particularly where the cover is already inadequate. The use of calcium chloride to facilitate high production rates has been found in excessive concentrations. This lowering of the pH of the concrete has severe implications on its durability (BRE 1982a, BRE 1982b). The corrosion of reinforcement, due to both of these effects, will exacerbate the problems. Concrete sections may become detached (Webb 1984), requiring partial demolition. Where a sandwich construction was used, ferrous fixings between the inner and outer leaves of concrete are liable to failure (Webb 1984).

Faults with panel joints The alternative methods of jointing panels is to either to seal with a mastic as with the Balency system, or to use an open or drained joint as with the Spacemeaker system. Both jointing methods were assumed to perform the dual role of weatherproofing and the accommodation of movement. Wide joint areas of mastic can fail due to embrittlement and subsequent lack of bond to the side panels (BRE 1978c). The drained joints can fail where either a defective vertical baffle, or horizontal flashing exists. Water leakage past the primary defence finds little resistance to further migration. A failure of the joint will increase the risk of corrosion of buried reinforcement. Figure 1.7(g) shows a horizontal section through an external wall panel joint in the Spacemaker system which incorporates an *in-situ* reinforced concrete column. Standing panels on neoprene seals and ropes may also have assisted the levelling process, although the cumulative effects of dimensional variability has given rise to variations in joint width.

Faults with claddings and facings

Where facing brick slips are laid in front of floor slabs, they can fail due to the shrinkage movement of the concrete and the slight expansion of the brickwork (Addleson 1982). Weathering of the exposed aggregate finishes may cause a loss of adhesion and subsequent failure, although this is usually a precursor to continuing failure of the concrete. Panel facings of mosaics and external tiling can become detached from their background. Differential movement and poor adhesion methods can be a cause.

Condensation problems

The provision of insulation within the panel construction enabled acceptable U-values to be achieved. These were seen as being part of the essential scientific ingredient of the systems. Very low values were possible. The thermal mass of the concrete meant that where high air temperatures could be generated rapidly, the likelihood of surface condensation increased. Where the insulation was not continuous through the structural joints between panels, cold bridges could occur (Addleson 1978), as shown in Fig 1.7(a). The positioning of steel fastenings to external panels, as represented in Fig. 1.7(f), places the fixings in a region where a risk of interstitial condensation exists.

Pattern staining; failure in the climate

Water runoff from the generally low-permeability external panels has produced severe pattern staining. This is common for many exteriors, and appears most marked on facades where light-coloured horizontal bands are provided. These commonly show colour differences between areas where water flow has cleaned surfaces, and those where dust disposition has occurred. Corners of tall buildings are vulnerable. Areas of discoloration of exposed aggregate panels can add to an overall staining of the building.

Verticality

The cumulative effects of dimensional variation between panels can cause progressive errors of verticality. These are most pronounced at corners and changes of direction. See Fig. 1.8.

Fig. 1.8 Inaccuracy in the corner of a multi-storey building. In this instance, the fault was caused by deterioration of the *in-situ* concrete in the corner columns which have lost their structural integrity.

The on-site adjustment of panel size was generally impossible, and where attempted would have profound structural and weathering implications. A reduction in joint widths to accommodate dimensional variation reduces the amount of movement that can be accommodated. Spalling of the concrete can occur. Variations in the verticality may also reduce the amount of panel-bearing surface and cause misalignment of movement joints through the structure.

Progressive collapse possible

Where inadequate panel fixings exist, either due to poor positioning or failure of the structural joints, instability of the structure can occur (Webb 1984). This can result in sudden failure, which may manifest itself as a progressive collapse, as in the infamous Ronan Point disaster.

Difficulty of determining structural stability

The technology of the systems is such that many of the fixings are covered by a concrete filling to the joints between panels. This makes inspection difficult, occasionally involving cutting away parts of the joints, which in turn can lead to a weakening of the structure. The position of any fixing metalwork may be unknown, since details of the 'as built' system may not be available. Corrosion of these fixings can cause reductions in lateral stability where they tie walls and floors together. In certain cases it may be necessary for the building to be temporarily supported until a safe demolition can be undertaken.

Fire safety

Reinforced concrete panels should provide good fire resistance. This can be seriously reduced where the cover to the reinforcement is inadequate, or non-existent. Movement of panels can open up joints between walls and floors to between 25–50 mm, so that a direct path for fire spread is created (Webb 1984).

General defects in non-traditional construction

The Housing Defects Act 1984

On 8 February 1983 the Minister for Housing and Construction announced a review of the condition of structural reinforced concrete construction in prefabricated houses built both in the inter-war period and post-war until the 1950s. Initial investigations were to be carried out into the Boot, Cornish Unit, Orlit, Unity, Wates and Woolaway houses.

Designations

Following this announcement a number of non-traditional house types were designated on 1 December 1984 under the Housing Act 1984. The Act placed a statutory obligation on local authorities to assist private owners of the designated properties to repair the structures. The assistance provided differentiates between private and publicly owned properties. This can create problems with the repair of semi-detached homes held in split ownership, especially where the proposed repair systems involve external modifications such as overcladding.

The scale of the problem

The scale of the problem is potentially enormous. For instance, a third of Birmingham Metropolitan Authority's 122,000 council-owned properties are constructed using non-traditional forms. Around the time of the Housing Defects Act the Labour-controlled Association of Metropolitan Authorities (AMA) produced a number of studies on defects in public housing. These presented evidence of the existence of a large scale of latent and other defects in the non-traditional forms of the 1940s and 1950s (AMA 1983a, 1983b), and the industrialised developments of the 1960s and 1970s (AMA 1984). They estimated at 1984 figures that the cost of rectifying the non-traditional stocks could run to £5000 million, and there could be a need for up to another £5000 million for the industrialised and system-built dwellings (AMA 1985).

A range of house types were designated under the Housing Defects Act 1984, including the Airey, Boot, Cornish, Parkinson, Orlit, and Wates houses. Designations were under the qualification of ineffective protection to embedded reinforcing steel in loadbearing parts of the building. In 1987 the Boswell House was designated under the Housing Act 1985, under the qualifying defects of excessive unburnt coal in the aggregates used for the concrete walls, floors or foundations, in addition to ineffective protection of the embedded steelwork. Smiths houses were also designated subsequent to the original 1984 designations.

Information sources

The largest source of information available for the analysis of defects in non-traditional buildings is the Building Research Establishment. A large number of specific reports and aids to identification of the non-traditional and systems forms of building have been produced. Considerable information has also been drawn together by the Birmingham TERN project*. The reader is advised to study these sources for the most complete and authoritative review of the condition of these properties. In the light of this, the following statements provide merely an overview of the problems.

* The Birmingham TERN Project (Training for Evaluation and Repair of Non-Traditional Buildings). A series of handbooks and videos appraising and solving some of the most troublesome defects which have beset non-traditional housing in recent years. Produced by and available from Birmingham Polytechnic.

Variability Keep in mind that if the construction of traditional housing varied enormously, even the non-traditional housing is regionally variable, and the findings of reports or other generalised comments may not be universally applicable. There may be exclusions or additions to regional or individual designs which influence the occurrence and symptoms of defects. Even with relatively highly structured systems there was a degree of variability in constructional form, and the surveyor must establish this in each circumstance. In some cases only samples representing a small proportion of the total houses built will have been reviewed.

Regional variations in Boswell houses Boswell houses, for example, have been found to be significantly more prone to deterioration in the Midlands than Merseyside because of the use in the Midlands of mining clinker in the *in-situ* concrete work.

Regional variations in Wates houses Take another example, the Wates house. The basis of design was a set of storey-height tray-shaped loadbearing panels. These were precast concrete delivered to site and craned into position for connection and incorporation of reinforcing steel at the joints. This is a prime area for the occurrence of cracking, especially at the corners of the building. There may be rainwater penetration.

A further problem is the rusting of the steel reinforcement, producing spalling of the precast panels (the designated defect under the Housing Defects Act 1984). This is likely above the window openings where the sectional dimension is small and likely to be wetted on two sides at least. Any variation in cover to the reinforcing steelwork is likely to be significant.

A ring beam runs around the building at first-floor level and provides the stabilising connection for the panels and primary fixing for the upper storey floor and wall panels. This may become displaced as the reinforcement to the joints corrodes.

Many of the Wates loadbearing panel system houses built in the north were commissioned by the National Coal Board (NCB). To maximise the use of NCB by-products clinker blocks were used extensively in the party walls and the linings to the outer walls. Concern over the long-term condition of these blocks should of course suggest careful assessment. Chandler identifies a further variation in the Wates design occurring in the South, where the ring beam at first floor is considerably deeper than normal. It carries a deep timber wall plate for the support of the first floor. There are other variations, such as the position of the head of the upper floor windows relative to the soffit boarding. Most significantly though, Chandler reports that there is a confusing variation in the party wall construction. Instead of the conventional twin-leaf blockwork reinforcing a beam and column system sandwiching an airspace, there is a precast and *in-situ*

Stability of the structure

Table 1.3 Common defects in non-traditional housing: prefabricated reinforced concrete

Components	Symptom	Typical causes	Aids to diagnosis
Walls	Spalling of external face	Corrosion of metal reinforcements due to lack of concrete cover presence of chloride accelerator carbonation	Visual inspection Government survey Sample and test Phenolphthalein test
	Cracking	Differential movement between *in-situ* and precast units	Visual examination
		Bowing of precast panels and cast *in-situ* panels Overloading. Foundation settlement, shrinkage	Bonding of tell-tales
	Falling out of panels Movement of panels	Insufficient ties between inner and outer leaves of external walls or panels; corrosion of ties	Endoscope survey
	Water penetration due to wind and rain	Movement of panels at joints Result of cracking	Visual inspection
	Cracking of render	Lack of bond to concrete surface Movement of substrate wall	Hammer test
	Condensation	Uneven. U-values in external walls, cold bridging due to use of panel and post construction Poor insulation	Visual inspection Endoscope survey
	Cold penetration	Poor thermal insulation standards	Infra-red meter
	Structural integrity suspect	Use of high alumina cement in beams and floor slabs	Sample and test for HAC Sample and test DTA *In-situ* UPV testing

combination wall. At the ground-floor level there is a precast concrete tray panel infilled with breeze which was sandwiched with a cavity on each side by twin leaf breeze blockwork. Refer to Fig. 1.4 for other inspection points.

General defects in PRC houses

The nature of defects in properties appears to differ with the constructional style, which is of course to be expected. The non-traditional housing of the 1940s and 1950s is mostly low-rise PRC (prefabricated reinforced concrete) construction. Table 1.3 summarises defects and their causes in PRC housing.

Corrosion of steel reinforcement

One of the general characterising problems of these structures has been the inadequate embedding of the steel reinforcement in the loadbearing concrete, a qualifying defect in 1984 Housing Defects Act designations. This may appear as patterned cracking in the external surfaces of the concrete walling or through render coatings. The corners of the building are a frequent victim because of relatively high exposure of the edges of the panels or the corner columns. The corrosion of the inadequately embedded steelwork may be exacerbated by the presence of chloride accelerators and by carbonation of the concrete with age. This latter problem advances more quickly with porous, poor-quality concrete.

Deterioration in concrete panels

There has also been deterioration occurring in the concrete loadbearing panels and claddings. This may be attributable to built-in defects initiated during the prefabrication or assembly stages. There is some evidence of poor training and quality control during the production of the prefabricated panels (BBC 1984). Quality through prefabrication was of course one of the key objectives of the systems. Clinker materials were frequently used as an aggregate to economise the production of these structures. The sulphates contained in these have produced severe problems with the concrete durability, and it is possible to find deteriorated clinker concrete that can be literally scooped out with the hand. There may also be questions of structural integrity of the buildings where high-alumina cement has been used in the fabrication of the structural elements (Urbanowicz 1986).

Differential movement

Problems with differential movement appear to be more likely to occur at the connection between *in-situ* and precast elements. This may lead to bowing or cracking of the precast or *in-situ* work, leading to secondary defects such as water penetration under driving-rain conditions. This has also been a problem caused by poorly detailed joints in otherwise stable structures. Another prime or contributory cause of movement or dislodgement of panels can be the omission of proper fixings and ties, or their substitution with inadequate alternatives (Urbanowicz 1987).

Long-term durability

Long-term durability problems have appeared with the reinforced concrete itself. These have occurred through carbonation or sulphate attack of the concrete, the latter originating in aggressive aggregates such as clinker. Some of the framed constructions have proven problematic because of poor durability in the external components such as window frames, and particularly the sills and other exterior woodwork such as flush-eaves fascias to flat roofs.

Flat-roof problems

There have also been myriad problems with the roofs themselves (AMA 1985), most of which were flat and covered with felt and chippings. In the 1950s it was common practice to construct unvented cold roof details, and interstitial condensation has been rife where vapour barriers were frequently omitted.

Table 1.4 summarises common defects and their causes in non-traditional housing.

Condensation

Condensation has also been severely problematic. The thermally heavyweight nature of the concrete walls and the high conductivity of plain concrete were poorly suited to the UK climate, and required additional insulation to improve the performance of the structure. The thermal insulation of these buildings differs greatly with the various designs, however.

The early solid-wall structures of the 1920s can suffer terrific rates of heat loss. The post and beam structures which emerged during the same era with partial cavities are still problematic, since the stiffening structure creates cold bridges through the walls. By concentrating heat loss they encourage patterned condensation.

The defects not specifically related to any constructional form such as overloading or soil movement still occur of course. The accommodation of foundation settlement, and particularly differential settlement, will of course appear as different symptoms in panelled construction compared with the small unit brick wallings.

General defects with the industrialised and system-built forms

The industrialised and system-built boom of the 1970s advanced the medium or high-rise design concepts. The use of novel construction techniques combined with the move to high-rise buildings has produced extra problems compared with the low-rise estates, colloquially termed 'the avenues' (BBC 1984). Some of these problems, such as the response of the

Table 1.4   Common defects in non-traditional housing (general)

Components	Symptom	Typical causes	Aids to diagnosis
Roofs	Cracking and water penetration	Degradation of roof covering such as asbestos cement, other corrugated roof panels, bituminous felts or metal sheeting	Visual examination with care
	Loss of heat through roof	Low-pitched roofs difficult to insulate	Visual inspection
	Loss of strength, bending of beams	Use of high-alumina cement joists in roof beams	Check for presence of HAC Sample and test DTA *In-situ* UPV tests
		Rotting of timber teams	Visual inspection
	Lateral movement	Inadequate wind bracing	Visual inspection
Windows doors and frames	General deterioration	Movement between main frame and opening frames	Visual inspection
	Corrosion of metal frames	Metal frames poorly maintained	Visual inspection
	Spalling of concrete cills	Corrosion of reinforcement due to lack of cover chloride accelerators carbonation	Covermeter survey Sample and test Phenolphthalein test
	Rotting of timber	Poor maintenance allowing moisture ingress	Check with moisture meter

European structures to the UK climate were unforeseen. A range of other deficiencies have been summarised by the AMA (1985), highlighting the hygrothermal performance of the structures. Table 1.5 summarises defects and their causes in high-rise buildings.

Rain penetration and condensation

Rain penetration and condensation have become almost legendary. Faults in the structure such as spalling of the concrete panelling or the claddings may arise from the quality and admixing of the concrete. Differential movement after completion and prefabrication and assembly inaccuracies have also been recorded. These are obvious at the relatively infrequent joints. This is an area of particular concern in the box forms using loadbearing panelling. The condition of the restraining fixings is also impractical to assess casually. See Fig.1.9

Table 1.5 Common defects in high-rise buildings

Components	Symptom	Typical causes	Aids to diagnosis
Reinforced concrete Walls Columns Nibs Corbels	Horizontal or vertical cracks	Corrosion of reinforcement due to use of calcium chloride accelerators; lack of concrete cover to reinforcement; porous concrete carbonation and loss in alkalinity	Visual inspection and: Sample and test for Cl. Covermeter survey Initial Surface Absorption test Phenolphthalein test
	Rust staining on concrete surface Spalling concrete	Corrosion process is exacerbated by exposed conditions	
Parapets Balconies	Fine hairline cracks on surface; colour change	Possible damage by fire	Sample and test concrete *In situ* UPV integrity testing
	Fine crazing of concrete surface	Rapid drying of concrete; over compaction	Examine crack pattern
		Possible alkali-aggregate reaction	Sample and analyse petrographically
	Vertical cracks along walls	Shrinkage cracking, no provision for movement joints	Visual inspection and measurement
	Diagonal cracks along walls	Differential settlement of foundations	Visual inspection crack gauges
Reinforced concrete	Vertical cracks	Overloading; inadequate reinforcement; design fault, thermal movement	Visual inspection and refer to structural engineer
Beams, slabs and lintels	Diagonal cracks at beam ends	Lack of shear reinforcement; overloading	Visual inspection and refer to structural engineer
	Excessive deflection	Inadequate design Insufficient depth	Visual inspection, refer to structural engineer
		Misplaced reinforcement	Covermeter survey

Table 1.5 *continued*

Components	Symptom	Typical causes	Aids to diagnosis
		Deterioration of concrete	Sample and test concrete
		Possible use of High Alumina Cement (HAC)	*In situ* testing using UPV NDT method
	Cracking of slab	Steel reinforcement at supports inadequate Excessive loading	Visual inspection Covermeter survey
Reinforced concrete	Random cracking in floors	Shrinkage of concrete; lack of movement joints	Inspection
		Concrete mix too rich	Sample and test
Slabs	Spalling of slab surface	Excessive traffic wear	Measurement of depth
		Poor quality concrete Frost attack, chemical attack	Sample and test Concrete core cutting and inspection
	Local settlement diagonal cracks	Poor compaction of subgrade Inadequate reinforcement Ground movement due to water erosion	Inspection. Taking concrete cores and excavation of soil for test
Precast concrete cladding	Rust stains on surface Spalling concrete Horizontal or vertical cracking	Corrosion of reinforcement due to use of calcium chloride accelerator lack of cover to concrete porous concrete (insufficient steel protection) carbonation of concrete surface	Visual inspection and: Sample and test Covermeter survey ISAT *in situ* test Phenolphthalein test
	Cracking (regular)	Differential movement between two layers of concrete in precast units – strong rich surface decorative mix surrounding a low-strength backing mix	Visual inspection and local excavation
	Movement of cladding panels	Lack of restraint due to incorrect fixings to frame or corrosion of fixings or absence of fixings	Visual examination Endoscope survey Radar NDT technique

Table 1.5 *continued*

Components	Symptom	Typical causes	Aids to diagnosis
	Rain penetration	Panel joints opened up Joints between frame and cladding open	Visual inspection and endoscope survey
		Hairline cracks permit rain to enter cavity exacerbated by high winds in exposed areas	Visual inspection
Brickwork cladding	Rain penetration	Differential movement of different materials causing cracking	Visual inspection
	Delamination	Poor bonding of brick slips to concrete faces, edge beams etc.	Visual inspection
	Cracking	Insufficient support in concrete nibs and corbels allowing transmission of full weight of panels to those below	Endoscope survey behind panel
	Cracking or bowing of panels	Lack of or corrosion of, wall ties between brickwork and concrete inner leaf	Cavity endoscope survey
		Eccentricity of load	Refer to structural engineer
GRC cladding	Bowing cracking and delamination of insulated sandwich panels	Excessive thermal movements due to heat build up in front of insulation	Hammer testing
		Inadequate fixing details	Special NDT techniques
Roofs	Water penetration	Differential movement on parapet walls Breakdown of flashing Cracking of membrane	Visual inspection Leak detector
	Cracking of mastic asphalt	Embrittlement with age	Sample and test
		Use of an unsuitable reflective paint	Sample and test
		Poor jointing techniques. No movement allowance	Visual inspection

Table 1.5 *continued*

Components	Symptom	Typical causes	Aids to diagnosis
		No protection in form of shingle or chippings	Visual inspection
		Use of underlying insulation causing heat build-up	Local excavation
Structural steelwork	Excessive deformations – out of plumb columns buckled beams	Due to poor erection, overloading, wrong grade of steel, inadequate bracing, slip or failure of support	Visual assessment Plumbing checking Consult structural engineer specialising in steelwork design
	Hairline cracks at welds and junctions	Defective welding practice Brittle fracture	Penetrant dye Ultrasonics X-ray radiography
	Slippage of joints	Overloading, incorrect type of fastening system, ineffective tightening of bolts	Visual examination Consult structural engineer
	Corrosion	Failure of protective system-(concrete) or paint system	Local excavation Tests on concrete (ISAT) Tests on paint thickness using NDT gauge
External staircase Link bridges and balconies	Spalling of concrete	Corrosion of reinforcement due to: lack of cover presence of chlorides carbonation	Covermeter survey Sample and test Phenolphthalein test
	Asphalt flooring cracked	Embrittlement due to age and exposure combined with possible impact damage in winter	Sample and test
	Water penetration	Failure of waterproofing membrane and cracked concrete floor screed	Visual inspection and leak detector
		Differential movement between floor units and edge beams excessive	Visual inspection and leak detector
		Poor detailing at parapets	Visual inspection

Fig. 1.9 Built-in inaccuracy between loadbearing concrete panels, obvious at the joints because of the intolerance of the orthogonal systems. The positioning faults may lie with fabrication inaccuracy, positioning of fixings or both. The tennis ball gives scale and also indicates edge damage to a panel.

2 Distortion and criteria

The criteria for determining the structural acceptability of movement and cracking in buildings usually focus on crack width. Cracking is symptomatic of a force or combination of forces greater than the capacity of the building or its component materials can bear. Generally the forces are tensile. Cracks should not be assumed to arise from a single causal defect, or even from a simple combination of defects occurring in sympathy.

Categorisation of building damage

Size and significance Categories of building damage and their relationship to the width of symptomatic cracks have traditionally been defined by the Building Research Establishment and are the generally accepted criteria. With the recent revision of Digest 251 (BRE 1981a) the classification by severity of the cracks has logically been removed, so minimising any prejudgement over the situations in which cracks appear. See Table 2.1.

Crack width categorisation and limitations These categories of crack width are used to rank building damage according to the probable severity of the fault and extent of remedial work required. It should always be borne in mind that these are relatively simple expressions which allow categorisation of damage only according to approximate crack width and any identifying features that may accompany them. They do not, of course, distinguish between different causes of faults. Nor do they directly translate to the degree of dislocation in the structure or to the extent of angular distortion accompanying any differential movement, both important features of residual stability.

Criticality It should obviously be necessary to assess the criticality of specific cracks according to their location in the building and their possible secondary consequences. In particular, consequential failures are not all equal, nor is the quality of design or construction. Variability in the state of repair of

Table 2.1 Classification of visible damage to walls with particular reference to ease of repair of plaster and brickwork or masonry. (Reproduced from Building Research Establishment Digest 251: British Crown copyright.)

Category of damage	Description of typical damage *Ease of repair in italic type*	Approximate crack width (mm)
0	Hairline cracks of less than about 0.1 mm width are classed as negligible	Up to 0.1[1]
1	*Fine cracks which can easily be treated during normal decoration.* Perhaps isolated slight fracturing in building. Cracks rarely visible in external brickwork.	Up to 1[1]
2	*Cracks easily filled.* *Redecoration probably required.* *Recurrent cracks can be masked by suitable linings.* Cracks not necessarily visible externally; *some external repointing may be required to ensure weathertightness.* Doors and windows may stick slightly.	Up to 5[1]
3	*The cracks require some opening up and can be patched by a mason.* *Repointing of external brickwork and possibly a small amount of brickwork to be replaced.* Doors and windows sticking. Service pipes may fracture. Weathertightness often impaired.	5 to 15[1] (or a number of cracks up to 3)
4	*Extensive repair work involving breaking out and replacing sections of walls, especially over doors and windows.* Window and door frames distorted, floor sloping and noticeable.[2] Walls leaning[1] or bulging noticeably, some loss of bearing in beams. Service pipes disrupted.	15 to 25[1] but also depends on number of cracks
5	*This requires a major repair job involving partial or complete rebuilding.* Beams lose bearing, walls lean badly and require shoring. Windows broken with distortion. Danger of instability.	usually greater than 25[1] but depends on number of cracks

Notes:
1 Crack width is one factor in assessing category of damage and should not be used on its own as a direct measure of it.
2 Local deviation of slope from the horizontal or vertical, of more than 1/100 will normally be clearly visible. Overall deviations in excess of 1/150 are undesirable.

buildings can be significantly indeterminate. If these limitations are acknowledged, they can be a powerful tool for comparative assessment.

In addition, it should also be acknowledged that a structurally insignificant crack can be quite unsightly. It may be unnerving for the owner, tenants and potential purchasers of a building.

Monitoring crack symptoms

Tensile forces

The occurrence of cracking in buildings is symptomatic of tensile or shear forces acting within the structure. The crack will appear at the weakest point or plane as the stresses exceed the tensile capacity of the materials. It may be a plain tensile fracture, or an indirect consequence such as cracking produced on the stressed skin of brickwork panelling as it bulges under compressive forces. Cracking does not always represent the total relief of inherent stress, so it is important to assess the dynamic stability. Frequently the cracking symptoms are physically and relationally remote from the causative defect.

Crack width analysis: cyclic movements

The most direct and hence the most useful way of assessing cracks is to monitor their dynamics. With thermal or moisture-related movements the edges of a crack may be moving relatively frequently and cyclically. Analysis of the width may show a pattern in variations which itself can be a strong indication of the possible cause. It is important to remember that the occurrence of thermal movement in a crack may not be the cause of the crack itself. Indeed, the discontinuity in the structure may now be allowing the thermal stresses to be relieved as movement. The same holds true for moisture-induced movements.

Timing and seasonal factors

Cracking that occurs early in the life of the building frequently involves the readjustment of the moisture contents of the various porous materials used for construction, combined with the loss of water used in the wet construction trades.

The rate of adjustment of the building and equilibria levels with the surrounding internal and external environments will depend on the season and exposure of the building faces, and on the nature of use of the building. Consequently the emergence of cracking associated with a predominance of these factors may occur in the springtime following construction during the wet season. There are also potential problems with the rapid drying out of the wet trades by additional heating of the property, and the rapid changes this produces can have marked effects.

Progressive movement Progressive movement which is an immediate threat to the stability of the building is sometimes obvious on first inspection. Where the building is currently stable but there is a question over its long-term performance, the use of a crack-measurement tool can be valuable.

Crack stability Where the crack is stable and the cause can be identified it may be possible to conclude that the residual stability of the building is satisfactory and shows no obvious signs of change. For the minimal cost and effort involved in making this analysis a lot of unnecessary disruption and cost is avoided.

Measurement tools

Tell-tales There are a variety of methods available. The range of tools that have been developed start with a simple glass tell-tale slip glued onto the surfaces adjacent to the crack. If movement occurs the glass will fracture. Periodic inspection for a sufficient period to assess the stability of the building fabric is simple. The applicability of this technique is limited. The surfaces require to be fairly level – any previous rotation of the building will prevent a simple tell-tale being fitted in the first place. In addition, the system is vulnerable to general or malicious damage and gives no indication of the degree of movement or its exact form. Also, as Melville and Gordon (1979) point out, it is quite likely to unnerve the building owner when it fractures.

Accuracy Modern tell-tales are graduated for temperature and allow fine measurements to be made with an estimable degree of accuracy. Other advantages are that they can be used at quoins, and are less alarming to the building owner. The accuracy of the perspex tell-tales is in the order of $\pm$ 0.5 mm (Richardson 1985a). See Fig. 2.1.

Markers An alternative to the clumsy glass tell-tale or the more sophisticated graduated tell-tale is the use of a set of markers and a strain gauge. These are less conspicuous and potentially more accurate (to $\pm$ 0.05 mm). This is usually sufficient for most monitoring purposes. The approach is the use of markers set into the structure, either non-ferrous screws or ball-bearings secured in epoxide resin. The use of screws is a relatively inaccurate and makeshift method. The precision with which the ball-bearings may be set into the building fabric allows a more precise set of measurements to be taken. The variations in their spacing can be measured using a micrometer, and this can be used to plot rotation or shear as well as tensile or compressive movement.

Fig. 2.1 Use of a commercial graduated tell-tale to monitor crack dynamics. Two tell-tales set at right-angles would have allowed any rotational movement to be measured. Poorly concealed previous cracking is indicative of a long-term and progressive problem.

Linear variable displacement transducers A further development suitable for taking very fine measurements to a high accuracy involves the use of linear variable displacement transducers (LVDT). These convert movements into an electrical signal which can be recorded at frequent intervals. The cost of the monitoring equipment and the LVDTs is relatively high and the potential accuracy of the system, around 1/100th of a millimetre either way, is infrequently warranted. However, it does allow hands-off monitoring of frequent and infrequent movement, and could be a useful tool for monitoring buildings where an

Fig. 2.2 Use of simple crack-monitoring tell-tales to determine relative movement in two dimensions.

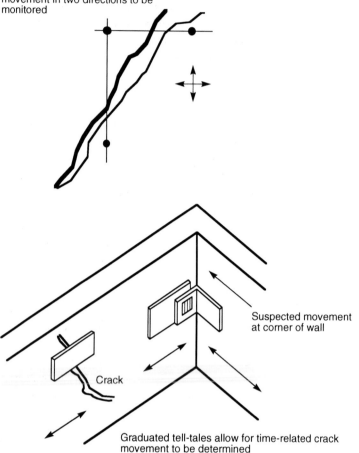

Fixing pins at 90 degrees will allow movement in two directions to be monitored

Suspected movement at corner of wall

Crack

Graduated tell-tales allow for time-related crack movement to be determined

early identification of minimal movement was necessary. The development of computerised data collection using portable electronic data hoarders, termed *squirrels,* makes measurement and analysis more practical. Be sure that this level of sophistication and accuracy (also cost) is warranted. It is easy to become swamped with data requiring an immense and possibly irrelevant exercise in statistical interpretation.

The relevance of weather

It will be important to note the weather conditions at the time of taking measurements, since this can obviously have a significant effect on the nature of movement in the building. If the measurements are all taken at the same time of day and in similar weather conditions they are unlikely to identify the trend of any thermal movement, for example. The frequency of measurement is governed by the accuracy of trends required, which in turn depends on the nature of movement(s) suspected.

Accuracy   The accuracy needed of the system will also be a factor in the choice of equipment. This will depend on the size of the crack and the scale of movement anticipated with the suspected causes. In instances of simple deflection it may be sufficient to use the traditional glass tell-tale. The calibrated version would be better, since it can give information about direction and degree of movement.

Tolerance   It is important that the tolerance of measurement is relatively insignificant compared against the fineness of movement expected. For example, a plastic gauge or tell-tale that measures to a tolerance of a similar scale to the expected dimensional movement is of limited use. It is likely to give erroneous and uninterpretable results. If this is not appreciated it can of course lead to a completely incorrect assessment of the defect. Otherwise, the uncertainty of errors undermines the certainty of measurement. Conversely, it is unnecessary to measure a coarse deflection with an extremely fine measurement system.

Vertical and horizontal applications   The measurements may be taken horizontally or vertically. A combination can give information about rotation of the crack or which direction(s) the movement is in. As well as measuring the movement of the structure to an appropriate accuracy, the analyst must be capable of distinguishing the movement traits associated with different symptoms and their causes. Both require consistency of application and thoroughness of analysis.

Unrestrained and cyclic movements   Unrestrained linear stresses will exhibit predominantly horizontal or vertical movement. Cyclical motion can be characterised only after recording movements over the relevant period of time and using an appropriate number of measurements.

Rotation and settlement   It will be valuable in cases of suspected rotation or localised sinking of the building to analyse both the top and bottom of a crack. This will allow an assessment to be made of the crack taper, and could of course be valuable in confirming hogging or sagging movements. Crack dynamics associated with plain settlement alone, for example, will generally minimise with time. If a stable state is reached they should stop moving altogether. Such a trend will only appear after measurements have been taken over a considerable period of time, and may be used to confirm the stability of relatively new building work.

   In the meantime it is to be expected that there will be pressure from the building occupier or owner to assess and rectify the problem rapidly. This should not be allowed to override prudent judgements.

Distortion criteria   The criteria for acceptability of distortion, or definition of failure, must be finely adjusted to reflect the function and type of building, its economics, and also the expectations of its users. Life safety is an over-riding concern, and where there is any doubt about the stability of a building, this must be an imperative. That the general public seems prepared to consider safety an absolute rather than relative measure, equating structural safety with absolute security, merely complicates the issue.

Types of movement

Deflection   Buildings rarely totally collapse, and even when they do there are usually some preliminary symptoms. The critical types of structural movement in buildings are the relative rather than wholesale movements. These produce deformation and dislocation of the structure which can significantly redistribute loadings. This can cause secondary failure in elements not designed for the new loads applied to them.

As knowledge of material properties has improved, the progressive adoption of components with smaller cross-sectional areas has contributed to the scale of deflections associated with normal use. Where the designer of buildings has not acknowledged this, there is the increased likelihood of cracking as mismatch in service occurs between rigid and flexible elements. The classic example is the use of the rigid partition bearing on a flexible floor.

Dominance and restraint   Where structurally dissimilar buildings are connected, the dominant structure may impose loads on the adjoining building if it tilts or otherwise moves. Relative mass and rigidity will determine the virtual restraint on movement. Particularly harmful can be the bounding of low flexible buildings by high rigid structures (Hodgkinson 1983). Where disconnection has been used to prevent this, there is the possibility of water ingress between the buildings at subsoil level and problems of weatherproofing above ground (Hodgkinson 1983).

Assessment of distortion   Assessment of distortion is not merely a matter of identification and cataloguing. Obviously these are important tasks that must be carried out thoroughly. However, the analysis and comparison of the designed and discovered structural systems is paramount. In all but the most simple and straightforward cases it will be necessary to consult a structural engineer.

Rotational and relative deformations

A fundamental distinction requires to be made between the types of deformation associated with relative vertical movement and those occurring with rotational failure of all or some of the structure.

Angular distortion in structures is the result of the tensile and shear components of rotational forces, and is the important criterion for the allowable deflection in loadbearing masonry. Masonry is a particularly important constructional form to assess, simply because it is a composition of numerous discrete and rigid units. The failure mode is strongly dependent on the relationship between the bricks and their jointing mortar, specifically their relative strengths and responses to movement. Masonry panel failure will be related to the nature of the surrounding structure. Composite structures will react to movement in a relatively complicated manner and may produce indistinct symptoms.

Sagging and hogging in masonry

The relative deflection of masonry walling is expressed as a ratio of the measured deflection ($\triangle$) to the relevant dimension of the panel, length (L) or height (H). This defines the change in shape of the section of wall as it sags or hogs.

The location and extent of any openings which provide zones of weakness in the wall will be critical to the response of the wall, and the maximum stresses that may be tolerated will depend on the restraint of the wall and the loading it is carrying.

Deformation limits for masonry

Where masonry is sagging, the relative deflection is usually expressed against the length of the panel, $\triangle/L$. The relative deflection may be also expressed using the height of the wall, for example with hogging deflection, $\triangle/H$ (Rainger 1983). The limits of acceptable relative deformation are related to the structural independence of the component, particularly its connectivity to the remaining structure. The limits of relative deformation for hogging of walls may be half those of sagging deflections (Rainger 1983).

Height and width of masonry panels

The relationship between the height and width of a panel will also influence the risk of cracking under distribution of movement, and hence the allowable relative deflection. Minimal values up to 1/300 for (sagging) acceptability in loadbearing walls and panels (Rainger 1983) may be adjusted to include a factor of safety such as 1.50, taking the limiting distortion to 1/450 (Hodgkinson 1983). This corresponds to 11 mm over a 5 m span. Hodgkinson (1983) notes that suggested limiting values are imprecise and range between 1:750 and 1:150.

Criteria for the measurement of distortion

The effects of loading and restraint of walls on stress capacity

Of real importance in the assessment of buildings is the judgement between progressive or stable deformations and cracking. This involves making the distinction between stable and dynamic causes. The state of the building must be assessed to evaluate whether it requires additional support and whether the extent of the distortion warrants action. These are difficult features to appraise and the implications are wider than merely the structural health of the building. It is imperative that expert knowledge is brought to bear on such problems, and will require the services of a structural engineer.

Distortion and stability

Distortion is important because of its effect on the residual stability of the wall. Static stability is dependent on the gravitational forces of the self-weight and imposed loads of the wall across its height. The response of an element will depend on the degree of restraint offered by its shape or any connected element of the building. Bear in mind that it may be the connectivity with the rest of the structure that is instrumental in the instability being created.

Effective width and relative distortion

For a gravitational force the assessment of the wall is clearly discussed by a number of experts (Richardson 1988). The basic rule for design and assessment is the $t/3$ (or middle third) criterion. This is an expression of the effective width of the wall and the relative distortion. Essentially, if any part of the wall overhangs the wall at ground (or foundation) level by more than one-third the thickness it is considered unstable. This may require total rebuilding of the wall, or there may be a possibility of stiffening it further.

Bookending, overhangs and openings

The actual significance of overhangs between one-third and one-sixth of the thickness of the wall will depend on the degree of restraint provided by the remaining structure, and the potential restraint available for the modification of the structure. For walls to provide restraint they must be used for support in their own plane, and consequently walls parallel to a failing wall cannot be used for support. This is a common reason why the domino effect of a series of walls tilting occurs in the bookend type of failure.

The degree of restraint posed by walls normal to the subject wall will depend on their inherent stiffness. As discussed earlier, a significant feature of this is the proportion and positioning of any openings. Essentially, well constructed restraining walls are likely to be sufficient to allow the building to be corrected without necessarily involving total rebuilding.

Other general distortion criteria

Stable distortions below about one-sixth of the thickness overhang may commonly be considered suitable for leaving as they are, but obviously this will depend on the specific circumstances. The visual impact of distortion may have positive and negative features and both must be considered.

Verticality and twist

Definition

The literal definition of verticality suggests that no error from that of a straight line is permitted. In this respect most/all buildings are not vertical, and yet few collapse. The need for guidelines with regard to the acceptability of verticality must be related to the type of construction and its structural form. Verticality is of particular concern for walls and columns, although the distortion may well have effects on the level of floors and roofs.

Verticality is a basic need

The need for vertical elements is a basic requirement of many simple structural forms. The resolution of forces into horizontal and vertical components is an essential prerequisite for simple structural analysis. Where this cannot occur due to vertical distortion, the building may exhibit instability. Any element can exhibit lack of verticality. In older buildings elements should only be assumed to be vertical following their measurement.

Tolerance related to building type

Older timber-framed structures are generally tolerant of vertical distortion, indeed this may be part of their quality of appearance. Conversely, panelled wall and floor systems are less tolerant, since the bearing surfaces may be small and cantilevered fixings are less effective. Masonry has evolved standards for new work, which can be applied to existing buildings for comparison. A value of $\pm$ 10 mm in any 2 m is suggested as being reasonable for new standards. This may be acceptable for low-rise construction, but would be unacceptable for tall buildings where structural considerations are paramount. The dimensional tolerances of each building type must be assessed in the light of the structural and constructional implications. This can allow severe twisting to occur to low-rise masonry buildings, as shown in Fig. 2.3.

Causes

The causes of verticality deficiency in walls can be due to similar reasons to those causing cracking and horizontal deficiency. The settlement of foundations due to differential loading, or changes in the bearing capacity of the soils, can cause walls to distort. Changes in the groundwater table due

Fig. 2.3 (a) Severe twisting and lack of verticality to a two-storey masonry building. The rate of verticality is seen to change in the area to the left of the window, giving rise to a bulge with staggered cracking. **(b)** Alternative view illustrates the complexity of the twisting throughout the length of the crack. Note how the crack passes through the doorway, a weak point in the wall.

to trees and drainage can also be factors. The nature of the loading mechanism from internal elements, particularly where eccentric loading occurs, can cause walls to bulge and distort. Existing older methods to restrain the movements, e.g. metal flats and tie bars, may merely serve to induce distortion in previously vertical elements. The lateral restraint provided by roofs and floors may have failed due to inadequate fixings, or material failure. Overloading of the roof may induce spreading, which can distort the verticality of the supporting wall. Lack of verticality and twist can occur where structural materials have deteriorated to an extent where their structural integrity is seriously affected. This is shown in Fig. 1.8.

Fig. 2.3 (b)

Perception of verticality The eye is extremely good at discriminating fine detail, since the visual acuity of the average eye can be 0.5 minutes of arc (Pritchard 1978). This would suggest that we are able to readily assess the verticality of buildings. Unfortunately this is not so, since the influence of other objects in the field of view, and their detail, can influence our perception of the surroundings. Things which appear vertical may not be, and vice versa. Care is therefore required in any subjective observation.

Stability of the structure

Measurement methods

A comparison of the possible measurement methods are shown in Table 2.2.

Access to the elements suspected of being out of vertical is required. Theodolites do enable measurements to be taken from a distance, although aiming targets must be close to the surfaces to be measured, and calculations are required. The simple plumb bob is ideal in windless environments, but requires attachment to the structure and is not suitable for tall buildings. Heavy weights attached to string work as effectively as proprietary equipment. Providing access is available, both internal and external measurements can be made. The 'Giraffe' and its more expensive brother the 'Laser Giraffe' are patented devices enabling the vertical profiles of walls to be determined by one person (Richardson 1988). This is not suitable for measurements above 10 metres, where autoplumbs, or vertically orientated theodolites, can be used. For many tall post-war buildings these were probably the methods used for the erection of the building. The accuracy of the readings obtained are related to the care in setting up the measuring device. Many measurements can be involved to determine the vertical profile of a wall, and regular checks on the orientation of the instrument are essential. For isolated checks over approximately 2 metres, spirit levels attached to straight extension pieces of timber or metal can be used. This can be useful for preliminary measurements. All of the optical measuring devices must be regularly serviced and checked for accuracy, since they are liable to impact damage and drift.

Permissible lack of verticality

Walls which are free-standing are liable to collapse when they are leaning more than their thickness. In this case the vertical component of the load in the wall is outside its base and rotation could occur. The point at which

Table 2.2 Comparison of methods of measuring verticality and twist of existing buildings.

Methods	Range	Personnel	Calculations	Notes
Spirit level	up to 2 m	1/2	No	Isolated measurements
Plumb bob	up to 10 mm	2	No	Liable to wind action
Giraffe	up to 10 m	1	No	Use with theodolite on windy days
Laser Giraffe	up to 10 m	1	No	Expensive No wind problems
Theodolite	100 m+	2	Yes	Skill in setting-up and reading
Autoplumb	100 m+	2	Yes	Skill in setting-up and reading

structural concern at the lack of verticality should begin is suggested to be when the wall is more than one-sixth of its thickness out of plumb (Richardson 1988). This must be set into a context which recognises the influence of the construction method, and the surrounding structure. The one-sixth rule applies to masonry walls of low-rise buildings. Masonry walls of one-third of their thickness out of plumb are likely to require rebuilding. The measurements of verticality should also be used to determine the underlying cause(s) for the deficiency. Framed and panelled buildings demand a different approach, since the bearing surfaces and the methods of fixing are very different from masonry. Reductions in bearing area will increase the risk of shear failure of the panel toe, and may increase the size of the structural joint. The vertical fixings, commonly metal, may be intolerant of dimensional variations. Therefore panel distortions of 40 mm may meet the one-sixth rule, but have inadequate lateral fixity. Structural investigation may be the only way of accurately assessing the criticality of lack of verticality of these building forms.

Static and dynamic verticality

Structures built out of plumb have an inherent deficiency. This may be due to poor workmanship and/or supervision, perhaps involving dimensional errors. This lack of verticality may be static, and present few structural problems. A series of measurements, as described earlier, over a period of time will enable monitoring of verticality. This would be appropriate where structural analysis has confirmed that the current distortion is producing no distress in the building. Any measurements will need to be accurate, and must take into account seasonal and loading characteristics of the building. Any associated cracking should be monitored over a similar period.

Timber-framed and other laterally well restrained buildings are very tolerant of lack of verticality. They will distort in harmony with moisture and temperature movements and follow patterns of foundation movement. Poorly connected concrete-framed buildings may be less tolerant, due to reduced bearing areas, and an inherent lack of tolerance of tensile stress.

Oscillating verticality

The movement of buildings due to wind action is essential for the relief of high tensile and compressive forces. This will mean associated changes in the verticality of the building. Because the wind action is more pronounced at high levels, they are the areas most likely to suffer lack of verticality.

Forced vibration tests, using the CEBTP (Centre Experimental de Recherches et d'Etudes du Batiment et des Travaux Publics, Paris) equipment, can determine the oscillating performance of the building. The influence of non-loadbearing elements in providing rigidity to tall buildings can be considered significant, although this will depend upon its construction. Occupants can be questioned as to the movement of the building: the threshold of perceivable acceleration is between 30 mm and 50 mm per second. Low-rise buildings, particularly those in built-up environments, can experience a wide range of fluctuating wind pressures. The risk of significant oscillation is small since the natural frequencies are relatively large. Although gable end walls can fail in high winds due to being sucked out (Buller 1988), the action of its oscillation may contribute significantly to its eventual failure.

Internal symptoms   The internal walls can be measured for verticality, in the same way as the external surfaces. Differences between the internal and external profiles would require further investigation of the wall construction. Some cracking at the vertical junction between the external and internal walls may be evident. Timber floors built into leaning walls are likely to become out of level, with associated cracking along the horizontal junction between the floor and the flank walls. Diagonal cracks appearing in non-loadbearing upper-floor solid partitions of tall slab and column buildings, may be due to excessive wind oscillation. A spreading roof may cause cracking of the plasterwork around the junction of the external wall and the ceiling.

3 The influence of structural form on failure symptoms

Loading and stability

External and internal forces

The full range of actions that influence building stability includes external and internal forces (Rainger 1983). External forces encompass the conventional dead, live and wind loadings applied to structures, but also involved are pollution, solar heat gain, and changes in humidity. Vibration of the building structure from vehicles or other processes can be categorised as external forces. The forces are generally variable and may occur cyclically and in combination, as may the response(s) of the building.

Internal forces produce the chemical and physical instabilities occurring within materials. These may be initiated or accelerated by external forces, but are differentiated by their relationship with the inherent durability and stability of the construction materials.

The coincident external and internal forces produce a combination of responses in the building structure and its constituent materials. Volumetric change in materials is common, and deformations of the building structure may follow fatigue or creep. The conflicting nature and gravity of the forces may exceed the traditionally applied performance criteria, culminating in an instability in the components or even the total structure. See Table 3.1.

Static and dynamic forces

Static forces comprise the permanent loads of self-weight and other structure(s). Nowadays they are mathematically well understood and in most circumstances reasonably predictable or estimable. In general, designers are sufficiently familiar with the concepts and assumptions of structural design to produce buildings in accordance with the current standards. Comparing the calculations and checks required for a modern structure with those of 30 years ago reveals a greater emphasis on theoretical analysis.

Overprovision and redundancy

In comparison with modern practice, the design and construction of early buildings is somewhat cumbersome, characterised by overprovision and a redundancy in materials. This is not totally disadvantageous, however, as

Table 3.1 Extrinsic and intrinsic forces affecting buildings (after Rainger 1983)

Type	Cause	Time and duration	Unrestrained volumetric changes
Extrinsic	External climatic temperature changes Solar radiation	Exceptional Intermittent Seasonal Diurnal	– Expansion Contraction
	Ambient humidity changes Wetting Drying Moisture content changes	Seasonal Exceptional Short term Initial Alternate wetting/drying	Shrinkage Expansion
	Loading Dead loads	At time of erection Permanent	Immediate: elastic deformations Progressive: creep
	Live loading (applied)	Permanent or intermittent	Elastic and non-elastic deformations
	Wind loading	Intermittent	Deformations
	Vibrations	Recurring	Oscillations
Intrinsic	Chemical changes Loss of volatiles	Initial	Contraction
	Corrosion and sulphate formation	Continuous Progressive	Expansion
	Physical changes ice formation loss of moisture	Intermittent Recurrent	Expansion Shrinkage

endorsed by the large number of surviving buildings. The response characteristics of the over-engineered buildings may be expected to differ from the closely engineered modern structures, as the accommodation of distortion or tolerances of overloading will generally be greater, if indeterminate. The nature of construction of old buildings may produce structures more tolerant of distortion

Deformation and creep   Static loads produce progressive deformation of structures and their loadbearing components from the time of construction. The extent of creeping deformation depends on the nature of the load and the response of the structural material. For example, the short-term loading of structural timber produces elastic deformations. The same loads applied continuously over a long term, however, produce excessive non-elastic creep.

Distortion

The amount of distortion produced as a structure creeps, and the sensitivity to overloading and other abuses, will be related to the margin of redundancy in the individual structural sections. In small buildings the scale of loadings often produces relatively insignificant deformations, but with larger buildings there may be real problems with the long-term dimensional stability of the structure. See Fig. 3.1.

Dynamic loads

Dynamic forces consist predominantly of wind loads and stresses developed in the use of the building by its occupants. However, in some areas of the country there is the additional possibility of seismic activity or mining subsidence. Ground movements may be unpredictable or appear seasonally. Soils with a high clay content and reclaimed or filled land can produce a dynamically unstable relationship with the foundations.

Stresses may be induced directly in the building structure from airborne vibrations such as sonic shock waves, accidental or intentional impacts by solid bodies (see Fig. 4.8) or explosions. Extreme dynamic forces may induce residual instability in surviving structures. This is a particular problem with buildings surviving earthquakes.

Predictability and design

The designer of buildings can only consider the conditions likely to occur, and for economic reasons alone cannot design for the unpredictable. This is somewhat of an over-simplification, however, as it is important to assess the potential consequence of the force. The definition of deficiency centres on expectations during the lifetime of the building, and clearly there may be instances of unforeseeable failure unconnected to conventional deficiency.

Structural instability

In addition to the possible structural instability occurring due to external forces, other instabilities originate from the thermal and moisture movement of materials. There may of course be chemical incompatibilities overlooked during design or construction, such as placing aluminium where it will be exposed to water running off concrete or render. The cement in the concrete will produce an alkaline rainwater solution, which then reacts with the aluminium and causes surface pitting or erosion.

There may also be inherent faults in the durability of materials. The response of the envelope to the external environment and other less tangible agents of decay and dysfunction such as neglect, perhaps even change of use, will affect the durability and stability of the structure.

Fig. 3.1 (a) Distortion in a timber lintel as creep occurs under long-term loading. **(b)** Note the response of the structure above the opening and at the bearings for the lintel.

Concealed defects   Some types of defects are generally more visible than others. These are usually the easiest identified and the ones concentrated on in texts. The majority of inspections of buildings occur after completion and occupation, and usually after decoration. Since the very objective of finishings is to conceal the structural elements, this makes access difficult and can present a mental as well as physical barrier to investigation. Nevertheless, a thorough analysis may require inspection of concealed construction for verification of the cause of a symptom of deficiency. Examples of instances where this may be appropriate include suspected foundation failure, or inspection of floor joists and roof spaces.

Duty to follow the trail   Clearly though, it is not appropriate to open up all constructions in all
of defects   instances and for all types of survey. The decision will depend on the brief for the inspection and the surveyor's perception of the actual situation. The purpose and extent of the survey will dictate whether you need to study the

Fig. 3.1 (b)

whole building or not. Regardless of this, the analyst has a duty to follow the trail of a suspected defect, for example suspected dry rot or subsidence. Where an opening up is not made it will be important that the surveyor can satisfactorily justify why this was not carried out.

The expense of opening up   With older buildings and other deteriorating structures it is often impractical to make more than an approximate assessment of the structural character or its stability without resorting to extensive exposure. Extensive means expensive also, of course. Much damage can easily be done to a building in breaking open to assess the constructional form and stability. Surface damage may also be created which may be impractical to remedy.

Understanding the structure

Nevertheless, plastering can cover a multitude of sins in new and old buildings alike, and exposure of structures can reveal numerous and conflicting alterations to the original design. A thorough understanding of the structure is essential if the building is to be properly assessed and it is important to establish the scale of residual stability of the building. Instances have been recorded where opening up has revealed the building to be critically dependent on a single timber beam for its structural stability.

Inspection of the roof space

It would generally be wise to inspect the roof space. If the roof is inaccessible from the building the potential problems that could be concealed may be justification alone for inspection. An access slot may have to be made, subject to the appropriate authorisation. With trussed roofs it is wise to check for adequate cross-bracing to stiffen the structure against lozenging or racking during high winds.

Whilst in the roof space check the state of the timbers and their connections, and follow up any symptoms of spreading of the roof such as may occur following re-covering with a heavier material, for example the replacement of slates with concrete tiles. Aside from stability, assess the ventilation and degree of insulation, also any services routed through the void. Also make sure that the insulation in the roof is not blocking the ventilation routes.

Do not neglect the covering. A large number of houses still exist with their original asbestos cement slates. These were designed with a 30-year life expectancy, and many houses are having their roof replaced. This has an implication for the content of the surveyor's report on a building. One of the functions of the surveyor is to make an assessment of anything that will cost money now or in the foreseeable future (Legrand 1989). The likelihood that the slates will require replacement should not be overlooked.

Buildings with poor roofs have often been treated with a felt covering applied to the outside of the existing slate covering. This not only renders the slates unsalvageable, but may restrict the air flow which had existed through the gaps around the slates. It is possible for condensation to occur within the stagnancy of the roof space. Roofs that have been treated in such a way are externally obvious, and should suggest an internal inspection.

Local knowledge

There may be little indication externally of the likelihood of concealed deficiency. It may be valuable to check the neighbourhood for symptoms in other buildings which confirm or refute suspicions. Local knowledge is valuable here, and can help immediately identify high-risk areas, such as those previously used for tipping or mining.

Buttressing and restraint in sub-structures

Basement structures are also particularly difficult to inspect and assess. The degree of restraint offered by internal basement cross-walls is difficult to determine. They may be buttressing the external walls against soil pressure (Colomb and Jones 1980) and themselves buckling through acquired deflection in more than one plane. The stability of the basement structure may be at least partially due to the self-weight of the superstructure above.

Post-tensioned structures

There can be similar problems with the partial failure or progressive demolition of post-tensioned structures. These rely on the loading above to deform the elements to their working positions. Removal of the self-weight loading from above will cause the elements to resume their hogging shape. If this occurs rapidly, the stress shock may cause explosive failure of the elements, with instantaneous collapse.

Limitations of structural commonality

Clearly then, it would be fallacious to assume more than a minimal commonality of loading conditions in buildings. Dynamic forces in particular will be situation-specific. The effects of these latter forces will depend to some extent on the building design, and in a broad sense on the immediate environment. They may vary from the design requirements throughout the life of the building, as may the strength of the elements of the construction.

The inspection of buildings

The analysis of the external structure is not an exercise to carry out in isolation of the remaining inspection of a building. There are usually a number of clues that can be picked up from the internal inspection of a building about the state of the external envelope, particularly its completeness and alignment. The analysis of the envelope is more easily done with these indicators in mind rather than in the reverse order.

The procedural order of inspecting the property is usually governed by the personal preference of the surveyor, and of course the specific form of construction. House inspections are obviously different to commercial and industrial property reviews. This book is not intended to cover the reasons for surveys, or the methods of inspection and report, only the technology of deficiency itself.

What is of paramount importance is that the inspection is systematic and there are no omissions. It is a tiring and exacting process, often conducted with interruption, and a procedure will help minimise the time and effort, and avoid unnecessary repetition. The conventional form of report will also flow more easily from a systematic order of inspection, assuming good-quality notes are recorded.

Procedural orders Some surveyors start with the inside of the building at ground level and work up to the roof, then finish with the outside. This prevents the external appearance of the building from influencing the judgement of the inside or biasing the search, and also leaves the dirtiest part of the survey until late in the process. If taken to the letter it also prevents the general description of the building from assisting in focusing the surveyor's attention internally. Obviously there must be a degree of overlap, since in inspecting particular forms of property the surveyor will be alerted to any potential faults characteristic of the type of property or the local area. Any information appearing during the external inspection must of course be rechecked against earlier observations for corroboration.

Others advocate starting in the roof space and working down through the property. Following the general description of the property, the roof is checked first from the outside and then the roof space is inspected. This can be a valuable area to start since the minimal restraint usually provided at the roof level will make any structural movement symptoms most obvious, and the structure is usually fully exposed internally. The state of repair may give some indication of the attention the property has received generally, also whether it was built well in the first place (Melville, Gordon and Boswood 1985). It can also indicate whether there are any chimney breasts that have been clipped off at their lower levels. Check whether the number of pots on the outside of the building correlates with the number of flues, and trace the routes of the flues through the structure. In the buildings of the 1800s the number of flues can be extensive, and virtually every room may have one.

Moving down through a house, for example, the rooms can then be inspected in turn, working systematically through the ceilings, walls and floors. The doors and windows and internal finishings are checked in turn, and the process repeated for subsequent rooms. The external works and drainage inspections are left until later, subject to their readiness for inspection by the builder.

Elemental or functional? Most survey techniques emphasise the elemental features of the building rather than the performance criteria and functions. For example, stability is assessed in terms of the state of the structural elements and their structural connectivity, rather than overtly the opposite. This is of course at odds with the adage of 'form following function' which may have been followed during the design process.

It is not appropriate to further discuss the mechanics of survey procedure here; that is the domain of a conventional surveying instruction book. The most important aspects of the order of the inspection are that there are no omissions, and that there are no preconceptions. If the inspection system is too rigid this itself produces risks. The authors recommend the text by Melville, Gordon and Boswood (1985) for excellent and very readable comment on this.

Acquired and built-in defects

The types of structural defect may be categorised as built-in or acquired distortions (Richardson 1985b). The symptomatic evidence will be characteristic of the individual building.

Acquired deflection in structures

Distortion

The principal causes of distortion in the structure of buildings are overloading or internally developed instabilities. These may be produced by inadequate attention to the design or construction of the building, the materials used for construction, or by abuse of the structure. Further, internally induced stresses arising from moisture and thermal movement can cause deformations as the elements elongate or shorten. This will greatly depend on the capacity for the accommodation of movement within the structure and the connectivity between elements. See Fig. 3.2

Cracking and relief of stresses

Differential or creep-related movements in the structure may be associated with the self-weight of the elements, the nature of imposed loads, or unfavourable interactions with the soil. Figure 3.3 shows distortion in a bay-fronted terraced house following subsidence.

The symptoms represent the relief of otherwise unresolved forces, which have produced damage within the structure. The alternative quiescent response is the containment of the forces within the structure. It should be appreciated that this is the normal response of correctly designed and constructed buildings. Buildings which exhibit no stress cracking do not necessarily have balanced internal forces.

Evidence of overstressing

Evidently, buildings with cracks have suffered excessive stressing at some time. There may be residual stresses producing a progressive fault, and this possibility must be considered carefully. On the other hand, the relieving effect of cracking or deflection may be complete and the fault stable. To further complicate the issue, the threshold at which stressing becomes

Fig. 3.2 overview of
(a) sagging arches and
(b) sagging lintels.

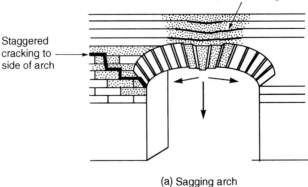

Centre of arch has sagged causing cracking to brickwork above

Staggered cracking to side of arch

(a) Sagging arch

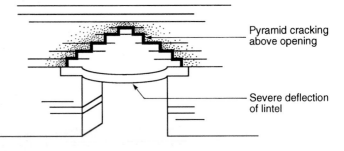

Pyramid cracking above opening

Severe deflection of lintel

(b) Sagging of lintel

excessive depends on the materials and form of construction. Indeed, the manifestation of any stress-relief symptoms will be a unique response of the constructional form. The structural and material relationships will be fundamental to the occurrence and resolution of stresses, and the appearance of faults.

Stability and progression

Firstly, it is obviously essential to correctly identify the related symptoms and isolate them from any red herrings. Recent cracks, for example, will have clear faces compared with older cracks, which will have dulled or become stained with pollution. In extreme cases this can provide a useful distinction between stable and progressive movement. Close inspection of painted walls, for example, can indicate whether the cracks were present when the coating was applied or have occurred subsequently.

Characteristic trends

Some types of defect produce characteristic trends in symptoms, such as

Fig. 3.3 Distortion and cracking relief in a stone building following subsidence. The bay movement is characteristic of its partial independence, and it is probably poorly founded. The building required demolition.

disjointing at the connections of walls with a deflection of the frontage. Evidently the age and style of the property can suggest to the experienced viewer possible mitigating features. Some system-built properties are renowned for particular types of defect, and the symptoms frequently follow a predictable and easily interpretable pattern. These features can be a useful guide, and trends of related symptoms may suggest further areas of investigation to corroborate or refute possible defects. It is important not to allow the suggestion of a possible defect to dictate to the objective analysis of all symptoms.

Generations of cracking and evolution in response

Crack shape and uniformity can provide valuable clues to the direction of movement and the nature of the forces at work. This is discussed later. The building may have a history of movement or distortion unknown to the surveyor, and cracking symptoms may be apparent from several generations of stress relief. The development of the symptoms may be in response to a progressive change in the structural nature of the building, rather than a change in the fundamental cause of the defect itself.

Location of stresses

Cracking and restraint

The location of stress is an important feature to assess in the analysis of symptoms. Stresses will produce cracking, bulging or differential movement at the weakest point in the local structure. There is also the possibility of cracking adjacent to zones of restraint, where the movement is not accommodated and the tensile forces exceed the strength of the material. See Fig. 3.4.

Fig. 3.4 Mid-span detail of row of garages. Expansive forces in the brickwork have been relieved by the cracking and lifting of the concrete roof slab at its centre.

Stress concentration   Stress concentration will be dependent on the form of construction. Any reduction in the thickness of the fabric produces a stress concentration which overloads the material locally. The connections between different materials will frequently form the concentration of movement-related stresses, particularly within a single elevation. Typical examples include a variety of claddings and curtain wallings to framed structures. A possible distortion pattern of the members in a curtain walling system are shown in Fig. 3.5.

All claddings require careful detailing to avoid damage under movement. Joint failure is likely. String courses in clad elevations will concentrate the

Fig. 3.5 Possible distortion patterns of a curtain walling system. (Reproduced by kind permission of R. Michael Rostron).

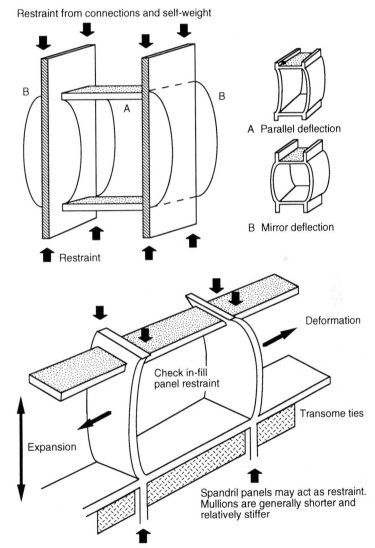

Restraint from connections and self-weight

B   B

A

A Parallel deflection

B Mirror deflection

Restraint

Deformation

Check in-fill panel restraint

Transome ties

Expansion

Spandril panels may act as restraint. Mullions are generally shorter and relatively stiffer

Fig. 3.6 Tapered tensile cracks to parapet wall. Restraint at the base of the parapet wall produces partial dimensional control. Expansion occurs in the exposed and relatively unrestrained head of the parapet. Cracking is similar to hogging failure.

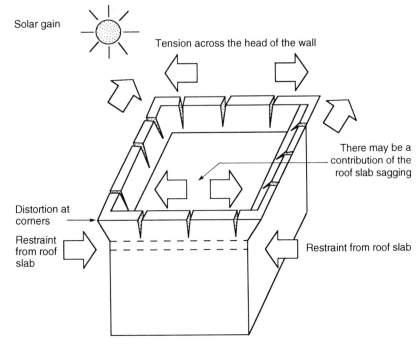

Solar gain

Tension across the head of the wall

There may be a contribution of the roof slab sagging

Distortion at corners

Restraint from roof slab

Restraint from roof slab

movement in the frame. Shear failure may occur at the junction between frame and cladding.

In the case of rigid brick cladding enclosed in a concrete frame the contrasting expansion and shrinkage produces tensile failure in the frame or bulging of enclosed brickwork as its free movement is restrained. An alternative effect may arise where unrestrained walling, such as a parapet, is built over a framed structure. The walling may fail in tension as the frame restrains movement, producing tapered cracks at the top. See Fig. 3.6

Cracking can also occur in the parapet wall as the supporting roof structure sags and produces tension in the upper edge of the parapet. This can easily be avoided by the inclusion of extra reinforcement in the upper (tensile) zone of the parapet, although this assumes the potential problem is predictable.

Connections between structures

Restraint and movement　Connections between two buildings of different rigidity or mass, or between an existing building and a new extension, produce stress concentration. Similar problems can occur at the connection of floors to a shear core. Multiple shear cores and rigid end walls exacerbate the problem greatly, since there is restraint in more than one location. Adequate and uninterrupted movement joints are particularly important in the structure spanning between pairs of shear cores. The height of adjoining buildings will be an important factor, since buildings of different height generally differ in mass also. Self-weight can be an important restraining feature and the short, lightweight building will be more likely to move. Movement will be relatively unrestrained and so occur freely.

Stress concentration   Changes in sectional thickness will concentrate stresses. These may be produced by changes in constructional form or where chases are inserted in the walls. Detachment cracking and differential movement will be concentrated at the junction between heavy or stiff components and the weaker or flexible elements. Adjoining structures of significantly different mass or rigidity can undergo differential settlement or movement. Internally induced forces in the dominant structure, for instance thermal or moisture movement, can produce symptoms in the minor building. Within a structure this can also be manifested as cracking adjacent to piers in walls, which act as localised stiff zones and form points of restraint for the concentration of stresses.

Differential movement   Differential movement can also be produced where damp-proof courses create slip planes. These transfer the forces through the structure poorly, consequently differential movement occurs across the plane of the damp-proof course. The degree of oversailing will require assessment to ensure that it is neither progressive nor unstable. This can usefully limit the extent of damage to the structure.

Butt ends of walls may be poorly connected. This was a common feature of Victorian terraced construction, where the frontage was constructed as a single facade across the previously erected party walls, indicated by the patterns of bulging. Bulging walls which appear to reattach between dwellings imply some degree of connectivity with the party walls.

Where butt walls are well connected, any movement in the direction normal to the span of the butting wall will not be accommodated. The wall will have little tolerance for such sideways movement and either will distort and crack around the junction or a simple cleavage will occur between the two walls, indicated by a vertical uniform crack in the internal corner.

Figure 3.7 illustrates the effects of connectivity and restraint.

Strained relationships   Differential movement occurs between zones of distinctly different structural context, such as sections of a building of different scale or mass or shape; between structural elements of differential scale and loading or restraint; and within thin laminated forms with variable stiffness or dimensional

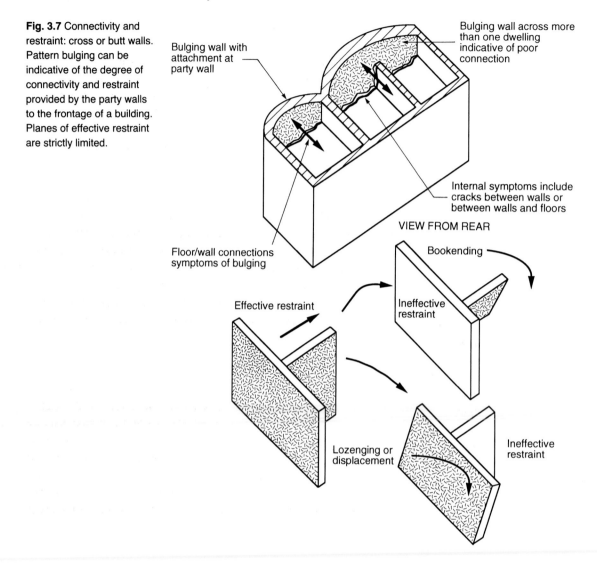

Fig. 3.7 Connectivity and restraint: cross or butt walls. Pattern bulging can be indicative of the degree of connectivity and restraint provided by the party walls to the frontage of a building. Planes of effective restraint are strictly limited.

stability and coefficient of deformation. There is a commonality of strained relationship.

Failure may be due to omission of the movement facility, either because of a lack of understanding somewhere within the design and construction chain, or because of ignorance of an incompatibility. The latent defects and other related legislative steps of the 1980s do not encourage the contractor to point out the error in design or suggest alternatives (Fernyhough 1989a, 1989b).

The analyst surveying the structure must establish the potential within the structure for restraint and free movement, and compare this with the potential differential responses of the structural and fabric materials.

Movement and structural form

Sources of movement

Sources of movement in buildings include external temperature and (solar) radiation effects. Others include thermal movements of concrete, moisture-related reversible and irreversible effects, elastic deflections under loading, and creep. With taller structures this may extend to wind effects including sway, and the relevance of mining or other seismic forces. The soil and building interaction is important and can lead to foundation failure.

Movement, restraint and relief

It is not always the case that lack of provision for movement or differential movement in buildings will lead to failure. In many instances induced stresses are restrained by the constructional form, or sheer weight of the building, and this can hide potential deficiencies in loading and distribution until failure criteria are met. The traditional masonry structure is less effective at concealing minor failures than some of the composite structures, but may have a degree of redundancy and other constructional characteristics that allow it to retain structural stability under partial failures, despite producing cracks that may unnerve the public.

Inertia and stability

Buildings with a high thermal inertia will respond to rapid fluctuations in a moderate manner, and for this reason the durability and relative dimensional and thermal stability of the older traditional forms of construction can be particularly advantageous. These possess a thermal stability and a minimal physical-movement response to rapid or inconsistent changes in environmental temperatures. This of course is not always beneficial and there may be distinct disadvantages in the areas of thermal performance and environmental response. Both of these are discussed later in some detail.

Accommodation of distortion

In addition to the thermal response nature of traditional forms of construction, these forms of building are usually of a soft construction. The materials, and in particular masonry structures, and the mortar jointing materials will accommodate a considerable degree of distortion before failure occurs. See Fig. 3.8.

Soft masonry may crack profusely, and where these cracks are minor there may be little detriment to the appearance of the wall. They may only be obvious on close inspection. In contrast, the brittle forms of brickwork bound using the harder and more resilient cement mortars have a reduced tolerance of movement. Cracking may be the sole relief of stress and so is usually more obvious. The masonry indicates greater instability from a similar type and extent of failure than the traditional soft forms.

Fig. 3.8(a) Accommodation of long-term and progressive distortion of a soft masonry structure. This example was exacerbated by the storage of motorbike spares on the upper floor.
(b) The paintwork and construction around openings reveals a history of distortion. Note how the door has been framed for compatiblity with the deformation in the surrounding wall.
(c) The plastic deformation of the soft masonry walling has produced a bulging in the front elevation.

The coefficient of expansion of materials is also an important feature which distinguishes the nature of thermal stability and response of the building to environment fluctuations. Traditional buildings are constructed from relatively few basic materials, all from similar sources and with approximately similar responses to thermal and moisture variations. Certainly the difference in response is less marked when taken in comparison with the modern forms of construction and materials.

The older buildings with soft constructions will probably have accommodated early settlement and irreversible expansion of the masonry materials well, and there may be little in the way of obvious external deficiency or symptoms. The exception to this is the traditional material timber, and there is ample evidence of the realisation and accommodation of

Fig. 3.8 (b)

differential movement characteristic of the older properties. Those which are prone to failure will probably have done so by now if the failure is to be by a plain expansive defect. Progressive distortion is obviously another matter. See Fig. 3.9.

Thermally lightweight buildings

The modern building construction form is generally thermally and structurally lightweight. This makes it significantly more efficient, and if

Fig. 3.8 (c)

properly designed it is more likely to load the ground and its supporting structure more evenly. Internal failure is not likely in such instances. However, capacity of the average building to accommodate ground pressures or other external or internal forces it was not designed for may consequently be more limited than its traditional counterpart. This is not a suitable area for further generalisation, but is a point worth bearing in mind in the context of analysis of the differential symptomatic response of buildings to apparently similar deficiencies.

Fig. 3.9 Coursed stone structure exhibiting considerable distortions without necessarily being in imminent danger of collapse.

Modern structural response

In general, the modern structures are structurally more responsive to the environment and the ground conditions, also to changes in use and loading and the relationship with adjacent buildings. Their temperature stability is in some notable instances more sensitive. The structure may be discontinuous and reliant on the interaction between the framed construction and a superimposed envelope bearing and transferring its own weight and imposed wind loads only. The change in emphasis on design produces a change in emphasis on failure.

Modern material response

The modern materials of construction are generally stronger than their predecessors and produce more efficient components. Advances in structural engineering and construction technology have allowed a much more efficient exploitation of their total loadbearing capacity. As a consequence, however, buildings may deflect more under safe loads. The more brittle and harder forms of construction are less tolerant of a marginal failure and partial distortion. The forms of construction are varied and widely ranging. Their intimate inspection is difficult, either because the construction is of a laminated and hence partially concealed form, or because the variety and scale itself may be unfamiliar to the analyst.

Material developments

Greater deflection requires greater design accommodations. The performance of the whole building relies on the interrelationships between the individual and hugely variable materials making up its structure. Intimate composites such as glass-reinforced cement (GRC) and laminated-contact forms such as insulated cladding have evolved.

If any development makes a significant impact on the design and construction of buildings, there is the risk it will influence the occurrence or type of defects also, even if only through uninformed misuse.

Design developments

During the changeover from one form of construction to another the state-of-the-art buildings will have exploited bastardised or embryonic forms of design. A number of failures have arisen in such contexts, generally attributed to design innovation. It is debatable whether this is sufficient justification or excuse. The benefits of advancement are readily identifiable and more widespread than the drawbacks. Also, if innovation is to be criticised, where do you start?

A feature of modern construction is therefore the inherent variability of materials and constructional combinations. Consequently so too are the types of movement and other common forms of failure.

Combining rigid and flexible forms of construction

Constructions combining rigid and flexible materials which are inadequately designed to accommodate the differing degrees of deflection can produce profuse problems. A common example is block internal walling built on top of *in-situ* reinforced concrete flooring. The two-way spanning floors which deflect relatively more than the single spanning floor forms pose greater potential deflection problems. To complicate this, the speculative production of buildings for letting frequently involves leaving the design and positioning of partitions until late in the process. The designer may have no knowledge of the type of use of the floor and hence the distribution of loadings and may have little choice other than to make general assumptions.

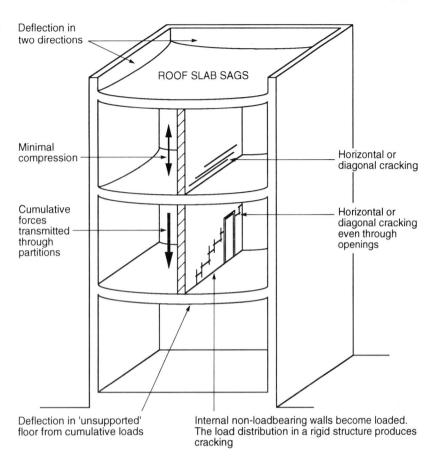

Fig. 3.10 Deflection of floors and distortion in internal walls in a multi-storey structure. Non-loadbearing rigid internal partitions distort and crack as deflection of the roof and floor slabs occurs and imposes loading on them. Lower floors are not subdivided and excessive deflection occurs from cumulative transferred loads.

Deflection in two directions

ROOF SLAB SAGS

Minimal compression

Cumulative forces transmitted through partitions

Horizontal or diagonal cracking

Horizontal or diagonal cracking even through openings

Deflection in 'unsupported' floor from cumulative loads

Internal non-loadbearing walls become loaded. The load distribution in a rigid structure produces cracking

Localised overloading of floors in use will produce excessive deflection which the relatively rigid partitions can neither tolerate nor accommodate without visibly cracking. Deflection of the floor occurs also as the concrete creeps.

Deflection and curvature

Internally, curvature deflections of floors or beams supporting rigid partitions can lead to tapered cracking in the lower sides of the partition as it bends and tensile forces are produced at the floor level. These may extend up as far as the neutral stress plane, normally situated approximately half way up a simple panel. There may also be horizontal cracking in the partition approximately one course above floor level. The mechanisms of these symptoms are discussed later in more detail.

The partial loading from the deflecting floors will be transferred to the walls, which can cause further problems. Where the subdivision of the floor spaces is not repetitive, lower-floor slabs without subdivisions below them may deflect more under the cumulative loadings transferred through the internal (non-loadbearing) walls. Cracking in the partitions deriving their support from the lower floors may be greater than at the upper floor levels. See Fig. 3.10.

Movement and deformation

Reasons for movement problems

Lenczner (1981) has noted a number of reasons for the increase in movement-related problems in buildings. He suggests that the use of higher working stresses for materials with a similar elastic modulus compared with the traditional materials is partly to blame, because this produces greater flexure. Also the materials are being used in larger spans and slimmer forms of construction, and taller buildings are exposed to greater dynamic loading. In addition to this the move to prefabricated jointed buildings, together with the emphasis on thermal insulation and heating, can exacerbate the movement.

Structural implications of movement

Movement of buildings in relation to their surroundings may have direct structural implications or merely produce a building that is aesthetically displeasing or unserviceable. Depending on the use this may be a fine distinction and either may be unacceptable.

Differential or relative movement

Differential movement is more destructive and problematic. It also produces a combination of symptoms that in themselves are problematic and difficult to analyse. Indeed, it is the mere variability in form that is one of the prime barriers to simple analysis.

Relative movement is likely to produce destructive damage at the interface between components or elements. This produces structural instability in the former case, and unserviceability in the latter. With box forms of construction, structural instability may be caused by failure of the infill components serving also as structural elements. Legislative steps to limit the potential consequences of these forms of constructional instability at the design stage have been implemented in multi-storey buildings constructed after 1973. The key concerns over the structural stability of the box forms of building include the quality of workmanship and supervision rather than the designs alone. In particular, there is recorded evidence of, and residual uncertainty over, the absence of critical structural connections between panels. Some of the poor workmanship associated with connections was related to over-optimistic design or fabrication concepts (BBC 1984).

Deformation and load transferral

The transfer of load between components is determined by their relative stiffness and the degree of restraint adjoining the elements present. The general form of relative movement occurs in a number of well-known areas such as differential movement between insulated and uninsulated forms of construction, deformation and differential movement within laminated constructions largely due to uneven temperature distribution and/or the variable response of different materials. As the non-traditional forms of construction have become traditional to the UK, this type of problem has become better understood.

Movement joints

The inclusion of movement joints can be a potential source of problems as well as a potential solution (Alexander and Lawson 1981).

Deformation of materials

The elastic deformation of materials is usually in direct relation to their Young's modulus. Young's modulus is a statement about the elasticity of a material. It relates the amount of physical or dimensional straining in a material to the level of force producing it. For an elastic material, within its range of elasticity, this is assumed to be a constant, and is represented by the ratio of stress to strain. The elastic deformation may finish at the stress limit and there will be disproportional straining with stress as the materials deform, the essence of ductility.

Elasticity, bending and deformation

Young's modulus of elasticity can be used in conjunction with the moment of inertia of an element and the area of the bending-moment diagram to calculate the anticipated deflection. This can be verified against the visible movement to estimate or verify the compression of columns under axial loading for example, or the sagging of beams under self-weight and imposed loads.

Shortening and bulging of columns

In addition to a reduction in length of a column under axial loading, it is to be expected that there would be a small change in sectional thickness as the element bulges in compression, in accordance with Poisson's ratio. If there is delamination or degradation of a reinforced concrete column, this bulging may be excessive and indicative of structural instability.

The external symptoms may include cracking of plaster finishes or breaking up of the surface of the column. Disruption may appear in lines parallel to the predominant reinforcement. This is to be distinguished from bulging of the column initiated by deterioration of the reinforcement, such as may occur where the cover to the steelwork is inadequate. Internal columns in benign environments are not likely to suffer this, however.

Shrinkage in elements

Shrinkage in the constructional materials may not be uniform. Perhaps there was non-uniform compaction of the concrete, producing a variable load capacity across the column. The element may be non-homogeneous, with the various components exhibiting different movement characteristics and loading response.

Bending straining

Columns may exhibit bending straining. Creep may vary throughout the section or between different columns also. Further, the loads may be eccentric, such as occurs at the eaves of a portal frame or a traditional pitched roof. This can produce spreading of the roof and damage to the perimeter walling, which is usually ill-equipped to restrain the movement. Where the structure does not bend, it may fracture when the stresses exceed the capacity of the materials. See Fig. 3.11.

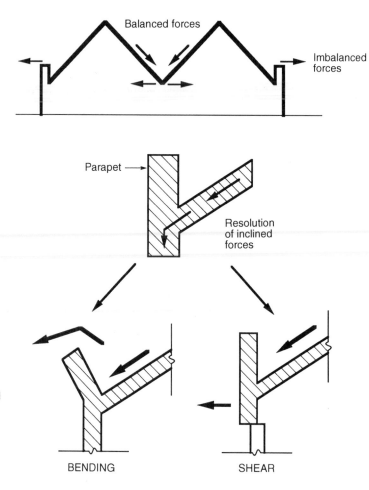

Fig. 3.11 Imbalanced forces at the eaves of pitched roof. Failure in the supporting wall may produce either shear cracking as the roof moves horizontally, or bending in the unrestrained parapet walling.

Flexural loading   In the case of beams bending conventionally under flexural loading, there will be a neutral plane above which the beam will shorten under compression, and below which it will lengthen marginally under tension.

Tension in elements   Materials undergoing tension will lengthen, and there may be some necking of the cross-section. Whether this is measurable will depend on the magnitude of the force and the capacity of the element.

Non-elastic behaviour   Many of the common constructional materials are not elastic, however. In concrete, the main brittle material used in construction, the elasticity modulus is dependent on the mix formulation and the aggregate strength. The modulus of elasticity also varies with age, and there is a contribution of creep. In fact there are a tremendous number of variables that may affect its elastic modulus.

Creep occurs where the loading on the concrete lying within the stress capacity causes a gradual deformation with progressive strain. This is obviously a long-term effect and is distinct from the short-term fluctuations in loading which generally produce elastic (reversible) deformations within working capacity. There is an additive effect of creep and shrinkage in concrete, however (Rainger 1983), and the total movement that may be experienced in a concrete structure is of serious proportions (Rainger 1983).

Visible and encapsulated straining   Materials such as concrete and masonry have a less distinct stress at which ductile failure occurs, and there is virtually no stress capacity in tension. In contrast to the traditional soft masonry constructions, their brittle nature produces cracking failure rather than a compliant deformation. The limit of strain capacity dictates the transition from encapsulated strain to visible straining, or relief through cracking of the brittle materials. Once again, there are different symptoms from similar causes.

Deformation in masonry   In masonry the elastic properties will depend on the softness of the jointing (lime or cement mortar generally speaking) and the compressive strength of the bricks. There is a wide range of elastic moduli between different types of brick (Alexander and Lawson 1981).

Timbers and elasticity   The elasticity modulus of timber is anisotropic and highly variable between species and across different sections of the wood. There is a great sensitivity in most timber to relative humidity and this will affect the creep properties also.

Restraint and connectivity

Distortion and restraint

The degree of restraint offered by the structure will dictate the symptomatic response to the movement forces. Corners and other changes in direction are important. Where a long wall with a short return expands or moves wholesale the shorter return will tend to rotate. This is particularly likely if it is tied at its remaining end. If the return is short (approximately 700 mm) it will probably crack, since it will not be long enough to accommodate the rotational movement. Longer walls will tend to rotate at the corner formed with the adjoining long wall. The long wall will tend to bulge outwards from the partial restraint offered by the return, producing tension in the skin of the wall. In extreme cases these tensile forces may produce vertical cracking. Figure 3.12 shows these possible distortion patterns.

Shrinkage, connectivity, and restraint

Drying shrinkage or other moisture-related movements of large span elements such as cast *in-situ* roof or floor slabs or beams will cause deflection of other parts of the building structure, as the conversion of stresses into movement is not restrained, but is nevertheless transferred through connectivity. Where the slab or beam bears on intermediate supports the degree of connectivity, its self-weight, and the frictional resistance to movement this creates, will be sufficient to drag the upper part of the wall with the slab as it moves. Unless there is some design provision for this movement, any intermediate supports to the slab or beam will be distorted. The symptoms in supporting walls running normal to the direction of shrinkage are likely to be horizontal cracks a few courses below the connection, produced as the wall rotates between courses. See also Fig. 10.1.

Distortion and floors

A potentially serious problem can arise with cantilevered floor and roof slabs. If these distort and deflect under loading any other components supported on them will deflect in turn. This may produce a deflection and cracking in walls and cladding panels deriving their support from the slab. Walls supported on the cantilevering floor or beam will fail by hogging, as the deflection causes compression at the base of the wall and tension in the top. Symptoms of cantilevering failure at the perimeter of the building from a deflection of a floor slab may be similar to a sinking of the building at its edge.

Fig. 3.12 Modes of failure and the influence of restraint. The relief of movement stresses occurs through cracking in short elevations of walling if there is no capacity for rotation, or through rotation of the shorter elevation if there is capacity for rotation. Bulging and vertical tensile cracking may also occur, usually in the elevation directly affected by the movement stresses.

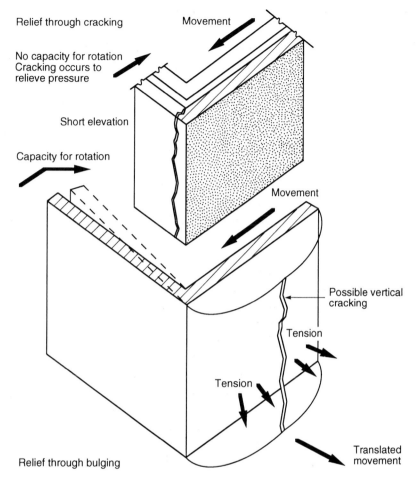

Relief through cracking

Movement

No capacity for rotation
Cracking occurs to relieve pressure

Short elevation

Capacity for rotation

Movement

Possible vertical cracking

Tension

Tension

Relief through bulging

Translated movement

Openings and stress concentration

The influence of openings on stress distribution

Other stress concentrations include openings in walls. Minor openings will produce local concentrations of stress. Cracking and differential movement will occur here because of the weakness in the structure. The transference of load around openings will mean that the walling immediately below the opening carries relatively little loading, whilst the walling separating openings carries a considerably higher loading. The redistribution of forces produces shear cracking and differential movement. There is an effect on the bonding of masonry walls. Weaknesses in the bonding are likely to occur at any structural discontinuity or where the loading is reduced surrounding windows. See Fig. 3.13.

This concentration of stresses into bands between openings where the construction is continuous occurs both horizontally and vertically. The

Fig. 3.13 An example of the stress concentrations around openings acting as a focus for vertical and diagonal cracking.

comments earlier about concentrations of loadings at changes in materials are also important. If there is variability in the materials within an elevation it is likely that any stresses will be relieved as cracking at the connection of materials and near the openings.

Openings and rigidity

Where there is a high proportion of openings in the wall there may be a significant loss of rigidity. The wall will bulge relatively easily and offer reduced resistance to cracking. This may have direct consequences for its response to loading, or may be a secondary feature contributing to the movement of the remaining structure because of the minimal restraint it offers. The slenderness of the wall will be contributory, as will be the relationship under loading between components in a laminated or cavity construction.

Exposure and stress location

Environment and exposure

Highly exposed locations in the building will produce extreme conditions. This will influence the deterioration of materials and the degree of movement associated with environmental changes. Parapets are a particular risk area because of the position of the building and the fact that they are exposed on both sides to the weather and so lack the usual buffer afforded to inner leaf construction. Figure 3.14 illustrates such deterioration of joints and failed material.

Fig. 3.14 Single brick parapet wall to garages showing early signs of joint and material deterioration characteristic of exposed brickwork, exacerbated by movement in the structure.

Stability of the structure

Prevailing winds   The direction of the prevailing wind will affect the weathering of the external envelope. This will cause cooling and water concentration on local facades, increasing the risk of weather-induced stress. The wind action will influence the concentration of water runoff from facades. This will load joints and influence the distribution of moisture movement stress.

Location effects   The siting of the building can be classified in accordance with an exposure index (BRE 1971b), and this should be related to the performance of the external materials. The improper selection of materials and components may result in premature failure or general deficiency. The siting of buildings within intimate groups of other buildings may give extremes of exposure or protection. These micro-effects may change within the life of the building, placing new and excessive loads on the construction.

Height and prominence   The weathering loads on structures can increase with height. As the wind loading increases, the effects of driving rain become more severe. Secondary effects of induced building movement due to extremes of temperature and oscillation may induce tensile and compressive loading. Although the water runoff rates will be reduced at high levels, roofs and parapets are vulnerable to dimensional changes. Lifting forces on roofs may reverse the normal loading stresses.

4 Instability: the role of external forces

Wind loading

Reasons for wind loading

The UK is situated in a temperate climate belt, and is subjected to a range of wind speeds due to the general movements of the climatic airstream. This is capable of sudden variations, as the gales during January 1986 and October 1987 demonstrated. In general terms, the prevailing wind is from the southwest. Since the UK is also an island, coastal and near coastal areas are subjected to sea-land breezes. When both of these factors are considered, it becomes essential to allow for the effects of wind on buildings. Local features around the buildings may be very different from the generality, and a failure to recognise their influence may create deficiencies in the building. Early building forms, erected on rule of thumb and experiential knowledge, would either collapse in the next severe wind, or resist until overloading occurs.

Historical data

Wind speed and direction have been measured over many years, and are the basis of design recommendations. The first British Standard Code of Practice relating to wind loads was issued in 1944. Since they are based on historical data, the guidelines are only as accurate as the last storm to be included. The action of gusting wind has been recognised as a major factor in damage to buildings, and new recommendations (BRE 1989a) have considered this effect.

Influence of location

Location will influence a range of externalities which act upon any building. Some small buildings may be completely sheltered from winds, and their probability of failure due to wind action will be very low. The sheltering of buildings can be due to natural features, artificial features, or surrounding buildings. The geographical location will also determine the intensity of wind loading. This is shown in Fig. 4.1 (BRE 1989a), although a range of modifying factors are needed to convert basic wind speed into wind loading. The diagram shows contours of reference basic hourly mean wind speed for the whole of the country. Local climatic differences can produce

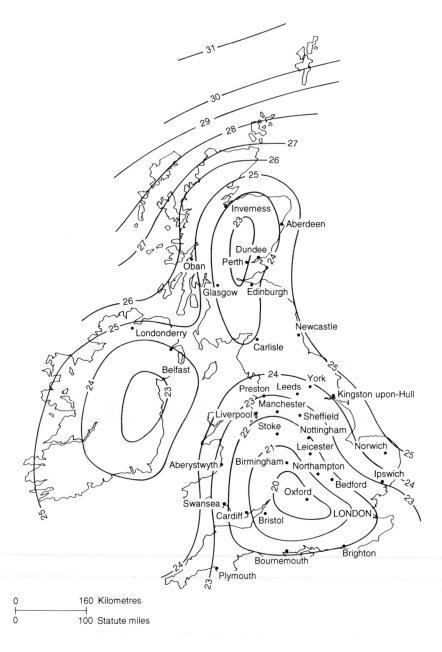

Fig. 4.1. Map of the United Kingdom showing the reference basic hourly mean wind speed in metres per second. (*Source*: Figure 1: BRE 1989a, (Crown Copyright 1989).

Fig. 4.2 (a) and **(b)** Airflow
patterns over simple building
for one wind direction.
(c) Wind-induced pressure
pattern from airflow shown
in **(a)** and **(b)**. (Reprinted
from Marchant E W 1989 by
kind permission of Elsevier
Science Publishers Ltd).

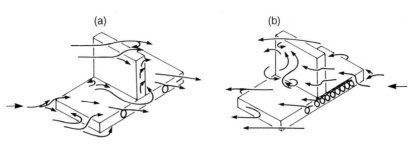

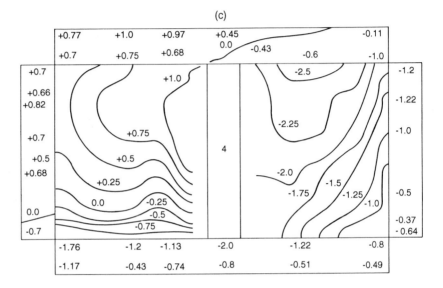

special problems, however, and careful examination will be required to
identify correctly the cause of the damage.

Influence of orientation

The wind patterns around buildings will create negative and positive
pressure regions. See Fig. 4.2. Corners and changes of direction can be
vulnerable. Where the orientation of the building presents large surface
areas to the prevailing wind, direct overloading can occur. This can be
accentuated by the movement of air around the building, causing negative
loading on other surfaces.

Both effects can cause defects, with the latter likely to detach items from
the building's surface. Aerodynamic modelling of buildings may be required
where they are likely to behave dynamically. These buildings can be defined
as being sufficiently stiff to prevent galloping instabilities, although wind
damage such as deflection cracking could occur (BRE 1989a). Framed
buildings with minimal structural cladding or loadbearing walls offer little
damping compared with masonry structures, and require specialist
investigation.

Stability of the structure

Influence of surroundings

The topography of the surroundings will modify the wind flow. Hills are obviously windy regions, but to assume that buildings in valleys are sheltered from wind-induced defects is an oversimplification. Wind patterns are recognised as being complex (BRE 1989a), and care when establishing causes of damage is required.

Built-up areas will slow down the general wind speed at low level. This may reduce the risks for buildings within the same height group, providing localised effects are not generated. Once above the surrounding buildings, the wind speed will continue to rise and therefore upper regions of taller buildings are vulnerable, regardless of their surroundings.

Influence of adjacent buildings

Buildings can affect the windflow around them. Tall buildings can cause pedestrian discomfort by diverting the high-level cool air down to ground level. They can also influence the wind pressure on adjacent structures. Passageways and external corridors between buildings can be subject to high wind speeds, creating negative pressures on wall and roof surfaces and affecting finishings and coverings. This was dramatically illustrated by the collapse of some cooling towers*.

Overhangs and projections are also liable to high loadings. The location of external doors in positions of high local wind speeds will cause increased wear and tear, and they may be severely damaged. The concentration of wind speed within groups of tall buildings is highly variable. Wind flow may alternate unpredictably between stagnation and turbulence.

Structural implications of wind loading

The maximum pressure exerted upon the external faces of buildings is related to the maximum wind speed, and in particular its gusting strength (BRE 1989a). This fluctuation in loading will cause a corresponding straining of the building. This may lead to defects in structural components and wear on fixings. The negative loading may be sufficiently large to cause the complete detachment of external features and elements. It is difficult to assess the wind loading resistance of flat roofs due to their construction, although they are particularly vulnerable to suction. When roofs are blown off framed buildings, this can lead to instability of the freestanding walls, which may then be blown over or otherwise collapse (Buller 1988). Masonry structures are considered to be well damped (BRE 1989a) and more resistant to wind loading. The taller framed buildings are likely to be less damped, giving greater amounts of movement. Cracking and movement of the structural frame may occur, particularly in the upper storeys.

* Collapse of Ferry Bridge power station cooling towers in November 1965. This was caused by wind pressures developed as a venturi effect occurred between the towers, sucking them in.

Non-structural implications of wind loading

Concrete non-loadbearing facing panels may be sufficiently massive to resist negative wind loading. The overall movement of the structure may cause some minor movement between the panels. This may lead to joint failure, and the ingress of water. The metal fixings may then be vulnerable to corrosion. These fixings are difficult to inspect, due to the height of the building and the lack of information concerning the construction method. Small-scale lapped roof coverings can move under wind action, causing the fixings, or the material to wear. Failure may lead to the units being blown away, or falling into guttering. Detached or partially detached renderings, or other thick applied finishes, may also be sucked off due to wind action. Gable ends appear particularly vulnerable since many were affected by the gale of October 1987 (Buller 1988). Windows may have their glass blown in, or under negative pressure effects, whole windows can be sucked out. Inadequate fixings or material failure can be typical causes. The regions of negative pressure around the external envelope may cause finishings to be sucked off.

Oscillation effects

Although many buildings move under the influence of wind, the movement is only considered critical when the structural performance of the buildings is threatened, or the movement is perceptibly uncomfortable for the occupants. This movement appears to be perceptible when the acceleration of the building is between 30 and 50 mm per second. Although the degree of flexibility of various structures has been approximately defined (BRE 1989a), where tall framed buildings are the subject of analysis, specialist advice should be obtained. This will usually involve external inspections of the structure. Many tall masonry buildings have been oscillating for many years, e.g. Church bell towers, and although they require regular inspection, many are performing adequately. Horizontal and vertical cracks in the masonry may open and close under the influence of wind, or varying loads.

Response of structural form

The response of the structural form to wind effects will increase with the wind loading. Very high winds have the potential to damage all above-ground structures, and in this respect all structures are vulnerable. Masonry structures, particularly low rise types, are well damped and are tolerant of oscillating loading effects from the wind (BRE 1989a). Lightly clad framed buildings are more flexible. Chimneys are special structures, but due to their shape and height, can experience oscillation. The massing of the building, its shape and height configuration, will channel wind around it. This can produce localised high loadings and/or oscillations. Buildings in exposed positions will require resistance to high winds and other hygrothermal effects.

Fig. 4.3 (a) Wind damage to the flat roof of a three-storey block of flats. The building is located at the edge of the estate and wind approaches across a large expanse of open ground. The roof was torn along the eaves, and large sections of the covering were blown off. **(b)** the damage was made worse by the fact that the bonding of the roof covering to the insulated substrate was poor and so provided a cleavage plane once the edge lifted.

Symptoms   A general deterioration of the structure may make the building vulnerable to wind-induced instabilities. Rain penetration can cause fixings to corrode severely. The attached elements may then fail due to moderate wind loading. Parts of badly weathered masonry may become detached. These can become a hazard to pedestrians and cause impact damage to the structure. Horizontal cracking, or the opening of joints on facades facing the wind flow, may indicate flexure of the building. These may also be seen on side walls. Small unit coverings may be missing, or blown into gutters. Ridge tiles are positioned in windy areas and are a good measure of wind loading. Few have positive fixings, and many rely on self weight to remain in place. Inspections following high winds, during the winter months, will enable some symptoms to be seen in relative isolation. Large areas of sheeted flat roof coverings can become detached in high winds. Examples of this are shown in Fig. 4.3.

Fig. 4.3 (b)

Snow loading

Historical data The national records of snowfall and its distribution are part of the work of the Meteorological Office. These records have been kept over many years and are used to determine snow loads on buildings. Although the exceptions are well remembered, snow does not normally lie for weeks or months. However, any significant snow-load failure is likely to cause major environmental discomfort to the building user. The build-up of snow depth that can occur during these periods will increase loading, in a manner which relates to the shapes of the lying snow. The standard design process (BRE 1988b) has used Canadian and Russian data, since their winters are more severe. Local weather stations will have extensive records of all meteorological data, and these can be consulted where symptoms of snow-related defects are suspected.

Influence of macro-climate The UK is situated in a temperate climate, and severe winters are rare. A measure of the national variability of snow loading is given in Fig. 4.4 and

Fig. 4.4 Basic snow load on the ground, at an assumed altitude of 100 metres with an annual probability of exceedance of 0.02 (*Source*: Figure 1: BRE 1988b, Crown Copyright 1988).

although these relate to snow on the ground 100 metres above sea level, they are useful to determine coarse regional variations. Design-life considerations for buildings traditionally disregarded global weather variation. There are fears now that global warming may produce a UK climate resembling that of the Mediterranean. Snow loadings may reduce also.

Influence of micro-climate   The specific locality of individual buildings may mean that the general snow-loading patterns do not apply. Shelter from adjacent buildings or other topographical features can reduce loads. The effects of buildings upon wind flow has been mentioned earlier, and winds can affect the disposition of snow layers on building surfaces. This may blow snow away, or produce drifts and thickening layers. The wind speed around the upper storeys of tall buildings will be faster and less turbulent, producing differences in snow loading (Mitchell 1976) to those areas in the lower, more turbulent windflows.

Building shape and configuration   Where wind can blow snow into corners and against walls, considerable depths are possible. The junctions between roofs and parapets, as well as structural walls are vulnerable. Flat roofs may allow the snow to remain lying until the thaw is complete. Normally, guttered pitched roofs allow the snow to slide off, when the frictional resistance between the snow and the roof is sufficiently low. This weight of descending snow may tear off weak guttering, and cause secondary impact damage to conservatories and other structures at low level. This is shown in Fig. 4.5.

Horizontal and vertical effects   Snow can accumulate at the edge of roofs and around the perimeter of buildings. The drifts will impose variable loads on the structure, which may be considerable. For general design purposes it is assumed that undrifted snow has a weight density of 1.47 kN/m^3 (BRE 1988b) and that this is likely to rise to 2.0 kN/m^3 for drifted snow. These loads can generate deflections which may lead to joint or material failure. Flat roofs are particularly susceptible. As progressive deflection occurs, the build-up of snow will increase. The structural implications of the increased loading can be serious. Walls are generally more resistant, although windows and doors are vulnerable. Outward-opening elements may become inoperative. Parapet walls may be pushed over.

Where highly textured facades exist, perhaps containing ledges and crevices, snow may be able to lie for some time, particularly where sheltered from the direct sun. The loading effect may be negligible, although staining can occur. Snow can be blown horizontally and this can mean that where no secondary barrier exists, snow can be driven through lapped coverings into roof spaces. Poorly fitting windows and doors, weepholes and defective construction around openings in walls and roofs allow snow to breach the primary defence.

Ice formation   The reduction in density which occurs when water turns to ice has the practical effect of causing unit quantity of water to expand. This expansion

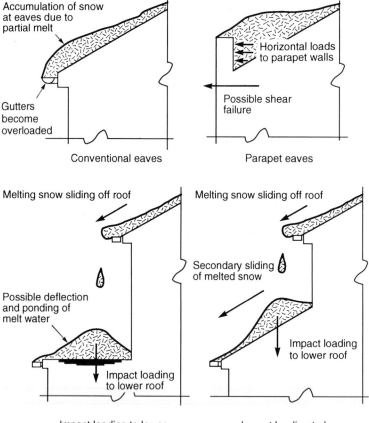

Fig. 4.5 Snow profile at pitched roof eaves, and risk of secondary impact damage to lower roof or structure. Accumulation of snow and risk of melt water ingress at parapet wall.

is disruptive to porous materials and a more detailed explanation is given in Chapter 8. Freezing cycles during the winter months, can allow successive snow layers to accumulate and allow frost attack to proceed. The local temperature will be related to the degree of solar exposure, and the wind chill factor. North-facing roofs with heavy solar shading may be at greater risk. The joints of lapped coverings can become areas where ice settles; it may also accumulate around the range of defects associated with this type of covering.

The influence of insulation

The temperature of the external envelope will depend on the level of thermal insulation and also the amount of heat in the building. Poorly insulated or overheated buildings will tend to be snow-free, or the snow loads may be very low. Problems with over-loading are more likely to occur where the structure is highly insulated and exposed, or under-heated.

The effect is useful as an approximate method for establishing the existence of roof insulation, since it can be seen from the ground. It also further complicates the basis of prediction of snow loads (Mitchell 1976). Where buildings are well insulated, wind action may become a significant factor in determining the amount of snow load.

The hygrothermal link

The effects of snow on buildings is complicated by its change of state when in contact with the building. Dry snow will load the building; wet snow can cause a range of defects associated with its hygrothermal performance. Any snow that enters the roof space can melt, causing damp penetration through walls and ceilings. The base layer of snow lying on pitched roofs can melt, and where the depth of snow is sufficient, water can directly enter the roof space. Accumulations of snow and ice on the felt underlay may overload it, causing bulging to occur. This may reduce its effectiveness, allowing water to move through poorly lapped or damaged joints. Drifted snow may be above the level of the DPC in solid walls, or those which are rendered, breaching the primary defence. This can also occur above flashings to parapet walls, or other walls around flat roofs.

Symptoms

There are many signs of defects associated with snow loading, and due to the link with hygrothermal performance many are similar to those associated with water entry. Guttering may be broken, due to load from thawing or drifted snow. This may render the guttering ineffective, causing rainwater spillage. Thawing snow may also cause secondary impact damage to surrounding structures. Trees adjacent to buildings can shed snow in a similar manner. Flat, or low-pitched roofs may have deflections caused by snow loading. This can cause ponding which generates more loading, causing more deflection, and so on. Joints to flexible sheeted coverings can be subject to frost attack when buried by snow layers. Melting snow blown into roof voids, or behind lapped covering units, can cause a range of dampness problems to the wetted interior. During long periods of cold weather this may soak roof timbers.

Drifted snow may bridge DPCs and enter walls though weepholes and poor joints to openings. Horizontal cracking or movement at the roof/parapet wall intersection point may be indicative of loading failure. Northern regions are more likely to suffer from severe snow loading, but local orientation, the height of the building above sea level, and adjacent buildings can also influence the amount of snow on any building. However, few existing buildings require structural modification.

Seismic shock

Mechanism

The UK is relatively free of earthquakes and the resulting effects associated with seismic shock. The few serious tremors that do occur are very newsworthy, but in global terms are considered small. One earthquake of 2.5 (approximately) on the Richter scale occurs once per month in the UK (British Geological Survey 1990). The movement of one section of the earth's crust relative to another is the simple description of inter-plate activity earthquakes. The UK is not close to plate boundaries and suffers from intra-plate earthquakes. These earth movements are complex, and are assumed to propagate a variety of wave forms. These movements can occur rapidly, sending wave patterns through the rock to the surface. At this point they impinge upon buildings. The waves can be classified (in a simple way) as Primary (**P**) when they are longitudinal, exerting a push-pull effect, and Secondary or Shear wave (**S**), when they are transverse and exert a side-to-side effect. They each travel at different speeds (P 8 km per second, and S 4.5 km per second) (Verney 1979), which also varies depending on the material that they pass through. Except where the building is close to the epicentre, their arrival times at the building are different.

The Richter scale

The use of definable scales to measure the intensity of earthquakes is essential. Observation of damaged buildings is useful, but must contain limited scientific rationale. The scale devised by Dr Charles Richter in 1935 (revised in 1977) (Verney 1979) and in common use, is open-ended. It measures the vibrational energy in a shock wave. Since it is a logarithmic scale, an increase from 1 to 2 on the Richter scale is a ten-fold increase in energy.

Historical data

The British Geological Survey has approximately 100 seismic measurement sites in the UK, and they provide reasonable coverage for the country. They constantly record and monitor any activity, and are a useful source of current and historical information. The earthquake on the Lleyn peninsula North Wales, on 19 July 1984, recorded 5.4 on the Richter scale and is the largest tremor this century (British Geological Survey 1990). This occurred at a depth of 24 km. After-shocks continued for several years. The tremor on 2 April 1990 centred close to Shrewsbury was at a depth of 14.3 km and recorded 5.1 on the Richter Scale. The data gathered by the survey are used to determine the probability of occurrence of future earthquakes. A 5.4 scale earthquake is likely to occur in the UK every 160 years.

Activity areas

Because the UK is situated in areas of intra-plate seismic activity, the locations of earthquakes are variable. Although there appears to be evidence of recent activity in the North Wales region, tremors could occur anywhere. The depth of the epicentre is another factor which must be considered, since deep seismic activity will be less damaging than shallow activity. The shock waves produced can be reflected from underlying rock surfaces, or can be refracted through them. This will affect the size of the activity area. The location of the epicentre is based on the difference in time between the arrival of the **P** and **S** waves at three or more measuring points. This enables accurate positioning to occur (Verney 1979).

Nature of loading   True vertical movement of the ground will apply compressive loads to the building. The resistance of compressive loads is a feature of most buildings, and if this were the only loading applied then many buildings would perform well. Unfortunately the vertical movements are dynamic, causing variations in compression load. A dynamic horizontal component is overlaid onto these loading effects. This is rather like shaking a jelly on a plate. Since the building may be less responsive to the earthquake-induced movements than the ground, at some point they may both be moving in the opposite direction. This will generate horizontal shearing forces within the building structure.

Duration of loading   The duration of seismic activity is another factor in the analysis of its intensity. Where times are extended a greater amount of structural damage can result. The peak effects of UK tremors are less than 10 seconds, although the build-up and after-shocks may last for hours or even years. They may not be continuous during this time, but have individual time scales of peak activity.

Structural effects   The structural effects of UK earthquakes, compared with those in other more active regions of the world, are minor. It is not reasonable to assume that a certain point on the Richter scale marks the start of structural damage; too many other factors are involved. Buildings will shake, and this can cause cracking to structural elements and finishes. In framed and panelled buildings, cracks of 45 degrees in external ground and lower-storey panel walls can occur (Whitley-Moran 1978). Shear failure of the lower storeys is more likely the taller the building. Long buildings may suffer the most severe damage in that part furthest away from the direction of the shock wave. The cracking failure of internal elements in parts close to the shock wave appears to absorb some of the load, reducing the amount of damage (Whitley-Moran 1978).

Non-structural effects   The shaking of the structure may cause loose or poorly fixed items to become detached. Roof tiles and slates are vulnerable. Projecting chimney stacks of masonry construction can fall over, causing secondary structural damage to the surrounding roof. Where already suffering from sulphate attack or other defects, the inherent resistance to load may already be reduced. Glass cracking, due to structural deformation, will only occur where the movement permitted by the glazing is exceeded.

Performance of building types   In the UK there is little need for specialised constructional forms or foundations to resist earthquakes, except for the most sensitive of building uses. Indeed, for framed buildings an adequate structural performance for wind loading appears to be a reasonable guide for earthquake resistance (Whitley-Moran 1978). The traditional loadbearing masonry construction is not good at resisting significant earthquake shocks. It is relatively inflexible and liable to cracking. Where buildings have a structural frame, they perform better, although an increase in height and an irregular plan form may reduce resistance. The ground and lower-floor column frames are at risk from shear distortion, the lozenge effect (Whitley-Moran 1978), particularly where lateral restraint is minimal. The traditional, and new variants of timber-framed construction can absorb much of the energy in the shock wave. They are able to distort to a considerable degree, providing that they are not contained within rigid construction forms. Rigid finishings are vulnerable to cracking.

Identification   The incidence of earthquake damage to buildings in the UK is relatively rare. It can manifest itself anywhere, and little warning is generally available. When inspecting buildings the local seismological recording station could be a source of information. The structural symptoms of earthquake shock, i.e. cracking finishings, structural cracking, and major distortion, are similar to those produced by a range of other effects, and these should only be eliminated following careful examination and investigation.

Vibration

Vibration mechanisms   Vibration is used to describe an oscillating movement back and forth. This is analogous to simple harmonic motion, and many of the associated terms have been used to describe the vibration of buildings. Structures have a natural or fundamental frequency,* related to their general dimensions and construction, which exerts a damping on the oscillations. Where the vibration has a similar frequency then resonance[†] of the structure could occur, giving rise to large displacements and therefore movements. Where movement of a building exists, this can cause distress to the structure and its finishings. The study of vibrations is specialised and has a scientific basis similar to sound and acoustics (Porges 1977).

Causes

The sources of external vibration commonly associated with buildings are usually outside the control of the building user/designer/owner. These may come from the ground, be due to traffic, railways, tunnels, and other underground sources. They can also be generated by wind action upon the building; further details of this are given in the section on wind loading. These can be termed transient vibrations (Beards 1983). The 'steady state' vibrations are normally associated with moving machinery sited inside the building. This particular type of vibration is outside the scope of this book.

Effects on loadbearing structures

Traditional construction methods appear to endure the long-term low-frequency vibrations of road traffic well. Where the structure has been subject to attack from other sources, the weakened areas may be displaced by the additional vibrations. Long-term vibrations may cause cracking of structural elements. Tall buildings which are poorly damped will exhibit greater amounts of movement. This can cause diagonal cracking in the walls of upper storeys. Where busy roads and railway lines are in close proximity to the building, traditional construction methods may not be structurally or environmentally suitable.

Influence of form and structure

The vibration-induced loading will be at a theoretical maximum where resonance of the structure occurs. This will depend on the construction, shape and configuration of the building. The degree of tolerance to vibration will be similar to that identified for the resistance of fluctuating wind loads (BRE 1989a). Tall buildings of loadbearing masonry construction are less resistant to vibrations than buildings with a structural frame.

Settlement and movement

The constant vibration of buildings due to external sources may cause settlement, since the building is acting as a crude compactor of the bearing

---

* Fundamental frequency: Where the transmitted force is limited only by the damping of the structure.
† Resonance: occurs when a body is excited by a periodic force at its natural frequency.

soils. Since this may proceed over many years, some measurable effect may occur. This may vary with the soil type and relative proximity to the source. Slight movements in the building may be structurally inconsequential, but can alarm the occupants.

Effects on non-loadbearing structures

The vibration of the building shell and structure, due to the above sources, can also cause vibration of the non-loadbearing structures within it, since the vibrations will be passed around the building in a manner dependent upon the degree of connectivity of the structure. These may create an obvious secondary acoustic effect within the building since they may have little damping. Movement caused by the vibrations will impose extra loads on the joints, and may cause distortion.

Symptoms

Buildings subject to ground vibration should be considered as suspect. In general the first symptom may very well be noise. Severe vibration can cause cracking in upper storeys. Where the structure is severely weathered, falling debris may be due to a combination of decay and vibration. Rigid cladding systems and glazing may also crack, due to vibration movement. Underground vibration from tunnels may not be immediately apparent, although a thorough site investigation should identify it.

Mining activity

Mechanism of mining subsidence

The basic scientific theory of mining subsidence has been refined several times since commissioners in Liège, Belgium in 1839 concluded that mineworkings deeper than 100 metres would not affect buildings at ground level. The recent theory considers the use of mining factors, site factors and structural factors to determine the amount and effect of subsidence due to mining (Shadbolt 1975). These are summarised in Table 4.1.

The mining factors will cause a range of movements at ground level, and these are related to the depth and position of the mineworkings. These effects can now be accurately predicted (NCB 1975, Whittaker and Reddish 1989). The prediction methods become less effective when the type of mining and its position are unknown. This has a direct relevance to the incidence of mining damage of existing buildings.

Locality effects

In coalmining areas the subsidence is likely to occur directly above the seam, providing that this is reasonably level. The width of the subsidence area is approximately half the depth to the workings, and this tends to

Table 4.1 Main process factors which influence the extent of mining subsidence. (*Source*: Bell F G (ed.) and Shadbolt C H 1975 *Site investigations in areas of mining subsidence*, Newnes-Butterworths)

Mining factors	Site factors	Structural factors
Vertical subsidence	Near-surface rocks	Size
Differential subsidence	Types and distribution of	Shape
Curvature (differential	engineering soils	Design of foundations
tilt)	Changes in hydrogeology	Type of superstructure
Horizontal	Faults in rock formation	Methods of construction
displacement		Quality of materials
Strain extension		Age
Strain compression		Standard of maintenance
Differential		and repair
displacement		Purpose of use of building

localise its effects. The front edge of the subsidence wave is at a point on the ground surface at an angle of 35 degrees normal to the seam. A representation is shown in Fig. 4.6.

As the wave passes along the ground surface, the ground slightly rises, tilts and lowers, then levels out at a depth equal to the total subsidence. During the tilting phase there is an initial extension (or tension) of the ground surface, a point of contraflexure, then a compression of the ground surface. This wave will hit, pass through and then leave behind buildings at the ground surface. The overlying rocks above the seam must deform in a similar manner, and in the case of sandstone this has caused cracks of 0.5 metre (Shadbolt 1975).

Rate of movement   The rate of ground movement will be related to the speed of working of the seam. The response of the overlying rock strata and the soils will also

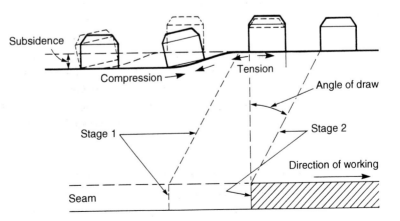

Fig. 4.6 Wave-like effect of mining subsidence (*Source*: F G Bell (ed.) 1975).

control the rate of movement. Rates are normally modest and allow a certain amount of stress distribution to occur within the subsiding building. This may not be the case where crown holes or pillar failure occurs (Bell 1975). These methods may be associated with mining other than coal, and originate from older workings.

Predictability

The subsidence from National Coal Board workings is very predictable, and is the subject of detailed records. These will cover much of the existing working areas, and records of older workings may also be available. The maximum amount of subsidence associated with seam working is 84–90 per cent of the extracted thickness. Where the mine is stowed, or backfilled, this amount can be halved (Shadbolt 1975). The records relating to very old mine workings become increasingly inaccurate and unavailable. Local authorities in mining areas can have expert knowledge and hold detailed records. Occasionally the only time that they are aware of an old mine is when the subsidence occurs.

Effects on structures

Subsidence will affect different structures in different ways. A uniform vertical displacement of buildings may not seriously affect their structure, although external connections to it may be severed. Mining subsidence may induce differential movement and this will create additional tensile and compressive loads. The tilting phase will place considerable loads on tall buildings. Any tensile and compressive loading of the structure may produce cracking in brittle materials. The degree of tolerance to mining subsidence varies between each structural form. An over-reliance on using the degree of movement of the structure as the only measure of damage, ignores this inherent variation. Rigid claddings or areas of glazing can fail due to minor dimensional changes of the structure. In the more flexible structures, including portal frames, severe damage may be limited to the rigid finishes. The degree of resilience of the structure may mean that after the passing of the subsidence wave, the structure is safe though suffering from some distortion. The structural integrity of the buildings should be assessed where the parameters for structural instability have been exceeded.

Effects on non-loadbearing structures

The external areas adjacent to buildings suffering from subsidence are likely to be displaced and loaded in a similar manner. Drives, roadways and car parks may become depressed, and entrances to buildings may become mis-aligned. Freestanding walls in external works may behave in a laterally unrestrained way. Underground service connections can suffer from similar effects.

Fig. 4.7 Mining subsidence of a pair of semi-detached bungalows. The new brick surround to the doorway on the right is the only part of the front elevation that is even nearly vertical.

Symptoms   Depressions in the ground around areas of mining are obvious symptoms. Rigid buildings may be distorted in a gross manner, as shown in Fig. 4.7.

The inherent strength in compression of certain loadbearing structures reduces the risk of compressive failure. Where they do occur the top of a building is vulnerable; here roof tiling may buckle, or be seen to overlap. A horizontal cracking of brickwork accompanied by vertical bulging may also occur. Some generalised symptoms are shown in Table 4.2.

Tensile cracking may penetrate the foundations and walls. The crack width will reduce with height, and the accompanying wall movement will affect the shape of openings. These are similar defects to those associated with foundation failure. Masonry walls can remain standing, even with a wide range of structural defects. These can collapse suddenly, and present a serious safety hazard where pedestrians or other properties are at risk. Severe symptoms could resemble those of explosion damage, although elements are more likely to collapse inwards.

Care is required when determining the cause of any subsidence, since it could be attributed to many other factors. The locality, and its history can be of assistance but this should not overly influence a rational appraisal of the possible causes.

Table 4.2 National Coal Board classification of subsidence damage

Change in length of structure (mm)	Class of damage	Description of typical damage
Up to 30	1. Very slight or negligible	Hair cracks in plaster. Perhaps isolated slight fracture in the building, not visible on outside
30–60	2. Slight	Several slight fractures showing inside the building. Doors and windows may stick slightly Repairs to decoration probably necessary
60–120	3. Appreciable	Slight fracture showing on outside of building (or on a main fracture). Doors and windows sticking, service pipes may fracture
120–180	4. Severe	Service pipes disrupted. Open fractures requiring rebonding and allowing weather into the structure. Window and door frames distorted; floors sloping noticeably; walls leaning or bulging noticeably. Some loss of bearing in beams. If compressive damage, overlapping of roof joints and lifting of brickwork with open horizontal fractures
>180	5. Very severe	As above, but worse, and requiring partial or complete rebuilding. Roof and floor beams lose bearing and need shoring up. Windows broken with distortion. Severe slopes on floors. If compressive damage, severe buckling and bulging of the roof and walls

Accidental impact damage

Forms of building attack

Buildings are subject to change from a wide range of different external agencies. Accidental impact to the structure can produce a range of defects which range from insignificant to structurally damaging. The accidental impact damage associated with people and vehicles can produce such a range of defects. Figure 4.8 shows the impact damage to a solid one brick wall by a motor vehicle, which suffered only moderate frontal damage. In order to be fully developed, this classification should involve an appraisal of the relationship between buildings and people. This is outside the scope of this text, which is concerned with physical causes and their effects on the building.

Fig. 4.8 The impact damage to a solid one brick wall caused by a motor vehicle which suffered only moderate frontal damage, much of which was probably caused by the roadsign post.

Impact on the structure

Casual impact of the structure by occupants and moving components will generate wear. This may be structural wear or fabric wear. Structural wear can occur at different rates for different locations within the building. The resistance of the structure to live loading will be flexural and the wear generated should have an insignificant effect. Major changes in this loading, as with a change of use, can have direct structural impact. Sudden impacts occurring within the building from live loads can be severe, and relate to the use of the building. Sudden impacts to the structure from outside, as in the case of motor vehicles, can be catastrophic. This is recognised as one of three broad types of accidental loading of buildings (Moore 1983). The vehicle may become lodged in the building, and offer temporary support (Mainstone 1978). This type of impact will usually only occur to the lower storeys of buildings.

Lightning

Lightning striking a building can cause a considerable amount of damage.

Stability of the structure

There will be an amount of mechanical damage from the energy in the strike, and there may also be additional damage to cabled systems within the building. In major strikes both types of damage may occur. The considerable forces released are capable of splitting masonry and punching through coverings. A simple reliance on the provision of a protection system may not ensure total protection, as demonstrated by the lightning strike on York Minster in July 1984 which caused considerable damage to the south transept. A fire risk is likely to be present with any lightning strike.

Impact on the fabric

Exposed structures are more vulnerable to impact damage. Where the impact loading is severe and localised the fabric can become punctured, leading to deterioration of the primary and secondary defences. Fracture shards or failed finishes may become detached, creating a risk of secondary impact damage. Impact on the fabric from within may cause similar effects, but the subjective deficiencies of appearance and decoration may assume greater significance.

Human attack

The human occupation of buildings brings with it the risk of accidental damage. Impact loading from falling and moving objects coming into contact with the structure can cause severe defects. Where buildings are accessible to the general public, the incidence of damage to claddings appears to increase (Anon 1986). See Fig. 4.9.

Contributory factors

The selection of materials is often made without reference to their resistance to vandalism. There are many possible causes for a poor resistance to accidental impact damage. The inability of the materials to accommodate impact loads may be due to their inherent deficiency. Brittle or inflexible materials may permit a rapid propagation of cracks. Ductile materials may have a low resistance to impact damage, yet may absorb large amounts of impact energy. If the deformation is severe, the material may have little residual flexibility.

Mechanical attack

Many accidental impact loadings are forms of mechanical attack. These may cause distortion where loading is insufficient to produce cracking or failure. External metal claddings can exhibit this effect, due to their elasticity. Where brittle materials are used in a layered form of construction, the impact may generate horizontal and vertical bond breaking between the interface layers. Direct localised mechanical impact may perforate several layers.

Fig. 4.9 Vandalism to six-storey curtain walling adjacent to a public car park has destroyed the primary defence layer of the building, exposing the fixings and the secondary defence layer.

Locality effects Major circulation areas within buildings are high-risk areas. External facades which are close to pathways, roads and drives, are vulnerable to vehicle and pedestrian attack. Traffic accidents may cause further impact damage, involving load shedding (some of which may be toxic), tree felling, or flying debris. The relative levels of traffic ways may allow parked cars to roll into walls. Shop windows and other glazed areas offer little resistance and are, by definition, sited on circulation routes. Where hard durable materials e.g., mosaic and ceramic tiles, are used adjacent to public areas they can also suffer from damage (Anon 1986)

Explosions

Resistance criteria The main structural considerations for buildings are the ability to sustain dead, live and imposed loads. Explosive loading is a specialised form of load, and for many buildings this is resisted by the inherent strength provided by allowing for a normal range of loadings. The risk of explosions in buildings is small, and the forces are such that major collapses are rare. The major types of explosions in buildings are gaseous (Mainstone 1976). In simple low-rise dwellings damage is normally limited to the source building, with secondary effects spreading to the surroundings. The risk of accidental loadings in non-residential non-industrial buildings is approximately three times greater than for a dwelling (Moore 1983).

Explosive loading

The nature of explosive load upon the inside of a building depends on many factors. The combustion of the explosive mixture in the room will proceed at the flame front, and in general this is not likely to move outwards at speeds greater than 10 metres per second (Mainstone 1976). This may increase in rooms which are structurally confined, or where the explosive mixture is dense and in an unconfined space. This development of the flame front will generate a rise in omnidirectional pressure, and this can expel the explosive mixture through vents, or even blow out weak parts of the structure. This is shown in Fig. 4.10. These effects will reduce the damage in the room. The peak pressures developed are highly variable and depend on the burning speed of the flame front. The peak pressures developed in the hall of the flat where the Ronan Point explosion occurred have been estimated at 80 kN/m^2. Obstacles and turbulence of the explosive gas mixture will increase the rate of combustion and the pressure developed.

Explosion-resistant buildings

The need to provide civilian buildings with significant explosion resistance is rare. Specialist requirements may apply to buildings of government agencies, rail terminus buildings and scientific installations. Buildings involved with the processing of explosives demand special treatment, and could be termed explosion resistant. In these cases it is essential that a suitable structural form is constructed to resist the possibility of widespread and progressive collapse of the building. The risk of flying debris causing secondary impact damage should also be considered.

Structural damage

An internal explosion will generate pressures on all surfaces. Where distances are small the pressure waves will arrive at most surfaces at the same time. This will apply a normal configuration of loading on floors, an abnormal loading configuration on ceilings, and lateral loads to walls. This may cause major structural damage to certain elements, particularly floors which are generally assumed to have little tensile load in their upper surfaces. Where venting of the explosion occurs the pressure patterns on surfaces will be significantly altered. Figure 4.10 shows the possible blowing

Fig. 4.10 The influence of weaknesses in the external envelope on the consequences of internal explosions.

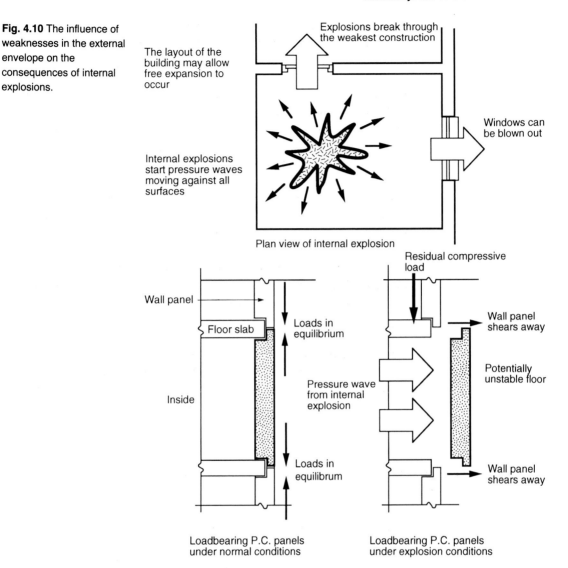

The layout of the building may allow free expansion to occur

Explosions break through the weakest construction

Windows can be blown out

Internal explosions start pressure waves moving against all surfaces

Plan view of internal explosion

Wall panel

Floor slab

Inside

Loads in equilibrium

Pressure wave from internal explosion

Loads in equilibrum

Loadbearing P.C. panels under normal conditions

Residual compressive load

Wall panel shears away

Potentially unstable floor

Wall panel shears away

Loadbearing P.C. panels under explosion conditions

out of a loadbearing storey-height precast concrete panel. This will reduce the general pressures on surfaces within the building, although causing an increase in the loading to the floor which it had supported. Pressure differences will become more marked in certain areas. Where a difference in pressure exists between two sides of a wall or floor, damage is likely. Where the venting or configuration of the room allows, the pressure difference may be very small since the explosive pressure is acting on both sides of the element at virtually the same time. The blowing out of weak areas, e.g., doors and windows, may reduce the pressure sufficiently to reduce the incidence of structural damage. Where confinement of explosions occurs then structural damage can be more likely since there is a slower decay of peak pressure (Mainstone 1976) and the duration of the explosion is effectively increased.

Accommodation factors

When a gaseous explosion can expand into free air, the generated pressures are much reduced. This will occur when the explosion has removed sections of the enclosing building. It can also occur to a limited degree when the layout of the internal spaces does not limit the explosion to one room. The layout of buildings is based on the needs of the occupants and use of the building, and this may not be commensurate with the minimisation of explosion risk. This may occur by default, where kitchens usually have windows, and may open into larger dining rooms through lightweight doors. Temporary structures in spaces (including furniture) can create turbulence which will increase the speed of the flame front (Mainstone 1976). This will increase the pressure developed.

External explosion sources

Gaseous material can enter the building from outside, either by ambient air movement, or through the ground. Methane from the ground, either naturally occurring or from landfill regions, can enter buildings through fissures (Grant 1989). Porous soils and services connections present a possible route into the building. There are more than 800 landfill sites in the UK where methane could be a problem, and depending on the pressure developed the gas could travel considerable distances. Basements may be watertight but not gastight, and methane may lie in them since it is heavier than air. Traffic accidents can cause spillage of volatiles, which may enter buildings through ventilation openings. Large-scale gas leaks can enter buildings in the same way. The growth in terrorism has meant that certain buildings are likely to suffer bomb attack. These sources of explosion are specialised, and demand military construction solutions.

Internal explosion sources

Any material that can burn can also explode. Finely divided material is also an explosion risk. Flour mills are vulnerable. Leaks from the storage of gaseous materials can permeate general spaces within buildings, and also enter voids in the construction. Where these are linked by interconnecting cavities the explosive mixture can move throughout the building. Leaking distribution pipework may flood ducts and basements.

Building tolerance

Brittle masonry materials are generally assumed to have little ductility, but under the instantaneous loading associated with explosions, they can exhibit some ductility (Mainstone 1976). This is influenced by any edge restraint provided. The lightweight areas of external and internal openings may be blown out by the force of explosions. This may cause secondary damage, but will assist in lowering the duration of the explosion and minimising the peak pressure in the explosion. Where large spaces are present, structural damage may be significantly reduced. Framed buildings containing large areas of glazing can provide good explosion venting. Solidly confined areas are more vulnerable to structural damage following internal explosions. Where the tying together of major elements is ineffective, the structural effects of an explosion may increase, causing them to flex and tear apart. This may leave them unstable since they may be deprived of end-bearing or lateral restraint. This may cause elements to fall in the opposite direction to the direction of the flame path. Ends and corners of buildings are more vulnerable to explosive forces since they have no surrounding structure to resist the applied loads.

Progressive collapse   A progressive collapse can occur following an internal explosion. With supporting elements blown away, there may be no structural support to elements above them. This is the type of collapse that occurred in part of the Ronan Point multi-storey block of flats in May 1968 following a domestic gas explosion. This was a Larsen–Nielsen system building of precast reinforced loadbearing panel construction. The inadequate jointing of the panels was seen as a critical factor in the collapse (Byrd 1981). This structure, typical of many industrialised buildings, possessed concrete inner walls and floors, and these confined the explosion, producing considerable pressures. Masonry structures are similarly resistant, although failure can develop around unit joints. The sections which remain may continue to offer some structural support, which can arrest progressive collapse. In multi-storey buildings where sufficient wall has been removed, the support to upper floors may allow them to fall under their own weight onto lower floors. This can cause overloading of elements potentially weakened by the explosion, and initiate a progressive collapse.

Symptoms   Major explosion damage can scatter debris over a large area, and effectively demolish buildings. Openings and other forms of lightweight construction may be blown out. Lightweight roof coverings can be blown off, and this can assist in relieving peak pressure. An outward movement of walls can cause horizontal and vertical cracking in patterns related to the shape of the explosive flame face. Explosions may cause scorch marks on decoration and finishes, and this may develop into severe fires. The notoriety of explosions will generate interest from supply companies, and since third parties may be involved the police may take an interest. Identification of the cause of explosions is a specialised area, and will take into account a wide range of factors. The defects analyst should identify the effects of the blast on the performance of the building.

5 Instability: structural form and deficiency

Inadequate loadbearing capacity

Mechanisms of deficiency

Inadequate loadbearing capacity in structures may arise through a poor design initially, or through the failure of the materials of construction or their connections during the lifetime of the building. The construction of the building offers a potential link between the two.

These deficiencies may be usefully categorised as built-in or acquired. The acquired deficiency arising after completion of the building may also be due to the influence of overloading in use, or poor internal load distribution. External forces may adversely affect the structure. There is quite likely to be a combination of contributory causes.

Imbalanced forces

The nature of response of the structure will depend on the form of loading, and in particular the direction and magnitude of forces. It will also depend on the structural form. Opposing and balanced forces do not necessarily introduce immediate instability unless their magnitude exceeds the loadbearing capacity of the particular element or the structural frame. Imbalanced or eccentric forces can create potential instabilities by producing non-uniform loading of the elements or the whole structure.

Creeping of the structure may be of significance in the long term. The limit of creep allowable before a building becomes unstable is sometimes difficult to predict. It depends on the flexibility of the materials and their individual susceptibility to creep. The tolerance of the joints between components and the structural interconnections within the frame will be important. Does the frame fail or continue to function after distortion? What is its relationship with the remainder of the envelope? For example, the tolerant designs and materials of the traditional soft forms of construction may accept marked plastic deformation, and are generally better than the rigid forms of construction at redistributing the internal forces. See Fig. 5.1.

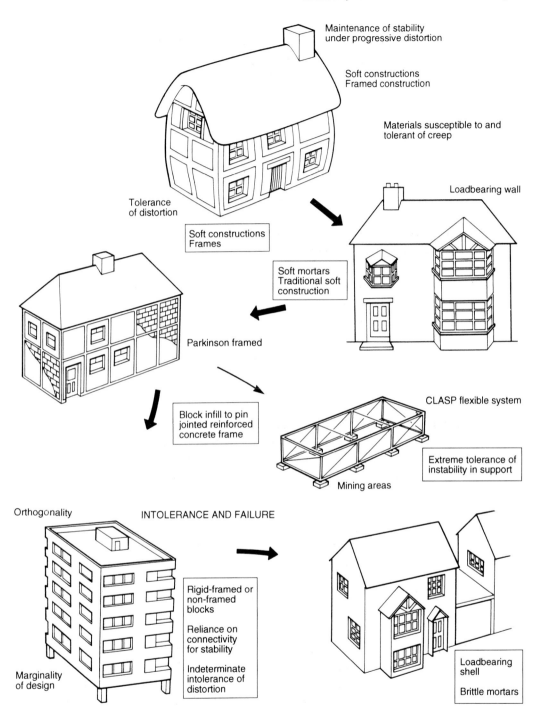

Maintenance of stability
under progressive distortion

Soft constructions
Framed construction

Materials susceptible to and
tolerant of creep

Tolerance
of distortion

Loadbearing wall

Soft constructions
Frames

Soft mortars
Traditional soft
construction

Parkinson framed

CLASP flexible system

Block infill to pin
jointed reinforced
concrete frame

Extreme tolerance of
instability in support

Mining areas

Orthogonality

INTOLERANCE AND FAILURE

Rigid-framed or
non-framed
blocks

Reliance on
connectivity
for stability

Indeterminate
intolerance of
distortion

Marginality
of design

Loadbearing
shell

Brittle mortars

Fig. 5.1 Tolerance of creep and distortion of timber framed and soft masonry
constructions contrasted with the development of the modern form and its characteristic
intolerance: failure may accompany distortion.

Fig. 5.2 The possible effects of the partial removal of chimneys. The structure as a whole relied on the chimney breast for buttressing. With this removed the building may suffer lozenging distortion. Partial retention of the chimney may exacerbate the problem by eccentrically loading the weakened walls, so encouraging racking of the roof and a progressive deflection in the gable wall.

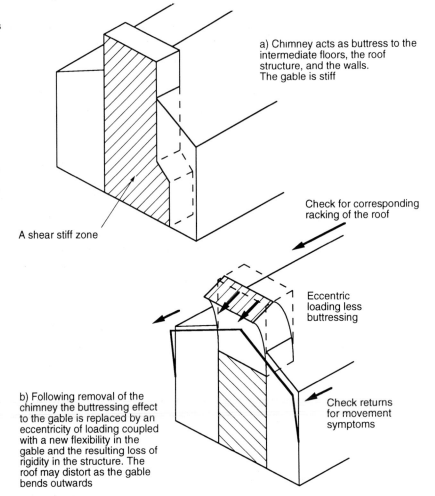

a) Chimney acts as buttress to the intermediate floors, the roof structure, and the walls. The gable is stiff

Check for corresponding racking of the roof

A shear stiff zone

Eccentric loading less buttressing

b) Following removal of the chimney the buttressing effect to the gable is replaced by an eccentricity of loading coupled with a new flexibility in the gable and the resulting loss of rigidity in the structure. The roof may distort as the gable bends outwards

Check returns for movement symptoms

Acquired deficiency A deficiency may be acquired in what was previously a stable structure. In making an assessment it is obviously necessary to understand the structural nature of the building, and the type and approximate magnitudes of the forces imposed upon it. It is particularly useful to identify any imbalances of forces or acquired loadings that were probably excluded from the design calculations. Alterations or additions to the building may strengthen or weaken the structure, and in doing so redefine the internal distribution of the loading. Differential settlement may be produced between an extension and the existing structure, resulting in the main building being eccentrically loaded.

Internally, the Victorian practice of using trussed partitions as loadbearing divisions is not always appreciated. It is possible to find that the seemingly innocuous addition of a doorway in a partition is the prime cause of a marked deflection to the centre of the building (Price 1984).

In more modern structures, the deflection of a trussed roof could easily impose some of the roof loading onto the non-loadbearing partitions. This may be transferred through to the upper floor-joists and produce sagging in the centre of the house or bowing of the lower floor non-loadbearing partitions. Where this is suspected, it is possible to carry out some simple tests to assess the stressing of the partitions. There will be no gap remaining at the head of the partition, and the structure will sing when struck (Benson *et al.* 1980). Benson *et al.* also mention some general distinctions of partitioning designed to be loadbearing. The walls will continue from floor to ceiling and will probably run normal to the floor joists. There may be other confirmatory telltales such as a beam below the partition to receive the load.

The fashion of removing chimneys in the 1960s has eliminated the shear core or gable buttress in many houses. Where the cosmetic alterations were only partial, it is possible to find the remaining upper portion of the chimney eccentrically loading the walls it originally buttressed. The remaining chimney structure may also be evident from an inspection of the roof space. See Fig. 5.2.

Modifications and deficiency

At the level of components, timber joists may have been notched, even cut through to make way for the installation of services. The re-roofing of slated roofs with concrete tiles will have altered the weight of the covering significantly. The roof structure may spread as the increased loading disturbs its established equilibrium. Other alterations may be concealed by re-finishing and the safety factor for the structure may be indeterminate.

The influence of constructional form on symptoms

The material and method of construction will determine the types of symptom that the structure will show under overloading. Bear in mind that deflection and overloading do not always accompany each other. Deflection produced by the creeping of structures may create visible distortion in some forms of construction without suggesting an imminent collapse, for instance in the traditional timber-framed buildings.

Rigidity, flexibility and tolerance

What distinguishes the tolerant and intolerant structure? Perhaps their very tolerance contributes to the pace and scale of distortion and deficiency? Overloaded rigid and brittle forms and materials of construction may give little advance warning of failure. The distress symptoms characteristic of the flexible and tolerant structures, such as diffuse cracking in brickwork or the creaking of timber buildings, are simply not there. Failure is sudden and may be catastrophic. The relief of stresses is not partial, and occurs only at the limit. But the characteristic of brittle materials to fail suddenly is only part of the problem. Generally, the creation of modern buildings relies more heavily on connections, often between materials with fundamentally different characteristics. Joints provide areas of concentrated stress. In the soft forms of construction characteristic of the pre-war eras, the design philosophy for buildings was largely orientated towards allowing, accommodating and concealing movement within the construction. The structures could undergo a considerable amount of distortion without losing their structural stability. The deep foundations required by the hard constructions to bypass seasonal movements in soils were not as necessary with the soft constructions (Benson 1984). The soft systems represent more of a dynamic equilibrium with the ground than the harder constructs. They could be considered tolerant.

Compare this type of performance with the modern structures, which may use larger units of construction, fabricated from the range of modern materials with variable dimensional stability, and the physical movement increases rapidly. The number of joints to accommodate this decreases with the increased size of components, and the softness or flexibility of the structure and its connections has been replaced with rigidity. Further, there has been a design shift which attempts to isolate the construction from its interrelationship with the soil. The hard forms of construction are reliant on stiffness either in the foundations or in the superstructure (Benson 1984), and can come into conflict with the ground rather than being in harmony with it. The appropriateness of design and quality of construction becomes critical. There is little margin for deficiency, little tolerance.

Consequently the significance of symptoms must be considered to be contextually dependent. Minor cracking in the loadbearing features of tolerant structures may indicate a new-found stability after the release of stresses. With intolerant structures the same apparent scale of cracking may represent a severe curtailment of stability. In the non-loadbearing components it may indicate secondary failure associated with primary deflection or distortion. This can be a valuable early warning.

Distortion and failure   It will be necessary to study the materials used for the loadbearing sections of the structure, and analyse the nature of the framework before deciding whether failure will accompany distortion. Flexible, tolerant structures may endure distortion without failure. Rigid, intolerant frames may accommodate little or no distortion before collapsing. The position and direction of cracking will also be important. Cracks usually cross through areas of weakness in the skin of the building. The formation of a crack is usually associated with tension, although compressive bulging of the wall may produce tensile cracking in the ballooning skin of the wall. Again, the context of the cracking is important. Shear can produce cracking, and of course there is no rule stating that forces and defects must occur in isolation! Evidence of eccentricity, shear forces or imbalanced multiple forces may be obvious where a crack is tapered or otherwise non-uniform, or perhaps shows indications of rotation or slipping. The range of crack symptoms and their possible causes are discussed at some length later in the book.

Load transferral and overload

In the case of infilled frames, there are distinct possibilities of differential movement, especially where the concrete frame encloses clay brickwork. Consider whether there are any symptoms of movement in the frame such as deflection and distortion. Where this movement is restrained by the enclosed brickwork, this can produce tensile cracking in the reinforced concrete frame itself. In other clad forms the occurrence of cracking in the frame may be concealed by the outer envelope. With the lightweight non-loadbearing claddings there are more likely to be movement symptoms in the cladding or its fixings than in the frame, since the magnitude of forces transferred will exceed the capacity of the envelope. Preliminary movement in the frame may produce distortion symptoms in the cladding visible from the outside of the building. Cracking in glazing is usually easily visible, and so can be load stressing although this may require polarising filters. Brick cladding may bulge prior to cracking. This should be visible by standing at the base of the building and looking up the outside surface. It will be difficult to be certain in the early stages.

Where there is transferral of loading to panels not designed to carry and distribute a load, the stability of the building will require very careful assessment. There may be fracture symptoms or distortion in components designed to be non-loadbearing that indicate clearly the current or previous redistribution of loads. Orthogonal designs, such as the large panel systems of the 1960s, may illustrate distortion clearly at their joints. It is important, however, to distinguish misalignment as a built-in defect from movement of the panel following fixing, an acquired defect. The built-in defect may be relatively stable and unlikely to move further; conversely it may be the root of consequential instability. The movement of a panel after fixing may represent omitted or dubious fixings. Alternatively, it may indicate flexibility where there should be rigidity under loadbearing conditions.

Tolerance and constructional form

The system-built structures of the 1960s are renowned for their inaccuracies, and on-site modifications or omissions. These possibilities occur with all buildings of course, but where there is evidence to suggest omission or sub-standard performance of the critical connections, concerns for the overall stability of the building should be much greater than in some other traditional forms of building.

The assessment of residual stability and realistic safety factors for buildings is usually a complicated process in existing properties. The effects of material degradation on the loadbearing capacity of elements and the quality of load transference at connections produces consequences that need inspired guesswork. The assessment of residual strength in systems buildings will be complicated, since the rigidity and degree of connectivity assumed during design will produce structures intolerant of distortion.

Loadbearing system buildings have also been found with fitting tolerances approaching the physical dimensions of their components. This will tend to cause overloading of the bearing surfaces, and in addition to the damage this causes directly there is also a reduced margin for movement of the structure under conventional distortion. The same holds true for the hogging distortion which occurs in concrete floor and roof slabs above compartment fires.

This potential criticality of the joints is compounded in ordinary use where the nonalignment of adjoining surfaces compromises the weather-proofing properties of the system. This allows water to drain or be driven into the joints, where any inadequately protected fixings will be at risk. Direct penetration of the envelope is possible, and water ingress behind the weatherline can easily become chronic. Brittle mastics installed in unservice-able or inaccessible locations make inspection and assessment difficult. Design assumptions about the site installation and inspection opportunities may have been optimistic. Repair will be expensive, perhaps even impracticable. Secondary deterioration in the panels may be more critical.

Recorded examples of large panel systems being discovered with no apparent fixing other than their self-weight, or corroded or substandard replacement fixings, suggest that the degree of instability at which a structure will predictably fall down or remain standing is unclear. Indeed, the exact stage of partial deterioration at which the buildings become structurally unstable is indistinct (BBC 1984).

Progressive collapse

Attempts to demolish these types of building have not always been as straightforward as assumed, considering potential progressive collapse is a frequent concern. There has been at least one recorded instance where demolition by explosive-induced progressive collapse was only partially successful. In other cases the indeterminacy of the structural stability means that demolition becomes a piecemeal exercise, possibly requiring the installation of expensive and time-consuming shoring to the lower sections of the building to prevent uncontrolled progressive collapse. The social and financial consequences of decanting or rehousing are also enormous. Assessing these instances of structural instability and placing the threshold of acceptability are the task for the specialist structural engineer. It will obviously not be a straightforward process.

The legacy of Ronan Point

The failure of Ronan Point by progressive collapse in 1968 precipitated the 1973 changes to the Codes of Practice to guard against progressive collapse. Here a relatively minor gas explosion on the 18th of 22 storeys of a system-built block of flats blew a loadbearing wall panel out. This was at least partially attributed to substandard fixings. The consequence of this was a loss of support to the floor. The collapse of floors pushed out the wall panels below until much of the corner of the building had collapsed in a domino effect. The relevance of Ronan Point twenty years on is the legacy of similar buildings it represents.

Structures and specialists

It is worth reiterating here that whenever you lack confidence in the accuracy of your appraisal of a condition, seek a second opinion. A structural engineer should be consulted for all but the most straightforward of structural defects.

Symptoms and constructional form

Overloading of mass brickwork produces diffuse cracking symptoms, or major cracks which are well defined and usually directional. The exact mechanism of collapse of a masonry wall will be related to the integrity of the whole building.

A timber-framed building will act as a composite whole due to the nature of the connections. In contrast, the sub-components of prefabricated buildings are more difficult to connect, and junctions may be liable to fail.

Portal frames give structural distortion to a greater degree and sudden failures. Rotational movement is considered to be a more destructive form of failure than shear, since it is commonly associated with the cracking of many or all of the structural components.

Crack width, shape and direction

Mostly these faults produce cracks as the structure inadequately responds to the movement expected of it. Eldridge (1976) classifies cracks according to the direction of travel (vertical, horizontal, diagonal) and the form of the crack (toothed, straight, or stepped). Also important is the width of the crack and its location in the building. Where cracks are of non-uniform width this can give valuable clues to the possible sources of the stresses. Note also whether the crack is varying during the day due to thermal influences, and whether the crack penetrates the whole element of the structure, is merely superficial, or in between the two.

Where cracks pass through the DPC there is a special type of problem. Alignment will also be an important indicator of causes of the crack. As a rough guide, compressive failures are associated with horizontal cracks, whilst tensile faults produce vertical and horizontal cracking. Bending produces diffuse cracks which are omnidirectional.

Instability in walls

Cavity walls are fundamentally reliant on the wall tie to give the necessary stability to their separate leaves. Cavities up to 3 inch (75 mm) width were frequently experimented with in the late 1800s, and below this figure the cavity size does not appear to be particularly (structurally) important. There is evidence that significantly above these values the leaves begin to act independently, and instability may arise as a built-in defect. The design threshold is currently 90 mm. Tall walls may suffer from excessive slenderness ratios, making them liable to bulging. This same effect is produced by the corrosion of the ties, since the wall loses a significant proportion of its effective thickness.

Stability of the structure

In addition to the loss of stability through the corrosion of ties, there can be additional disruptive problems as the ties expand on rusting. The resultant downward force must usually be within the middle third of the width of the wall for stability. Exceptions are with the resilient forms of construction. The horizontal components of floor loadings exacerbate any instability.

Slenderness and stability

With the movement towards slim, tall structural elements there is a corresponding increase in slenderness ratios. Meanwhile the physical tolerance within the feasibility limits is being extended. Working tolerances obviously must be sharpened to similar tolerances if the structure is to perform satisfactorily.

The traditional one-brick-thick masonry walling has proved to be a reasonably stable wall structure, despite its waterproofing problems. If the masonry envelope is adequately restrained and the loading is reasonably uniform, then the light loads characteristic of domestic-scale buildings can easily be tolerated.

The modern design criteria for brickwork have been developed around concepts of required thickness in relation to height and length. These are rarely appropriate for the assessment of old walling, since the residual material strength is difficult to determine without samples, and the methods of construction were even less ideal than nowadays.

The nature of the jointing is a critical component of the overall strength and stability of walling. In particular, the older, soft forms of construction are relatively tolerant of seasonal distortions in the ground conditions. At the domestic scale, they rarely have the equivalent of modern foundations, which makes their renovation and alteration relatively straightforward. It also means that the walls bear directly on the soil and their stability depends on a more intimate relationship. Meanwhile, the changeover from the soft lime mortars to the harder cement mortars which appeared after the First World War radically influenced the dynamic behaviour of brickwork.

Cavity walls

There are some instances of the use of non-metallic ties in older buildings, but these are rare. They are commonly based on masonry materials, for example, glazed bricks and cross headers. These have caused problems with cavity bridging leading to damp penetration (Hinks and Cook 1988e).

Internal disruption

Disruptive aggregates from natural sources are uncommon in the UK, although the Scottish dolomite has been recorded as being problematic. Partially burnt or unburnt colliery waste or other artificially modified aggregates can contain quantities of sulphates and other disruptive chemicals which react with the cement gel. Although the disruption mechanism is the same as sulphate attack, it tends to occur in the internal zones of the building rather than external. The occurrence of the aggregates is distinctly regional and associated with the aggregate sources. Boswell houses in the Wolverhampton/Birmingham area of 1920s construction suffer sulphate attack due to the clinker content, whereas those in Liverpool do not.

Inconsistencies in mortar and brick strength

Unless there is a sufficiently high cement content in the mortar to produce excessive shrinkage, the mismatch between mortar and brick strength is likely to produce a variety of symptoms from the same cause. Movement cracking in brickwork bound with very strong mortar may produce vertical cracking through alternate courses of bricks. In walls where the mortar strength is more closely matched to that of the brick, or of a lower strength, there is more likely to be a toothed crack pattern which follows the joints.

Weak mortars may be more susceptible to sulphate attack than the stronger mixes. Within reason there will also be a general improvement in the durability of the wall, but the corresponding reduction in the flexibility of the wall will restrict its accommodation of moisture and thermal movement. The air-entrained mixes may be more resistant to frost attack, but may adhere less strongly and allow greater rain penetration.

Inadequate restraint of elements

Restraint through connectivity

Cross-walls and connections between gable and elevation walling provide self-restraint to movement through their connectivity. Problems occur where these and other structural connections fail. Rotting timber joist ends and structural timber connections can lead to the imposition of imbalanced forces on walls coincidental with their loss of bracing. There is also the possibility of wall-ties placed in direct contact with joists transferring the forces of deflecting floors to the outer skin.

In traditional roofs, lateral restraint was provided by purlins, binders and ridge plates. The rigidity and connectivity of the traditional three-dimensional structure supported the weight of the covering and was self-restraining. Where this has broken down due to the defects associated with the timber, the formerly well-braced structure reverts to a two-dimensional form, and lateral movement can occur. These faults may occur either through progressive deterioration of the timber frame and/or through overloading, snow loading or high wind. In the traditional forms wind bracing was almost incidental compared with the critical contribution it makes to the rigidity of modern trussed forms.

Bracing and restraint The need to provide lateral restraint for the trussed rafter roof was not recognised initially (Munday 1989). There is a relatively small allowable movement out-of-plumb before an unbraced trussed roof loses its structural integrity, perhaps as marginal as 40mm horizontally (Munday 1989).

Other overloading problems may be due to re-roofing with heavier materials. Buildings should be compared with similar buildings in the area for the practicality and consequence of re-roofing, also to identify foreseeable cost for the building owner.

Sagging in roofs Another feature of movement in pitched roof framing is the outward movement which accompanies sagging. This may be produced as the critical frame components rot, particularly the tensile ceiling joists, or simply where the roof is overloaded. External symptoms may be a tent-like bowing of the ridge between rigid gables, which may carry greater load, and an outward movement of the top courses of masonry at the eaves, which are weak, and relatively lightly loaded. Confirmation of this may be provided by high-level horizontal cracking in the upper rooms of the building.

It is important to remember that pitched roof structures do not usually impose a simple vertical load on a wall. If there is deflection and failure of the ceiling (tie) joists the roof will sag under the net force, and the loading will be transferred to the walls.

Long-term creep In addition to lateral instability in the roof it should be appreciated that buildings can undergo a long-term creep, or plastic deformation. Whilst the degree of movement may not be sufficient in itself to cause instability, there may be a secondary effect as distortion and the transferral of loads to non-loadbearing partitions occurs. Here the importance of understanding the structure becomes evident. In some of the Victorian town houses the internal timber partitioning is loadbearing and contributes to the overall stability of the building. Their indiscriminate removal may produce such faults in the general structure. Studs may be identified as loadbearing by a number of tell-tales. For instance, they may sing when hit, there will be no gap at the head and there may be a beam running underneath.

Distortion of non-loadbearing partitions Where non-loadbearing partitions have structural loads thrust upon them, they are likely to distort. If they are flexible, then an indication may be given by sticking doors. This type of symptom may arise from vertical loading and also from distortions in the external walls. As the external frame lozenges it distorts the internal partitioning which acts as a bracing. This produces a racking stress in the internal partitioning, leading to distortion in the openings. In such instances the external walls are likely to be out-of-plumb.

It is inappropriate to assume all non-loadbearing partitions will not carry load. It is common for them to have loads imposed on them; it is the degree of overloading and their reaction to it that is important. Price (1984), for instance, points out that non-loadbearing party walls frequently transfer a greater load to the ground than other loadbearing walls.

Deflection of the support to non-loadbearing rigid partitions may produce horizontal cracking at the floor level or between the lower courses, or a toothed crack running upwards at an angle. This is a similar effect to the archway cracking formed by sagging lintels and may pass through the partition, perhaps originating from weaknesses such as doorways.

Partitions and floors If the supporting floor deflects transversely to the line of the partition, there may be wholesale downward movement, which could produce a uniform horizontal opening at the top of the partition and vertical cracking where it abuts adjoining stable walls. In extreme cases a tensile crack may appear on the underside of the floor in line with the partition.

Shear resistance and the chimney The contribution of chimneys to the stability of a structure commonly extends to providing a shear-resistant core for the entire building. Consequently, where chimneys are being removed this may place excessive loading on the surrounding construction coincident with a reduction in stiffness.

Where chimneys are retained they usually impose high loading density on the foundations. Any weakness in the soil or differential movement between the chimney and the rest of the building will manifest itself in cracking. Shear cracking may occur through the fireplace in a chimney as it sinks, since this is the weakest point.

Differential settlement in walls Another example of differential settlement in walls is due to highly fenestrated elevations. The diversion and concentration of loads through the masonry piers between the openings can produce localised overloading of continuous foundations.

Lightly loaded walls Bowing and general movement in parapets or lightly loaded walls can be produced by leaning chimneys as they undergo sulphate attack, or by foundation settlement. Clearly it is important that the defects analyst can identify the chain of events producing a series of secondary faults, and trace it back to the fundamental fault. It should be remembered that with low-rise masonry construction there is frequently a large margin of redundancy, perhaps accompanied by a low self-weight. Differential loadings may be accommodated with minimal symptoms. Faults may take longer to emerge and may also be more serious. Note, however, that the relatively weak lime mortars display cracking symptoms differently from the modern high-strength cement mortar mixes (Pryke 1980).

Structural displacement

Displacement of all or part of the structure usually manifests itself in the external walls. The symptoms usually occur at some plane of weakness such as at a damp-proof course which gives little resistance to movement. The traditional bitumen felt DPC is liable to extrusion and general distortion under such loads, and may often present symptoms of internal forces affecting structures.

Where the loading restraint is relatively low, such as in parapets, or where the wall is subject to side forces at floor slab connections, structural displacement can occur relatively easily.

Parapet movement

Parapet movement can occur because a number of forces are at work. These are normally related to the physical properties of the constructional materials. Moisture or thermal expansion movement causes dimensional instability in the parapet. With fresh bricks, irreversible moisture movement can occur.

Reversible moisture and thermal movements happen readily since the parapet is usually lightly loaded and is relatively exposed to the weather, so offering little resistance to its effects. Frost attack of the brickwork or mortar may be a problem because of the high exposure. Other causes include sulphate attack. The symptom is usually physical damage to the brickwork and oversailing of the parapet associated with the expansion component of cyclic movement. Cracking may occur to the brickwork. See Fig. 5.3.

Eldridge (1976) comments that it is not always necessary to rebuild the parapet; it depends on how severe the cracking and oversailing are, and whether frost or sulphate attack is responsible.

Distortion and the DPC

The problem of extrusion of bitumen DPCs arises from slight movement of the wall at the plane of weakness, or frequently of a solid concrete floor. In the stages whilst the distortion is moderate there is often extrusion of the bitumen – this can be considerable and may produce a beaded effect along the edge.

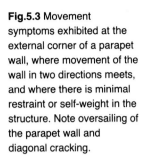

Fig.5.3 Movement symptoms exhibited at the external corner of a parapet wall, where movement of the wall in two directions meets, and where there is minimal restraint or self-weight in the structure. Note oversailing of the parapet wall and diagonal cracking.

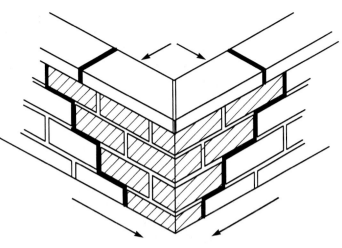

Isometric view of the diagonal cracking associated with oversailing

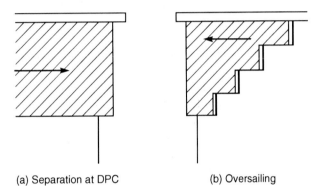

(a) Separation at DPC   (b) Oversailing

Floors and movement

It will be necessary to establish whether the extrusion of the DPC is accompanied by movements in the floor or walls. Floor movement may be exhibited by the surface lifting or arching or by the appearance of cracks as the internal stresses are released. Such floor expansion may be a primary symptom of an underlying fault such as expansion of the base. Problems have arisen frequently from the degradation of colliery shale, for example.

Lateral and longitudinal instability

Lateral stability

The cross-construction of walling or the presence of returns is vital to lateral stability. A two-dimensional structure may be unstable, whereas stiffened by a third dimension it may be relatively stable. Lateral instability is one of the most frequent forms of structural instability in masonry structures. There are a range of symptoms of lateral instability such as bulging or cracking in walls.

Restraint and stability from returns

In domestic-scale property the moderate length of walls and the provision of return walls usually provides some degree of restraint. This will be provided by the returns and there may be additional restraint offered by the connections of the floor joists, depending on their direction of span. There should be connections between the joists and the walls parallel to them, although this is not always so. In such instances, where the joists are providing restraint, the lateral instability of the wall will be concentrated in the upper level above the floor levels, and leaning or bulging of the wall is likely to commence and be most severe. This is frequent with the terraced forms of construction, where the joists will span between the front and rear walls running parallel to the party walls.

Floors and walls

Where the joists are parallel to a wall but untied, instability may produce bulging across the whole height. This is frequent with shallow gable walls where the floor joists have been run between the front and rear elevations. It may be accentuated at the ridge of the gable where there is minimal restraint from the roof structure. In instances of sufficiently high-level restraint the gable will perhaps bulge only partially. This will be greatest around any openings in the wall, since these represent weaknesses in the skin.

Cross-constructions

Cross-wall construction relies for its stability on the structural properties of the floor and roof. Connections between these elements and the walls are critical if satisfactory structural behaviour is to occur. Inadequate fixings, timber rot and differential movement can cause this link to fail. See Fig. 5.4.

Floor joists which run parallel to the cross-wall must be tied in if they are to offer lateral restraint. Joists which span perpendicular to the wall may subject it to a sideways thrust due to moisture or thermal movement, or expansion in a fire. Walls may bulge or lean outwards up to 40 mm out-of-plumb and still not exhibit cracking, even on the returns.

Terraces and the need for continuity

Where terraced buildings are partially demolished the remaining buildings may be deprived of the lateral restraint provided by the original buildings. This may cause progressive leaning of the party cross-walls. Distortion and cracking may be evident on elevations and coupled with bulging of the exposed gable.

Fig. 5.4 Incomplete connection of reinforcement in *in-situ* cross-wall construction means that the structural integrity of the building is below design expectations.

Cracking at junctions   In addition to vertical cracking at the joints between the cross-wall and external elevation, there is likely to be corresponding horizontal cracking at the junction between the ceiling and external wall (Melville, Gordon and Boswood 1985). The width of cracking between the frontage and the party walls will increase towards the roof, and may be visible from within the roof structure at the eaves. Apart from the immediate structural implications, there is of course the secondary deficiency in the fire resistance of the party wall. Incidentally, it is to be expected that the sound insulation will be poor.

There may also be gaps between the floor and the skirting as the front wall separates from the internal structure.

Lateral instability and floor joists

Lateral instability may be produced by the floor joists also. Where the joists become overloaded or deterioration leads to excessive deflection, the rotational extension of the end bearings will transfer to the supports. Joists built into the walls will push out the wall locally on extending. The mechanism is based around the rotation of the joist and the transferral of the load from the designated surface bearing to an edge bearing on the inside of the wall. This will create an inclined line of thrust into the wall and the bulging may appear below the line of the floor joists. In comparison, any eccentric loading from parapet detailing, such as from a cornice, will produce an increased instability because of eccentricity.

Lateral and vertical restraint in timber-framed buildings

Timber-framed constructions clad externally in masonry may receive lateral and vertical restraint from the tied masonry although the design and construction of the timber frame and the masonry cladding are considered as autonomous. In practice this has been questioned, on the basis that the wall tie connectivity produces a linked structure. Relative movement between the two leaves induces stresses in the brickwork. As gradual settlement and creep of the timber structure occur, the roof may load the external walls at the eaves.

Lateral restraint and the pin-jointed structure

Pin-jointed structural steel frames are commonly laterally restrained by the outer skin construction and roof purlins. This can often be clearly seen in modern high-tech buildings. Where the jointing between the beam and column elements of steel framing is done using the early basic jointing, the rotational effects of movement of the frame can cause their failure at the joints.

Impact damage

Impact damage to the structure may generate horizontal forces in excess of those associated with wind or building use. Explosions will tend to reduce the time scale of the applied force, which brings with it severe loadings on elements.

Acquired instability of frontages

Where the roof starts to sag due to the unresolved forces in its internal structural system, the net lateral forces cause outward movement of the walls at the eaves. This is most likely to happen at the centre of long elevations where the resistance effect of returns is minimal. This effect will

will be modified by any effective resistance of cross-wall construction. Commonly, however, the junction between the elevation and the cross-wall fails in tension and produces a vertical crack which is likely to increase in width with height. This should not be confused with a similar internal vertical cracking between unbonded or partially bonded, cross or party walls and external walls. The fault can be easily verified by a detailed inspection of the interface between the internal and external walls.

Sagging and load transferral

There may also be some tying effect from floors, and for this reason bulging is most likely to occur in the upper storey. Obviously where there is any interference with the structural system of the roof or building this is likely to occur. See Fig 1.1. An additional symptom of this type of failure might be a sagging ridge line. This may be a precursor to more serious cracking in the walls. As the load of the roof is transferred to the walls, the sagging of the roof structure may also impose some overloading on the gable walls. The sagging ridge pushes the gables out at its ends. This system of failure is particular to high-collared traditional roofs with separate ridge plates used to produce a three-dimensional system. The failure of two-dimensional trussed systems is likely to be concentrated in individual trusses and hence localised.

The contribution of distortion

The likelihood of the gable wall distorting and allowing the roof to move is increased if the return and internal cross-walls provide minimal restraint, either because they are not tied, or if they have a large proportion of their area as openings. It should not be forgotten that the wall may have been built out-of-plumb rather than acquiring its distortion.

Gables and restraint

This omission or failure of a continuous vertical joint between the elevation or gable and the internal cross-wall is an important fault. It is possible in old terraced buildings, since the party walls and the elevation may not have been built together; rather the elevation may have been built as a single wall without any bonded connections.

It is the degree of restraint provided by the connected walls and in particular the cross-construction that gives large expanses of brickwork their stability, particularly under dynamic loadings such as the wind and foundation settlement, however minimal.

Instability in the gable wall

Where the gable wall has not been effectively tied into the roof structure there exists the possibility that it can be pulled away under the action of high winds and explosions. The progressive movement of the gable will also

Fig 5.5 Tapered cracks indicate rotational failure in the gable wall. The cracks typically pass through openings in the wall. The fault shown passes throughout the complete height of the wall and is a case of suspected subsidence.

allow the roof structure to lozenge. This is common in early 1960s trussed-roof constructions, which were frequently built with minimal bracing. As the gable wall distorts, tapered cracks may appear at the connection with the returns. These will be widest at the top. An example is shown in Fig. 5.5.

Rigidity and the internal structure

The resistance and stiffness of the internal partitions and cross-walls was exploited by the Victorians in the construction of timber internal partitions. This was a different approach from the current practice of producing non-loadbearing partitions. They were diagonally braced and designed to partially support the loading from the floors above, around mid-span. Any interference with these partitions, such as their removal or the insertion of extra openings, can produce a sagging throughout the internal structure. This will be evident from distortions to floor surfaces.

Thermal movement and distortion

Similar symptoms may be produced by the thermal stresses of expanding roof slabs. This will produce the greatest physical deflection across the longest dimension. Distortion may occur other than in the frontages of the building or the gable returns. Where brickwork fails under these loads this is likely to occur at a high level where the restraint and self-weight are minimal, and may produce horizontal cracking between courses. Although the expansion and contraction forces may be cyclic, where the contraction effect is restrained, the net movement will be in one direction only.

Bulging, ledges and instability

Where movement occurs over the entire roof there may be some reduction in ledge bearing. Roof components may move independently of each other and the total net movement may increase with time. There may be progressive movement away from bearing surfaces causing a potential support instability. Further symptomatic evidence of this type of movement occurs with component roofing, by patterned cracking of the ridge and bulging of the roof surface. Internally, the finishings over the connections between the roof and walls may be disturbed.

The rooftop will expand to a greater extent than its underside. This will cause the roof to bow upwards and outwards in a similar fashion to a ballooning wall. The extent and nature of roof movement is to some extent dependent on the position of any insulation layer or finish to the roof. This movement may be accompanied by horizontal cracking in internal walls abutting the roof structure, and will follow a cyclic pattern. Where the walls are non-loadbearing this may be acceptable. Expansive thermal movement of a roof's upper surface may also trigger thermal pumping.

Tying

Where buildings rely on being tied to the floor structure this must be adequate and not rely on either end-bearing or poorly fixed joist hangers. Corrosion of fixing metals can occur. Concrete anchor blocks cast into the floor can be pulled out (Richardson 1988). Metal flats found in older properties may have been put in during construction, and not necessarily as a response to structural failure. These have been commonly inserted to offer lateral restraint to walls which run parallel with the main floor joists but were inadequately tied together. The bulging symptoms of this type of failure are not confined to the weaker parts of the elevation at roof level, but may occur at any or all floor levels.

There is a general commonality with the shape and direction of vertical cracking, which is likely to concentrate in the centres of the elevation.

Where the traditional metal tie is stronger than the bulging forces this can distort the rear wall to match the front wall. They move in parallel. Effectively the bulging of the front wall pulls the rear wall with it. Where the tie has been anchored into a corner or a shear-resistant element, this type of failure is less likely (Richardson 1985). A contributory factor to this type of failure may be excessively slender wall structures and any unresolved forces from the roof. The effective shear resistance of the end walls may have been reduced where partial demolition of terraced buildings has occurred without the provision of new shear (buttressing) walls or one-dimensional shoring has been applied. Buttressing may also be provided by the addition of extensions, and in some cases this may have been a design intention.

Ballooning   Where a structural resistance is provided which is opposed to vertical expansion of masonry, there exists the possibility of outward bulging both in the vertical and horizontal plane. This ballooning of the frontage will apply tensile and compressive forces to the structure, as well as jeopardising the structural integrity of floors and walls. In many cases acquired lateral instability may produce horizontal and vertical displacement of walls. Where this is the result of sulphate attack, the characteristic vertical expansion of the outer leaf (or the entire solid wall) may be resolved into the ballooning of the wall. The degree of vertical and horizontal restraint will greatly influence the reaction of the structure to the unresolved forces. Mid-terraced properties may have effective resistance to movement from the neighbouring properties, and the weight of concrete roofs may be sufficient to provide vertical restraint (Richardson 1985).

Where ballooning does occur, the bulging of the wall will be punctuated by any effective restraint offered by party walls, possibly producing a repetitive series of waves. Any potential weakness around openings in the walls will encourage tensile cracking of the external skin. The reduced cross-section of the wall will concentrate the stresses which produce the bulging.

Roof structures   Thermal movement occurring in roof slabs will usually be horizontal. In walls it may be horizontal or vertical. When the roof slab expands against an effective restraint such as an adjoining rigid building there is likely to be distortion such as bulging or rippling. Relatively unrestrained expansive movement of the roof structure is likely to cause failure in the walling at a high level. Similar symptoms may be produced by shrinkage in concrete roof slabs. This is usually accompanied by tell-tale cracking at connections where the tensile strength of the joint is easily overcome. The shrinkage along the longest dimension of an element will usually be the most significant.

Bookending

Where the ridge of the roof is parallel to the frontage of the building, there will be only minimal restraint from the roof at the gable end. This will be from the ridge plate and any purlins. Indeed, in cases where the roof itself is unstable the transferral of these loads will produce or exacerbate instability in the gable end. The wall may lean in either direction as the roof lozenges.

It is important to make a distinction between failure of the stability of the roof and the distortion of the gable end as it leans, and that produced by a lozenging of the whole frontage. In severe cases there may be some evidence in the end wall to help determine the true cause. The distortion may be uniform across its height or extreme only at the roof. This is not sufficient evidence in itself, however, and some internal inspection of the structure will also be needed.

Where movement joints are omitted in long runs of masonry walling, the expansion and contraction of the walling may produce a gradual extension of the elevation at a high level. The initial expansion of the wall occurs at the high level because of the minimal restraint of self-weight. It may produces vertical cracking on contraction. There is minimal thermal movement close to the ground due to a heat sink effect and the natural restraint offered by the subsoil. Any filling of these cracks by loose materials or repointing allows the wall to undergo further net expansion, and there is a progressive increase in the length of the building. With long elevations, this can produce a change in elevational appearance from that of a rectangle to a trapezium. Effective restraint at one end of the building will produce a lozenging effect, where rectangular structural areas distort at the top of the walling. There may be some disruption to roof coverings, as the differential movement of the wall is carried through to the roof finish.

A similar effect can be produced in lightweight steel-framed buildings with long elevations. The restraint produced by the connection of the structure at ground level is not matched at the eaves. Lightweight cladding provides virtually no restraint, and the structure will extend either symmetrically or as the shape and stiffness allows (Hodgkinson 1983).

An 'M' roof may also push a parapet or eaves wall out of plumb, as the structure sags and expands laterally. Minor distortions across a number of roof elements may act cumulatively to produce considerable movements at the restraining walls.

Openings

Where a large opening exists in the wall, there may be little lateral restraint offered by the bounding piers, allowing lozenging to occur. The relatively slender piers and abutments may also lack the required bounding restraint. This is common in large doorways, suitable for vehicle delivery. These are generally progressive modes of failure, since they are dependent on a gradual loading cycle.

The analysis of these and many other defects is impractical, without recourse to detailed examination of the building structure. Much of this will be hidden by the finishes, and/or nature of the construction. In certain cases the exposure of the elements to assess their degree of deficiency may compromise the remaining structural integrity of the building, in itself leading to failure.

Chimneys

Chimneys can suffer a number of structural defects. In most cases the symptoms are similar, a leaning distortion. The sulphate attack of the mortar joints in the chimney from inherent sulphates or migration from unlined flues is usually concentrated in the exposed faces of the stack above roof level. This produces differential expansion and pushes the stack into a leaning posture.

Where the chimney stack is sited on an external wall, internally induced leaning eccentricity may occur. This can be due to the wall suffering from distortion. If the chimney is stiff and incorporated into the external end wall, there is added restraint to the deflection of the wall, as the chimney effectively buttresses the structure. If the chimney is of minimal construction the deflection of the wall may be inadequately restrained, and the chimney may follow the distorted shape of the wall.

A wall leaning throughout its whole height may carry the chimney in the same direction. The eccentric loading induced will have implications for the wall's stability. An external wall containing a chimney stack and which is partially restrained at eaves level, by a hipped roof for example, may bulge below the eaves. This can produce a point of contraflexure at the eaves. The chimney will deflect over the roof above the eaves.

Instability of untied elements

Tying and stability

The tying in of the walls using the floor joists can improve the resistance to bulging (Benson *et al.* 1980). Where timbers are tied into the external structure problems can arise due to timber rot. The performance and durability of many buildings so constructed is evidence that the approach is not totally reprehensible.

Structural continuity

Timber frames bring with them special problems with regard to structural continuity, since the walls and floors are tied together. This is a three-dimensional rather than the traditional two-dimensional structure. If they are untied then the whole structure may be deficient. This is in contrast to the traditional structure where elements may remain standing when others have collapsed.

Instability of untied elements

Most untied elements rely on gravity fixing bearing onto masonry support walls. The dead and live loading of the floor provides the holding force. Where the span of the floor is parallel to the support walls they may not receive direct bearing. There may be built-in cambers in the beams, and deflections may alter with loading.

Intermediate floors and stability

Precast concrete slab floors similar to those used in certain industrialised building systems need to be tied to the walls in order to provide structural integrity of the entire building (Larsen-Nielsen). The hollow pot construction floors are commonly held together with a structural screed, incorporating steel reinforcement. Where omission or inaccuracies in the steel reinforcement occur the structural integrity of the floor may be lost, as it breaks down into individual loadbearing components. These are liable to excessive deflection, producing cracking to the floor. Secondary effects may occur within the perimeter structure and underlying partitions, as they become excessively loaded.

Roof slabs and stability

Severe thermal stresses affect roofs, because of their total exposure. There is usually little lateral restraint from the surrounding parapet structure, and they are relatively free to move. The degree of movement may be compounded by the position of the insulation. Cold roof detailing allows the concrete slab to reach a similar temperature to the roof surface. Warm roof details tend to be buffered against such extremes, and in these instances there is usually greater temperature-associated damage to the waterproof covering. In addition to the expansive problems of poorly restrained roof slabs, drying shrinkage of the concrete in the initial life of the building can produce excessive physical shrinkage. With large-span roofs the physical movement of the slab pulling the lightly loaded walls inwards can be significant.

Timber roofs and stability

In contrast, timber roof structures, owing to their natural resilience and lightness, are more prone to loading movement. Their reduced coefficient of expansion makes them more likely to exhibit larger moisture-related movements. This is particularly so where roof coverings have failed, and may be accompanied by water ponding due to deflection.

The connection between untied elements and their surroundings is usually the scene of symptomatic cracking or bulging. This can produce secondary waterproofing problems in the case of roofing details.

Interaction and instability

Other movement problems arise from reliance for support and stability and rigidity on elements of structure. Consider the example of a loadbearing solid construction partition built on a flexible floor surface. This is still common practice, and causes cracking problems with the walls in a number of ways depending on the openings and the nature of movement of the flexible floor.

Interaction of partitions and floors

Internally the problems of inadequacy of floor design or misuse may be illustrated by cracking partitions. This normally occurs when the floor used as the support for an inflexible partition deflects. It is usual for this to occur to some extent, but when it is excessive, a horizontal crack will appear at or near the base of the internal partition (Eldridge 1976). The type of aggregate used in the floor structure can make a significant effect on the amount of shrinkage. It is suggested that this can vary by four times. Workmanship may also be significant. These cracks are visible in any applied plasterwork, but may be obscured by covermouldings and skirting boards.

An alternative source of cracking is the shrinkage of the internal wall structure. Where this is built of concrete blocks or calcium silicate bricks, then the irreversible shrinkage in the early stages of the life of the building will tend to produce cracks at the juncture between the internal partitions and the external walls. This is related to the moisture changes of the material.

Roofs and instability

The expansion of a roof truss as it sags under its own weight and that of the coverings can push out the walls of the building. Larger irreversible movements which can also occur with time include creep and sagging (deflection) of the roof timbers. Timber performs well in short-term responses to loading; however, over long periods there is a tendency to creep. This produces sagging of roof structures, and outward pressure on walls at the eaves (Benson *et al.* 1980), even under relatively modest loads. The possible distortion patterns are shown in Fig. 5.6.

Where roofs are under-designed, or have become so because of replacement of one covering with a heavier alternative, sideways forces can

Fig. 5.6(a) The effects of the roof structure sagging may be visible at the gable ends of the building and across the ridge. The support from the gables will minimise sagging at the ends of a conventionally ridged roof, consequently the spreading is likely to be greatest at the centre of the building. **(b)** The spreading of the roof at ceiling level will produce an outward force on the walls at the point where they can provide least restraint from tying into the structure or self-weight. The rotational force on the loadbearing inner leaf will tend to produce a horizontal tensile crack, larger on the inside of the wall than the outside.

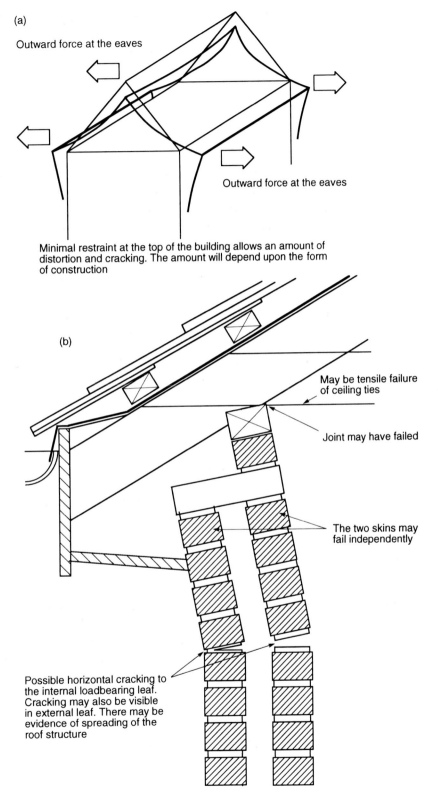

(a)

Outward force at the eaves

Outward force at the eaves

Minimal restraint at the top of the building allows an amount of distortion and cracking. The amount will depend upon the form of construction

(b)

May be tensile failure of ceiling ties

Joint may have failed

The two skins may fail independently

Possible horizontal cracking to the internal loadbearing leaf. Cracking may also be visible in external leaf. There may be evidence of spreading of the roof structure develop at the eaves. The reduction in structural performance of the timbers may also be as an indirect effect of fungal or beetle attack (Eldridge 1976). This will tend to push the upper courses of the wall outwards as a spreading effect occurs. This may produce horizontal cracking, most obvious internally where the larger dimensional movement occurs. It is suggested that this occurs mostly in the top three or four courses externally and may appear as only one crack in the internal leaf. The cause is similar to that occurring with the bulging of walls at intermediate floor levels.

Openings and instability

Where the support system above an opening is inadequate, compressive and tensile forces will develop in the area above the opening. Deflection of the lintel can cause cracking in brittle masonry materials. Compressive loads are normally concentrated at the top of the lintel, unless the deflection becomes so severe that hogging occurs. Timber bressummers over long spans are subject to the problems of time-related strain and can deflect significantly. Where the flashing and weephole provision above the lintel is inadequate, the risk of fungal and insect attack will increase. There may be damage to the edges of surface cracks where large compressive forces have been at work. Cracking above openings can mark the boundary of the maximum practical tensile stress in the walling material. This may be useful when considering the instability of several openings.

Cracking and structural movement

Slip planes and the DPC

Where there is any continuous plane running through the structure, there exists the possibility of the loss of structural linkage and continuity. The provision of movement joints is an example of this, although the need for continuity reinforcement is recognised. Its provision contradicts the rationale of the joint, but is seen as a structural compromise.

The DPC will need to satisfy the separate criteria of resisting the passage of moisture, which requires continuity, and ensuring that the structure is not compromised. The traditional method of using engineering bricks solves both problems, although the need for alternatives has changed the acceptance of this as an economic method. Modern plastic based materials can have low coefficients of friction, although the textured bearing surfaces may increase this. Bitumen felts may distort and compress under the action of compressive load and high temperatures. The provision of DPCs in buildings may concurrently be providing an inferior movement joint, at the wrong place, designed in the wrong way.

Total building movement

The wholesale movement of entire buildings is unusual. Structures which accommodate a substantial live loading component may act in a dynamic way. This movement may approximate simple harmonic motion, with little residual movement of the building. This oscillation may cause cracks to open and close in harmony with the changing loads. These may appear insignificant under a no-load condition. Where a residual movement occurs this may be seen at structurally weak slip planes.

Structures which are resisting lateral loads by reliance on dead weight may be subject to a creeping movement. Buildings which have retaining walls to one or two sides can be vulnerable. The overburden load from the soils will change with its moisture content, and this can induce differing strains in the supporting structure. Where this contains marked cleavage planes at low level movement can occur.

Overloading of the structure

Change of use

Where buildings have undergone a change of use without due consideration of the structural implications of any increase in loading, there exists the possibility of the overloading of the structure. The new dead and live loadings may be significant increases on the original design, although in many cases this may have been based on *ad hoc* rules. In some cases an over-design has been included, which appears to allow buildings to perform satisfactorily. This process is using up the allowable stress margins between working and failure stresses. As the failure stress is approached the structure becomes less stable and when exceeded may fail. Variable loading cycles, e.g., wind, may take a safe stress and increase it above the safety margin. In older deteriorated buildings the margin is difficult to determine. Consult a structural engineer.

Distortion of frontages

The distortion of frontages due to acquired lateral instability may be induced by such overloading. The bending reaction of floors to overloading produces an extension of the beam or slab and this transmits a horizontal component of the resultant force into the wall.

Openings and load distribution

Bresummers can distort due to overloading caused by changes of use of the building, or due to the deterioration of the timber itself, perhaps accentuated by damp penetration. There is an inherent mismatch between the bresummer material and the masonry, and the movement properties of the bresummer may induce cracking in the inflexible surrounding masonry. This is particularly problematic where relieving arches are omitted.

Bonding timbers and bulging of walls

Other instabilities occurred in older structures built around the time of the Industrial Revolution, incorporated bonding timbers. It is common for bonding timbers to be faced with snap header construction producing a natural cleavage plane for the bulging of the wall, when the timbers rot and compress. Such bulging will be localised, and may occur frequently across the height of the building elevation. Snap headers can be found used in concealed walling, as well as used for facing effects. The instability can be structurally severe.

Where the wall has integrity across its thickness, similar failure in timbers inserted internally, for decorative or reinforcing purposes, may occur. This can produce predominantly horizontal cracking between timber and timber-free areas in the walls. The fault may be repeated over an entire elevation, producing a wave effect that may be made more obvious where the fault has been present for some time, due to the action of weathering.

Bonding timbers were frequently used in houses built before 1900 (Melville, Gordon and Bosford 1985). The occurrence of rot is dependent on the cover provided to the timbers by the solid wall. The exposure of the wall may be such that sufficient dampness can penetrate to the timbers. This may allow them to become repeatedly or chronically damp, leading to rot. The rotting of the bonding timbers may be a secondary effect of water entry onto the wall from a number of causes.

The isolated pier and overloading

Sections of loadbearing walls can be considered as isolated piers, where concentrations of loading can occur. Symptomatic of this is vertical cracking of the masonry as the overloaded pier sinks. This is a similar problem to that encountered by the differential loading of the ground under chimneys and basements. Symptoms include the localised sagging of integral floors where joists derive support from the wall. The problem is accentuated where walls contain a large number of openings. The transference of loads around openings concentrates the stress into the masonry columns. This can produce a number of symptoms dependent upon the relationship between the pier and the remaining structure, including shear failure. Failure of individual piers may be associated with shear failure.

6 Instability in the structure: the role of substructure

Foundation failure

Symptom and cause

Failures in foundations may take many symptomatic forms, and it is important to be able to make a primary distinction between the possible causes by analysis of the effects on the structure. This may be confirmed by exposure of the foundations, and their specific analysis. See Fig. 6.1.

Settlement and cracking

It is common for fractures to occur within masonry as the forces in the structure are redistributed when the foundations settle. Probably all buildings settle to some extent, although in the majority of cases the structure withstands the forces this sets up and there is no apparent distress. Where failure occurs it can produce dramatic effects in the structure and be very costly to repair. Indeed, it is quite possible for the cost of repair to be unjustifiable and it may exceed the value of the building intact. The symptom is most frequently cracking as tensile forces in the superstructure are relieved and the building settles or differentially moves. There may be localised bulging distortion of the walling, but it is not very likely that the structure will bulge in the manner associated with lateral instability.

Foundation depth

It is important to appreciate the range of foundation types and degree of provision that exist in buildings. Houses with foundations at shallow depth are little better than houses where the walls are built directly onto the soil, but many have survived. They fall far short of our current regulations derived from experience, but may not be deficient. The current regulations pay little heed to the individual circumstances and are a blanket set of rules. It is unclear whether they are based on worst-case or average scenarios.

Structure and rigidity

Foundations of lesser dimensions than the Building Regulations recommend now are not necessarily deficient. However, in such cases there is probably some contribution to the overall strength and resistance to movement of the structure from the weight and bracing rigidity of the superstructure. There

Fig. 6.1 Evidence of foundation settlement beneath tenement frontage. Significant distortion at window emphasised by unsympathetic remedial works.

can be very tight margins of safety in these foundations, and the introduction of any instability produces severe effects.

The contribution of soil   It is important to assess the type of soil, since interaction between the foundation and the soil is critical. This was shown dramatically in the mid-1970s drought when buildings on clay soils with shallow and minimal foundations suffered heave and failure after many years trouble-free service.

It is easy to become involved in the riddle of deficiency and predictable and reasonable expectations, particularly where the building has exceeded its original design-life expectancy. Pragmatically, there is failure and immediate deficiency in the performance of its functions. Analysis is important in the correction of the fault. Whether wholesale alterations to all properties with the benefit of hindsight would have been wise or not is to question the very mode of development of the Building Regulations.

Types of symptom   Differential movement across the frontage may occur where the piers between openings carry a disproportionate amount of load and effectively produce point loading on the foundations. In marginal foundations or soil conditions there is the possibility of localised failure and excessive settlement. This will be indicated by the differential movement of the piers and the sagging of the brickwork in the piers.

Differential settlement   It is to be expected that a minor settlement will occur in all foundations during the first five years or so of a building's life. Foundations of differential loading and at different depths will tend to settle differentially. Adjacent buildings may settle differentially if one has a basement construction, since the loading and depth of founding will vary. In a terrace this may produce a saw-tooth effect in the roofs if there is significant alternation between depths of foundations to party walls and basements or flank walls (Melville, Gordon and Boswood 1985). If settlement of the building is initial settlement it may become stable.

The lightweight structure of bay windows as components of the frontage often led to them being founded at a shallow depth. Differential movement will appear at the connection between the bay window and the remaining walling in the form of a shear crack. See Fig. 6.2.

Soil overload

Overload   The ground may become overloaded. Where the total loading of buildings changes then this must increase the load transferred to the foundations. This may be non-uniform and produce differential settlement, since bearing capacities of soils are determined from soils before the construction of buildings. See Figs 6.3 and 6.4.

Heaving movements   The moisture content of soils and its migratory path is likely to change with time and this may produce reductions in long-term strength of the soil. An

Fig. 6.2 Differential settlement of bay window producing rotational or shear movement.

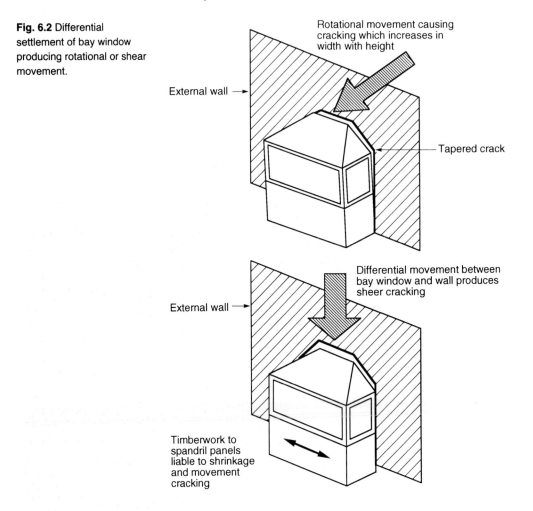

Rotational movement causing cracking which increases in width with height

External wall →

Tapered crack

External wall →

Differential movement between bay window and wall produces sheer cracking

Timberwork to spandril panels liable to shrinkage and movement cracking

extreme example of this dynamic reaction is the response of clay soil heave and shrinkage and the resolution of fissures. With granular soils, fines may be washed away.

Initial settlement
However, there is a degree of stability established following initial settlement and compaction, which if disturbed by the insensitive addition of further loading (e.g., extensions or adjacent buildings) may lead to differential settlement (see later).

Basements present a particular problem in this respect. The loadbearing capacity of soils will vary with depth, and so there is likely to be relative settlement between shallow foundations and basements. Where the design or loading configuration accentuates this difference, differential settlement is likely to occur in the structure (Richardson 1985d). This may manifest

Fig. 6.3 Gable end vertical cracking following the line of the rigid chimney structure.

itself as distortion and cracking in the main walls (a basic settlement crack), or uneven floor levels where the floor has followed the settlement. Particularly obvious will be settlement in terraces, which can produce uneven roof lines. Other symptoms include binding windows and doors.

Fig. 6.4(a) Rotational failure in a wall producing a characteristic diagonal cracking, tapering from the base of the wall. **(b)** There is significant evidence of earlier cracking which may be due to progressive soil collapse. The fault passes around the corner of the wall and through an opening.

Load transference failure

Soil adjustment Soil instability can be a primary source of structural failure. The relationship between the soil and the building imposed on it may be complex, and there will be an initial period whilst the soil readjusts. This will usually involve some degree of settlement (Melville and Gordon 1979), which may or may not be problematic for the building. The range of problems and uncertainties involved in founding buildings is bound to produce a large variety of faults.

The failure of the ground/foundation interaction commonly produces settlement, although upward thrusts are real possibilities also. The significance of settlement depends on the nature of the structure and its tolerance of such movements. A uniform settlement may be virtually unnoticeable; a differential settlement, on the other hand, may produce marked cracking and disruption to the superstructure. Settlement damage resembles a number of other foundation faults, and it is advisable to consider the alternative causes (Melville and Gordon 1979).

Structural failure can occur as a result of vertical or lateral movement.

Stability of the structure

Loadbearing capacity　　Failure of the soil may occur as a result of the loadbearing capacity of the soil being inadequate or changing during the life of the building. It is not infrequent to find buildings founded on inferior ground, and the risk of this is increasing. Up until the exceptional weather during the 1976 drought, and despite the earlier problems of 1946, the problems of clay foundations were little appreciated. Domestic foundations in the 1930s were commonly little more than 300 mm deep and a vast proportion that survived intact are evidence of their suitability at the time.

　　Other soil-related problems arise from existing locations of ponds, wells, quarries, etc. (Melville and Gordon 1979). The nature of local subsoils may be evident by surrounding conditions (Melville and Gordon 1979).

Subsidence　　Subsidence may occur either as a uniform progressive effect under consolidation of the soil or as a sudden effect such as can occur with mining settlements. Alternatively, soil recovery in the form of heave or lateral slip may occur under the building. In clay soils it is reasonably well known that problems with cyclic changes in soil moisture content can produce heave effects which disrupt buildings constructed on conventional foundations.

External forces　　Defects in the performance or stability of the soil may also occur as the result of local outside forces or actions. For instance, adjacent buildings imposing loads on the soil will tend to cause deflection and long-term settlement that may adversely affect the subject building (Hodgkinson 1983). Indeed, multiple compartment buildings may produce such an effect themselves (Hodgkinson 1983), as major divisions occur in the structure. See Fig. 6.5.

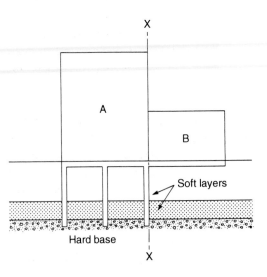

Fig. 6.5 Unless the structure is jointed at XX, block B tries to settle and tilt to right, cracking along XX. (Reproduced by permission of the Architectural Press).

Soil reaction It is possible for the reverse effect to occur if the loading on the building is released rapidly by the demolition or removal otherwise of adjacent building loads. This is likely to be particularly relevant in areas where multi-storey buildings, imposing larger loads on the soil through buoyancy foundation construction, are removed or the stresses on the soil are altered.

The relationship between the soil and the building is critical to stability. Too frequently rigidity is associated with structural soundness, whilst patently this is not the case. Examples: the use of CLASP in mining situations, the alternatives to earthquake resistant construction, flexibility to respond to wind also as a side-issue.

The analyst must not assume a particular type of foundation to exist. Building is such a diverse art that only in obvious cases such as inflexible design systems can it be taken for granted.

Foundation forms Melville and Gordon (1979) report that a great number of buildings constructed up to the end of the 19th century were given very little in the way of foundations other than a couple of courses of footings laid on the ground.

It was common for the 17th- and 18th-century buildings to be highly fenestrated, producing regular loads at piers between window openings (Melville and Gordon 1979). This overloads the soil and causes localised settlement. Similar effects occur with Victorian housing where the loads of the house are transferred by a beam across the bay window to a pier between the bay and front door (Melville and Gordon 1979). Most of these buildings which have survived are evidence of at least minimal stability. Overloading of the foundations may arise through change of use, or structural alterations, however. It is common therefore to find foundations overloaded because of inattention to their limits. This is not surprising if their form is uncertain.

Combinations of movements within the soil will need to be considered and the construction designed accordingly. The effects of differential movement within the structure are important also (see Fig. 6.5).

Instability Common defects with the stability of the structure as a whole exhibit themselves as foundation failure. This movement may be serious and require preventative and/or restorative action, and assessment should be made with a view to identifying whether the cracks are consistent with foundation failure. This can then give a clue as to the cause.

The severity and significance of cracks is important to assess, since a building with cracks may not be in danger of collapse, or even be significantly defective. On the other hand, it may be that the cracks are the symptoms of a progressive problem requiring preventative actions (BRE 1981a).

Wholesale ground movement

In 1971/72 Bickerdike *et al.* were warning of a likely increase in problems with houses as low-risk building land became scarce. Land previously rejected as being too risky is now being developed (Denton 1989), bringing with it the greater chance of soil- and stability-related defects. These potential defects can result from a low bearing capacity, but poor ground stability is both more critical and difficult to design for.

Subsidence generally

When mining subsidence or seismic activity (such as a geological fault, collapse of fill, clay shrinkage, or vibrations) reduces the loadbearing capacity of the ground under the foundations, dramatic and very sudden damage to the structure can result. A similar effect can be produced by sulphate attack of the concrete in the foundations.

Subsidence can produce a variety of symptoms. The external walls can become out of plumb, there may be movement at the DPC level, and internal walls may crack in an apparently random manner depending on the exact circumstances of failure and the method of construction (which dictates the structural response). Cracks can be quite wide (25 mm), but need not be.

Extent of settlement

The range of expected settlement can be extensive; Hodgkinson (1983) comments that in Mexico City settlements in the order of storey heights have occurred. Recommendations for damage limits for loadbearing walls and panels are available, and give an indication of the scale of the problem.

Differential ground movement

The DPC threshold

Across the structure, between structures or connected sub-divisions of structures, it is common for settlement-related faults to produce cracks that bridge the DPC. This is often a useful indication of problems in the ground, since problems with the bulk of the structure itself (e.g., thermal movement, moisture movement rarely pass through the DPC).

Tilting, swelling and hogging

Tilting of the structure can occur as the foundations or soils allow differential settlement or other movement. Soil swelling in the centre of the building may cause a hogging failure. Such cracks are likely to run vertically, becoming wider towards the top. See Fig. 6.6. The roof slates/tiles may spread also.

Differential settlement   Differential movements tend to occur for reasons of the soil structure (uneven soil types, sloping strata, or pockets of poor fill) (Hodgkinson 1983). Tall buildings which impose high loadings or have unevenly loaded adjacent ground may tilt (Hodgkinson 1983).

Differential settlement is an acute problem with subdivided structures (Hodgkinson 1983). With structures bearing on different strata, problems may arise if a piled subdivision to a sound base is attached to a relatively shallow, conventionally founded building on a marginal soil. The relative settlement between the two structures causes a cracking potential at the juncture, (see Fig. 6.5). Long-term articulation will necessitate a special flexible joint, which itself poses waterproofing problems (Hodgkinson 1983). Connections between subdivisions made before large settlements will cause problems.

Weak pockets   In cases where the building has settled differentially, it will be necessary to establish what the soil conditions are underneath. It is possible that the soil strata are inclined (Hodgkinson 1983), or that there is a localised pocket of low bearing-capacity soil or fill. The position of the weakness relative to the building will determine the type of symptom.

Clay soils

Reversible movement   Clay soil movement may be reversible, and any cracks must be filled with compressible fillers unless the structure has been modified to avoid the problem recurring.

Clay heave, hogging and symptoms   Clay soil heave produces hogging in the centre of a building, which tends to be indicated by single vertical cracks in external walls. The crack is often repeated in opposite elevations and also across solid floors and roof slabs. Alternatively, with a pitched roof there may be an opening up of the slates and tiles as the top of the building is pulled apart (see Fig 6.6).

Clay shrinkage, sagging and symptoms   The similar clay soil drying out in the centre of a building can produce a crack showing the converse symptoms. It will be necessary to review the recent climatic conditions and the nature of the site when appraising these types of cracks. Note that buildings which have their perimeter surrounding paved may have little direct water penetration below the foundations.

At the corner of buildings, cracking can occur with clay soils as shrinkage occurs and the foundations become unsupported. As they drop to the new

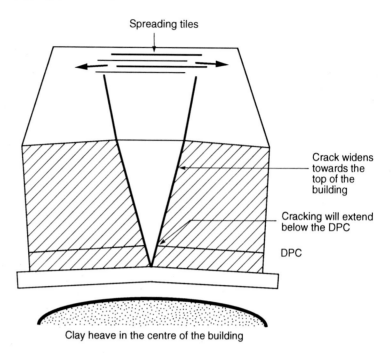

Fig. 6.6 Hogging failure in a simple masonry building. Single tapered cracks along the centre of a building may indicate swelling of the soil in the centre of the building or shrinkage of the soil around the perimeter. The tapered crack and spreading of the tiles indicates the building is failing by hogging.

Spreading tiles

Crack widens towards the top of the building

Cracking will extend below the DPC

DPC

Clay heave in the centre of the building

soil level a diagonal crack forms where the lower part of the wall rotates. In brickwork a characteristic stepped crack will appear, in concrete the crack may be relatively straight and closely follow the radius. This often appears or passes through a particular weakness in the wall, such as a door or window opening that cannot resist the tensile stresses as the corner rotates. With a corner crack the crack will be widest at the corner itself. See Fig. 6.7.

Seasonal and other factors

The fault is likely to occur after prolonged dry periods. The susceptibility of the building will depend on the nature of the soil, the depth of the foundations and the type of construction. Tree proximity places a significant extra emphasis on the soil's water content and may exacerbate the problem, making it worse or quicker to appear (more sensitive to dry weather). See p. 190.

More heave problems

Converse symptoms may appear to those associated with the corners of buildings if the clay soil is heaving. Here the soil is increasing in moisture content and bulking. This produces an upward pressure on the foundations and lifts part of the building. As with the shrinkage problem at the corners of buildings, the soil at the perimeter of the building is more susceptible to changing moisture contents than that under the centre of the building (the exceptions sometimes occurring with tree root damage), and so a corner of

Fig. 6.7 Corner cracking of a simple building failing as the supporting clay soil shrinks. The crack tapers and increases in width with the radius of rotation. Any openings in the walls will produce weaknesses which will concentrate the cracking. Since such types of defect originate at foundation level the crack will pass through the DPC. The crack width is likely to vary with the seasons.

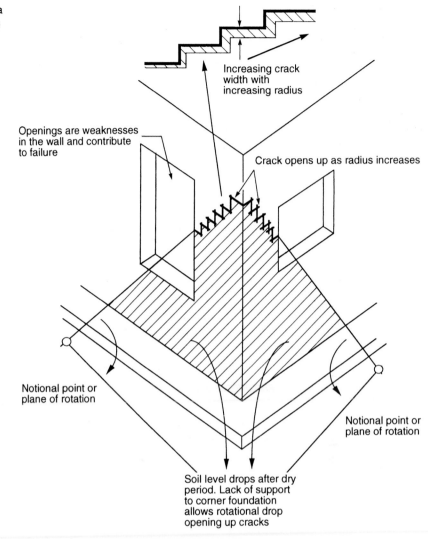

Increasing crack width with increasing radius

Openings are weaknesses in the wall and contribute to failure

Crack opens up as radius increases

Notional point or plane of rotation

Notional point or plane of rotation

Soil level drops after dry period. Lack of support to corner foundation allows rotational drop opening up cracks

the building, or even a side of the building, is lifted. This produces diagonal cracking, again often through points of weakness such as window or door openings (stress concentration points). The cracks will be wider towards the ground.

Moisture variability   These faults often arise with clay soils which have had trees cleared shortly before building (see p. 190), or were exceptionally dry when building took place. This soil then increases in volume with increase in water content. Where this fault arises in the centre of the building the type of crack that will appear is as shown in Fig. 6.6.

The role of trees in foundation failure

Tree planting

Stability problems can arise from the presence of established trees (Desch and Desch 1970), particularly after their removal (Cutler and Richardson 1981, BRE 1985a, Eldridge 1976), or in time from imprudent planting anew (Melville and Gordon 1979, BRE 1985a). Actions on adjacent land and beyond the control of the owner or developer can also produce stability problems for buildings. Checking for these and any other historical relevances will be an important preliminary step in the assessment of a structural defect.

Problems with tree planting

Melville and Gordon (1979) report that a lot of trees planted in the post-war booms were unsuitable for their location and their rapid growth has caused problems for buildings (Desch and Desch 1970). Cutler and Richardson (1981) report that prior to 1970 the safe distances for planting had been based on too little information, and to be fair they were qualified as being a rough guide at the time (BRS 1965a). Aldous (1979), however, comments that the role of trees in damaging buildings, and subsidence more generally, did tend to become overstated during and after the mid-1970s drought. Disproportionate insurance petitions (BRE 1981a) followed, fuelled by the recent introduction of subsidence cover to many domestic policies in 1971 (Hill 1982).

Factors affecting risks

In assessing the potential of existing nearby trees to damage buildings, account should be taken of the species of tree, and their size, number, and condition (Cutler and Richardson 1981), all of which may vary their particular water requirements. In particular, the area of leaf in the canopy will influence the transpiration rate. Pruning back the canopy will obviously only encourage growth unless the roots are pruned also (Benson 1984).

In addition to this the building structure itself will influence the response to soil movements. The harder the structure, the less compliant it is, and so the greater the likelihood, all else equal, that significant movement will produce visible cracks. The soft constructions, designed to accommodate movement, may absorb some of the soil movement and the risk of serious or obvious damage will be reduced.

Root drainage

Root drainage depends on the type of tree, but can realistically affect soil to depths of between 2.0 and 2.5 m, and critical movement zones of 3.0–5.0 m should not be discounted (Hill 1982). See Fig. 6.8. Poplars and oak, for example, root deeply in clay soils (Cutler and Richardson 1981). The critical movement zone on grassed sites can extend to 1.0 m (Hill 1982).

Fig. 6.8 (a) Zone of shrinkage and swelling in clay soil due to seasonal variations in moisture content where trees and vegetation are absent. **(b)** Effects of nearby trees on critical movement zone. (Reproduced by permission of W F Hill).

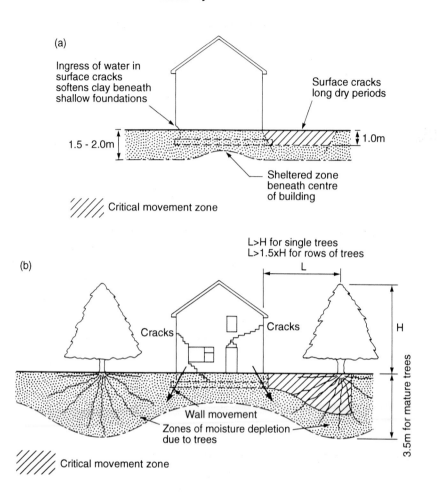

(a)

Ingress of water in surface cracks softens clay beneath shallow foundations

Surface cracks long dry periods

1.5 - 2.0m

1.0m

Sheltered zone beneath centre of building

Critical movement zone

L>H for single trees
L>1.5xH for rows of trees

(b)

Cracks

Cracks

L

H

Wall movement
Zones of moisture depletion due to trees

Critical movement zone

3.5m for mature trees

Root radii The placement of the building relative to the root radii (BRE 1985a), the amount of water available in the soil, and its shrinkage potential under drying (BRE 1985a, Cutler and Richardson 1981) will determine the site risk. Where the tree is close to the structure, it will be important before assuming the mechanism of failure to assess the relationship between the tree, the building and the available water supplies. The affinity for water will influence tree root development strongly, and a tree in between a building and a plentiful supply of water will make clear root development towards the water (Benson 1984). This does not eliminate the clay shrinkage problem, however, but the mechanism involves the drier clay below the tree drawing moisture from the subsoil below the building to maintain equilibrium. Differential subsidence can still occur in the structure even without tree roots below it (Benson 1984).

Hill (1982) suggests certain types of building can be expected to be particularly vulnerable to the differential movements of clay soils. These

include lightweight, low-rise structures with shallow strip foundations, particularly bungalows. Long walls and/or those with large openings have a poor stiffness, and are vulnerable to movement.

Alterations to buildings can influence the response of the structure to trees to disadvantage as well as advantage. Extensions to buildings made in a dissimilar constructional form, such as mixing soft and hard construction concepts, or with varying depths of foundation are likely to produce a composite building which responds to the influence of trees and other soil movements differentially.

It is also worth remembering that the dynamics of the tree and building system can retard as well as advance. Partially underpinning a structure to overcome the problems of clay shrinkage whilst a tree approaches maturity is likely to encourage relative movement between different depths of foundation during the heave following death or removal of the tree. The same holds for trees that have reached maturity. Further, Benson (1984) notes that the stepped underpinning of a structure can influence the migration of moisture in the subsoil and encourage additional settlements. The water collects at the deepest point of the underpinning, conventionally nearest the tree to best balance the effects of moisture depletion, and this attracts the tree roots. See Fig. 6.9.

Species and risks Species such as the poplar and oak are renowned for their root range (BRE 1985a, Desch and Desch 1970); Cutler and Richardson (1981) also highlight the plane, which was commonly planted as a street tree and so is commonly close to houses. Although they discount shrubs as a serious threat alone, Desch and Desch (1970) reviews a wide range of trees and shrubs responsible for substantial settlement damage. In prolonged dry periods

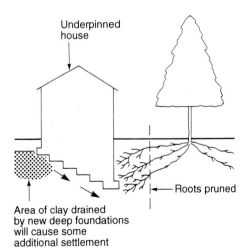

Fig 6.9 If the tree is allowed to grow freely, without root pruning, it will eventually grow down to the lowest point of underpinning, where the water has collected. (*Source*: John Benson 1984)

trees growing in shrinkable clays and drawing water from them can exacerbate natural shrinkage tendencies (Cutler and Richardson 1981), particularly in the highly or very highly shrinkable samples. Cutler and Richardson (1981) and Driscoll (1983) give details of analysing clay shrinkage characteristics.

The radius of impact potential problems varies with the tree, and in cases of groups or rows of trees this distance is increased (Hill 1982). BRE (1981a) cites the oak, poplar and willow as notorious risks to house foundations built within distances equal to their height on very highly and highly shrinkable clay soils. In other circumstances with less reactive soils it recommends some concessions in distance. Cutler and Richardson (1981) comment that 90 per cent of reported damage incidents occurred within a distance of the tree height. Desch (Desch and Desch 1970) reports not personally encountering any damage from coniferous species. Some support for this is given by BRE (1985a) which, although not discounting coniferous trees, does state they extract less moisture than the deciduous broad-leaved species.

Do not assume that the nearest tree is necessarily the source of any roots found to be causing damage (Cutler and Richardson 1981). Desch, for instance, recorded finding oak roots at 65 ft and causing damage, whilst a suspected row of poplars sat at 55 ft (Desch and Desch 1970). Climatic conditions which increase the transpiration rate of trees will enhance the threat from trees (BRE 1985a), so producing a seasonal trend for defect appearance in this country (BRE 1977a). Desch and Desch (1970) comments that the most likely period to discover cracks in structures that are attributable to existing trees is at the end of an unusually dry summer or the start of the autumn.

Established trees Potential problems with tree-planted soils may not manifest themselves for long periods, until exceptional weather occurs, such as the abnormally dry summers of 1975 and 1976 (also 1946/7) (BRE 1977a). This drought caused a large number of problems, often for the first time, for buildings founded on the shrinkable heavy clays present in the South East (Hill 1982), but also produced useful experience (BRE 1985a). Citing figures reported in 1979, Cutler and Richardson (1981) remark that nearly a quarter of all subsidence claims implicated trees, and of these 88 per cent linked clay soils and tree roots (Reece 1979*). Mature trees will often only severely affect the soil in extreme conditions, and their removal may trigger more problems as the soils recover (BRE 1981a).

---

* Figures of 2285 claims in a total of 10,684, and of the 2285 claims, 88 per cent implicating clay soils and tree roots, the remaining 12 per cent due to trees combined with other factors.

Stability of the structure

Preservation orders Developing trees responsible for structural faults may produce worse damage in the future, so there may be a case for their removal (BRE 1981a). In certain circumstances it may be possible to rescind a tree preservation order, but the implications should be carefully considered (Desch and Desch 1970).

Drains and trees Another form of tree damage can occur when drains are blocked and fail as a result of root interference (Cutler and Richardson 1981). Where this produces localised wetting and cavitation of the soil, settlement can occur (Cutler and Richardson 1981). In contrast, sudden wetting can produce a heave effect (Hill 1982). Hill (1982) notes that the incidence of direct physical damage by trees is likely to be rare and localised.

Heave Removing established trees can increase groundwater as the roots of felled trees no longer drain the soil (Hill 1982). This causes a localised heave effect in susceptible soils at least as far as the root radius (Hill 1982). Heave below the building can produce cracking and hogging failure. For instance, Bickerdike *et al.* (1971/72) recorded a case of a design fault in a house built on London clay soil. Here they identified the problems as being the result of removal of trees from the site only two months prior to commencing building work in dry weather. Hill (1982) notes that heave problems can occur unpredictably over several years. Common recommendations for cleared-site recovery periods range from at least one year to five, although research findings published by the BRE (Cheney 1988) show measurable long-term heave in clay soil 32 years after completion of the building (BRE 1981a, BRS 1965a). Cheney (1988) comments: 'If the problem of building heave arises it is virtually impossible to prevent swelling taking place . . . It is not sufficient to delay construction for "a few years" after tree removal at a new site.'

Long-term heave Many textbooks note the problems with a variety of prevailing trees, and recommendations for the time adjacent land cleared of trees should be left to lie fallow abound. Long-term research reported by the BRE suggests that on clay soils the effects on the soil of the removal of trees can be measured up to 25 years after the removal occurred. The subject of trees and clay soils requires careful analysis.

Symptomatic cracks In the case discussed by Bickerdike *et al.*, the result was that the previously well-dried soil swelled as the moisture content and level of the water table rose, lifting the house and causing a corner stepped crack characteristic of

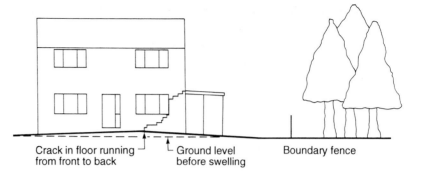

Fig. 6.10 It is important to distinguish between differential settlement of two parts of a building at their joint and the possible influence of tree roots on clay soils. Bickerdike, Allen, Rich and Partners, O'Brien T 1971/72 (Reproduced by permission of Building (Publishers) Ltd.)

Crack in floor running from front to back — Ground level before swelling — Boundary fence

drying shrinkage movement (BRE 1977a). The crack, which would widen towards the top (BRE 1981a), opened up as differential movement occurred between the house and a corner attached to the garage, as in Fig. 6.10. On first sight Bickerdike *et al.* (1971/72) noted the similarity to the rotation between a sinking garage and the house.

Soil recovery

Illustrating the significance of root position and the effects of heave, Melville and Gordon (1979) offer a circumstance of soil recovery after tree clearing which may produce cracks that become narrower towards the top. See Fig. 6.11. Similar underlying causes produce contrasting symptoms, and comparing Figs. 6.10 and 6.11 emphasises the importance of relative location of the tree(s) and the building to the damage caused (Cutler and Richardson 1981).

Bickerdike *et al.* (1971/72) had rejected the common conclusion that remaining trees had drained the soil under the corner foundation, and thus caused the crack. To come to this conclusion it is obviously necessary for the surveyor to know whether trees have been removed prior to building, since the apparent symptoms may be similar. BRE (1981a) illustrates a crack pattern similar to that assessed by Bickerdike *et al.* (1971/72), but due to drainage of shrinkable clay.

Fig. 6.11 Soil recovery. Although the reverse of settlement, movement due to soil recovery bears a marked resemblance in the pattern of cracks formed to differential settlement in the ends of a wall. There could be confusion between the two if no account is taken of the site conditions. As the ground swells parts of a building are forced upwards and the cracks which form diminish in width as the height increases. (I A Melville, I A Gordon 1979).

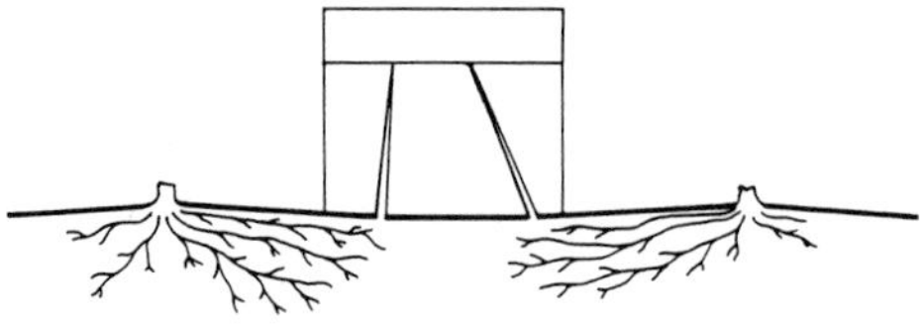

Stability of the structure

Reversible movement In addition to the indirect effects of trees on buildings, operating through the moisture sensitivity of clay soils, there is a possibility with lightly loaded walls that large tree roots may physically uplift the wall. This is unlikely to occur in buildings; rather it is limited to the lightweight structure and loading of garden or boundary walls. This type of movement should be distinguished from reversible soil effects.

Secondary symptoms Frequently the damage to the soil will be indicated by secondary symptoms, such as cracked pavements close to trees or crazed patterns on dried ground (Melville and Gordon 1979).

7 Instability in the fabric

Permeable envelopes

Development of the clad form

Cladding is usually distinguished from conventional solid or cavity walling on the basis of its functions. There are usually obvious constructional differences although these may be concealed by the exterior design appearance of the building. It is essential that the cladding can maintain its structural integrity whilst accommodating the movements of the structure. Where this has not occurred the cladding may become defective. Examples of deficiency exist across a range of systems and materials.

Scope and revised understanding

The term cladding relates to clothing, and the overlaying of materials. For the purpose of distinction in bylaws, cladding was defined as an enclosing facing which did not have a common action under loading with the substructure (Gilder 1989). The revision of understanding of cladding now has led to a performance-orientated definition concerning hygrothermal and structural function.

There are arguments of definition over the transition from vertical or near-vertical cladding to roofing. The distinction in terms of occurrence and identification of apparent defects is less pressing, although categorisation may clearly be of relevance in legal issues.

Loadbearing and non-loadbearing distinctions

Claddings may further be distinguished as loadbearing or non-loadbearing, and this has proved a constant source of dispute. Such classifications can easily overlook some of the most important cladding roles, and most likely sources of defects. A principal requirement is the resistance to imposed wind loads. The cladding and its supporting framework must not deflect excessively, nor should it compromise its weather-tightness properties. Obviously the fixings to the main structure will be critical. The forces may be positive or negative.

Very rarely does the modern external cladding skin contribute to the loadbearing capacity of the structure, and this is a particular distinction. One exception is the stressed skin structure.

Stability of the structure

Serviceability records   The actual number of cladding failures is relatively low taken in the context of the amount of cladding used (Gilder 1989). Indeed, there appears to be a relatively good serviceability record. The cost implications of failure in use are extreme though, and the serviceability record must be set against the degree of difficulty in carrying out inspections and maintenance. Hidden connections and lack of simple access to the rear of claddings further complicates inspection, meaning that failures are usually discovered at advanced stages.

Failures and fixings   Failures in the weathershielding of buildings generally form the bulk of building defects (Tietz 1989). They are by no means restricted to cladding. There are recorded incidents of cladding falling from buildings where the fixings have corroded or fatigued, and where detailing has been deficient (Dore 1989). It is obviously important that fixings are durable, but also that they can accommodate differential movement between the envelope and the building. It is essential that any structural secondary framing can accommodate all the forces transmitted through it to the structure. In tall buildings sway may be a significant factor (Hunton and Martin 1989).

Recorded faults with stone claddings   With the traditional heavyweight forms of cladding, such as slabs of Portland stone, granite and marble, the self-weight of the cladding makes a significant contribution to its stability. Their inertia also helps them resist the rapidly changing wind forces. The tying back with fixings was largely unremarkable, except during a period where the fixing of all four sides of panels was common practice. This prevented relief of secondary stresses arising from thermal or moisture movement in the panels, and damage could occur (Green 1989).

Problems with permeability   Rusting of concealed steelwork has been recorded in buildings constructed with the permeable envelope in contact with the steel frame (Gilder 1989). The problems with permeability in brick cladding is largely dependent on the mortar and the method of pointing.

The use of permeable materials enjoyed particular popularity in the 1980s (Green 1989). There are, however, significant cost and weight implications of choosing a brick cladding. Also, pressure to maximise lettable space has led to the replacement of the traditional thick panels with thin, sometimes laminated, forms of insulated wall construction. The general concept of a unitised or panelled construction has emerged. There have also been concerns regarding the position of the weatherline, which is now nearer the surface of the building envelope (Gilder 1989). Consequently, modern permeable claddings contrast less sharply with lightweight framed and glazed panel systems and the developments in structural silicone glazing.

Whilst retaining traditional monumental appearance, the connections to the building, and the relationship with the cladding material has changed. In particular, the use of rigid panelling fixed to relatively flexible steel structures has become recognised in the USA as requiring careful design (Dore 1989). There has been at least one recorded instance in the USA where complete re-cladding has been required following distortion in thin-sliced stone panels (Dore 1989).

Recorded faults with precast concrete claddings

Precast concrete cladding symbolises the industrialised building era. The structural nature and condition of these systems varies widely, principally between loadbearing or non-loadbearing envelopes.

Focusing on non-loadbearing concrete claddings, the tying to the structure is critical to prevent problems occurring with differential movement as the frame initially shrinks, followed by long-term creep. For instance, 3 mm creep in a panel dimension of 3 m is quite possible.

The possibility exists of acquired loading as the frame shortens vertically. The compression transfers structural loading onto the non-loadbearing cladding. Symptomatic failures include bowing. Visible cracking may occur as the outer skin is tensioned, possibly culminating in the complete failure of panels as the nibs connecting them with the frame shear under the imposed loads. Examples of possible defects are shown in Fig. 7.1.

Recorded faults with brick infill claddings

The brick-clad structures of the 1970s also suffer problems due to differential movement. A common cause is the counterstress between the expanding clay brickwork and *in-situ* concrete frame as it undergoes drying shrinkage. Where there is inadequate provision for expansion, the brick panels may buckle and dislodge, producing failure in the weakest and thinnest parts of the brick cladding. This commonly appears at the slips between floor levels. The adequacy of fixing can be an issue in itself. The provision of deeply recessed joints to brick claddings can further reduce weathering performance. These have all been confirmed as significant sources of problems in the past (Gilder 1989).

Conflicting forces

An alternative mode of failure can occur with the concrete supporting nibs supporting the brickwork being pulled out from the remainder of the frame by the distorting brickwork as it is compressed. Even where the expansion of the outer leaf is significant, the concrete inner leaf may show few or no signs of distress.

Fig. 7.1 Representation of defects associated with precast concrete cladding panels.

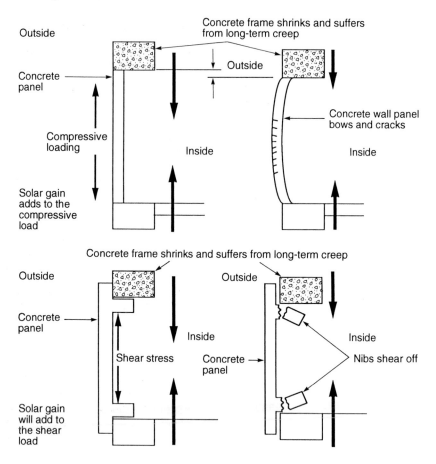

Recorded faults with rendering

The incidence of reported rendering defects appears to be increasing (BRE 1983b). Some common faults are summarised in Table 7.1. This escalation is for several reasons: the increase in rendered building stock; an increased awareness of the problem; and an extension of the legal liability for inherent defects.

Workmanship and material characteristics

The quality of workmanship when rendering has always been critical. With the advent of Portland cement approximately 160 years ago came the opportunity to produce strong durable renders. Until that time lime renders were used. The properties of both types of rendering are quite different. In contrast to the lime-based renders the cement-based coatings are characterised by their brittleness and poor tolerance of substrate movement. This shows up where renders pass over substrates with markedly different movement characteristics, as shown in Fig. 7.2. The newer single-coat types contain a range of resins and polymers which give the material improved strength and movement tolerance.

Table 7.1 Defects associated with rendering

Defect	Cause	Remedy
Cracking (random)	Where cracks are related to cracks in the wall	Remedy defect in wall. Remove, and replace with suitable mix
	Shrinkage of rich mix. Excessive trowelling	Where surface is sound decorate with filler type paint. Where render is loose the options are: i) Patch repair ii) Total removal iii) Consider alternative external finish iv) Apply weather cladding over the rendering
	Sulphate attack from background (see horizontal cracking)	Consider providing rain screen cladding Re-rendering of minor areas may be possible if the defect source can be eradicated
Cracking (horizontal)	Sulphate attack. Where general the source may be a clay brick background. Where localised the source may be the mortar	Arrest the leaching of sulphates by the use of rain screens Minor areas may be re-rendered when the source has been eliminated Re-rendering of large areas on a suitable waterproof lathing
Areas detached in random pattern	Under burnt clay bricks	Remove isolated bricks Where damage is extensive rain screens can be considered
Flaking	Cement content of top coats greater than undercoats	Where isolated, damaged areas can be removed, reface with less cement-rich mix
	Poor mechanical key	Remove, and replace onto suitably roughened background
	Shrinkage of the background	Remove and replace, where the risk of major shrinkage has reduced
Blistering of paint film	Water in the rendering	Remove defective paint, remedy the cause of the moisture in the render (which may be due to interstitial condensation)

Table 7.1 continued

Defect	Cause	Remedy
blistering (cont'd)		Repaint with a more permeable paint film
(Where the back of the film is soft)	Saponification	Remove defective paint, which may be large areas, and replace with alkali-resisting permeable paint

Stucco finishings   Stucco, which dates from around 1790, is another specialist finish with particular cracking problems. Where stucco or render coats develop defects these may lead to secondary faults. The coatings may become cracked throughout their depth, allowing ingress and trapping of water to produce long-term dampness in the substrate. The consequent damage may be more severe than if the substrate was totally exposed.

Fig. 7.2 Failure pattern of rendering over a composite background to a bay window. Weathering damage to the exposed window head is evident.

Salts in renders The dissolved salts and minerals in the mixing water can affect the durability of render coatings. This may be exacerbated by polluted or acidic rainwater. The salt content may retard curing of the render, and leave it liable to frost attack in winter working. This may emerge as an immediate fault or leave the render susceptible to premature failure.

Background and bonding Defects in the rendered substrate may be chemically or physically induced. There may be failure of the mechanical bond, or differential movement between the coating and the background. The causes of differential movement may be moisture or temperature related. In addition to this, the tendency of cementitious material to shrink may combine to stress the inter-face bond. The loss of adhesion may also be due to the rendering being applied to saturated walls (BRE 1985c). In such circumstances frost and cold weather can also cause weakness in the rendered coat (Taylor 1983a).

Appearance Poor surface texture is a common cause of appearance defects. Plain sheets of cement render present an appearance of utilitarian dullness. They are likely to craze and develop a patchy appearance. In contrast, where the surface irregularities are deep this prevents the surface self-cleansing. Discoloration can be rapid and very unsightly.

Cracking symptoms Extensive cracking will lead to breaking away of sections of the render, and will allow rain penetration of the primary defence. Any water penetration between the render and background or between separate render coats will accelerate the deterioration. Examples of cracking patterns are shown in Fig. 7.3.

Shrinkage and sulphate cracking Shrinkage cracks occur during hydration, and variations in the cement content can create areas of render with differential shrinkage characteristics. Problems with sulphate attack may arise from within the render itself or the background, especially brickwork. Cracking associated with sulphates in the brickwork may be random or predominantly horizontal. Random cracking is caused by a uniform distribution of sulphates in the brickwork; horizontal patterns are produced where the sulphate source is the mortar joint.

Advanced sulphate attack can cause detachment of large sections of rendering, or can be symptomatic of differential movement.

Recorded faults with mosaic cladding This is an area of envelope construction where the likelihood of long-term successful adhesion of a tiled surface depends on strict quality control

Fig. 7.3 Crack patterns associated with defective rendering.

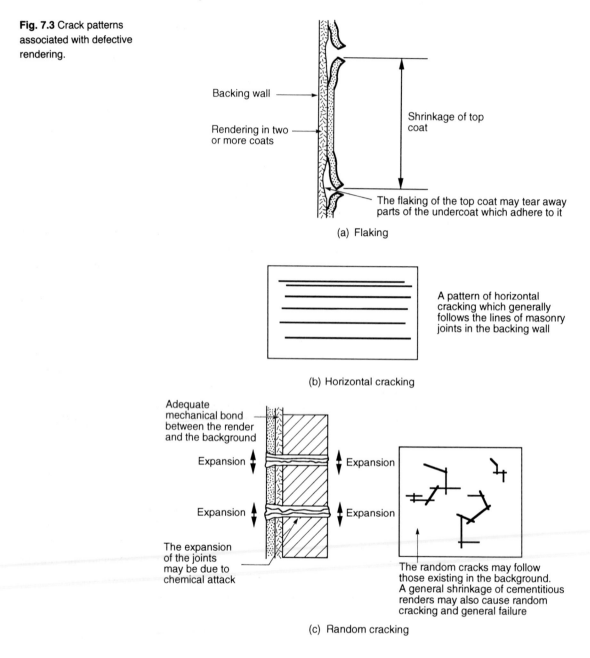

Backing wall

Rendering in two or more coats

Shrinkage of top coat

The flaking of the top coat may tear away parts of the undercoat which adhere to it

(a) Flaking

A pattern of horizontal cracking which generally follows the lines of masonry joints in the backing wall

(b) Horizontal cracking

Adequate mechanical bond between the render and the background

Expansion

Expansion

Expansion

Expansion

The expansion of the joints may be due to chemical attack

The random cracks may follow those existing in the background. A general shrinkage of cementitious renders may also cause random cracking and general failure

(c) Random cracking

during construction, and the types of mixture and their application. Indeed, Ransom (1981) warns that failure is more probable than success. In a number of ways the detachment or security of tiled surfaces depends on similar criteria to that of rendered skins. Both are usually fixed to the building without reliance on mechanical fixings.

Mosaics used to clad the edge beam detail of a panelled frame building can suffer the same problems with compression failure caused by creep or shrinkage-induced bowing of the panels. Bulging and wholesale detachment of sheets of mosaic are possible since the resistance of the adhesive to the compressive stresses is inadequate.

Failure and detachment Failure and detachment of tiling or mosaics can be entirely due to incorrect choice of adhesive or application method. Debonding fluids used in the casting of the framework and improperly removed may act as a release agent for the finished cladding.

The type of tile will directly influence the quality of bond. Pressed tiles have a relatively shallow ribbing on the rear side compared with that possible by extrusion, and the fixing key to the adhesive or mortar is correspondingly reduced. Failure is characteristically different. Pressed tiles detach from the fixing material. Extruded tiles tend to bring the backing adhesive with them. The correct application of the adhesive is very important for the security of the tile, but has significant cost implications. Patchy adhesion causes localised stresses and encourages early failure.

Bond quality Quality of bond will also be influenced by the absorbency of the background. A similar condition occurs with plaster and rendering. Tiled surfaces frequently suffer stresses at the interface between the tiling and the background, caused by differential expansion coefficients. This is particularly problematic with concrete backgrounds where the irreversible shrinkage of the concrete will cause high compressive stresses. The tiles will crack or become detached from the wall. Such problems are amplified where the bedding material is used as levelling screed for unevenness in the background.

Frost and sulphate attack Frost attack of the bonding material is possible, particularly where partial failure or long-term porosity of the mortar joints between the units allows water ingress. The pressure exerted by the freezing of water, and the reduction in strength of the bonding material following cyclic freezing rapidly decreases the quality of attachment. Failure is either as a direct result of the freeze–thaw cycles or as a result of the weakened resistance of the connection to differential movements between the structure and the cladding.

Stability of the structure

In addition to the risk of frost attack there is a distinct possibility of sulphate attack to tiled or mosaic claddings applied to highly sulphated brickwork. Ingress of damp will lead to dissolution of the sulphates and the tiling is forced off.

Background movement and failure

The movement of tiled panels is not necessarily indicative of a primary fixing failure. Any movement in the background is mirrored at the surface, so giving an early indication of the failure or cracking of the background. This is shown in Fig. 7.4. The intolerance of the rendering to background movement may therefore be a relative advantage over many of the other choices for the cladding of modern buildings.

Fig. 7.4 Deterioration of factory-applied finish to precast concrete panel indicative of background failure around an expansion joint and at ground level.

Impermeable envelopes

Definition   An alternative method of creating a weathertight envelope is to use impermeable materials. A generalised summary of absorption and impermeability is shown in Table 7.2. The differences between impermeability and absorption will mean that high surface absorption may be associated with materials of low impermeability. Where pore structures are small, total absorptions are likely to be low, although it is only where a developed pore structure passes through the material that permeability exists.

Historical evidence   The use of impermeable materials for the external envelope of buildings can be traced back to the use of hides, leather and leaves. Modern examples include the use of glass, metals and plastics. All of these are subject to defects associated with the weathering of the material. Many early forms of impermeable envelope were associated with the introduction of new materials and methods, e.g., curtain walling and vitrified glass. These could be considered as the practical development phase of some of these methods. The risk of failures occurring to finished buildings appeared to be acceptable. The lighter materials were used as the non-loadbearing external envelope to framed structures.

Relative impermeability   The molecular scale of the moisture will also be important when considering the relative impermeability of materials and systems. Simple measurement of absorption and impermeability may fail to account for the combined effect of both. Water droplets and flowing water may pass over the material

Table 7.2 A generalised summary of the relationship between surface texture, pore size and structure on absorption and impermeability

| | Surface | | Pore size | | Pore structure | |
	Smooth	Textured	Small	Large	Linked	Unlinked
Abs. (High)	No	Yes	No	Yes	Yes	Yes
Abs. (Low)	Yes	No	Yes	No	Yes	Yes
Imp. (High)	Yes	Yes	Yes	Yes	Yes	No
Imp. (Low)	Yes	No	Yes	No	No	Yes

Abs. = Absorption
Imp. = Impermeability

surface whereas water vapour may migrate through it due to the difference in general molecular size. This helps to explain why impermeable claddings may also be poor at allowing a hygrothermal equilibrium to exist across them.

Dissimilar materials   The use of a range of materials for the external envelope of a building will mean that a difference in absorption patterns will be created. This will have implications for the general surface temperature and cause concentrations of water runoff. Slight differences in permeability can influence the pattern staining of facades.

Material impermeability   Material impermeability directly influences the impermeability of the envelope. The time taken for water to migrate from the outside to the inside of the envelope will generally be extended by the use of impermeable materials. However, materials which under normal conditions do not allow water to reach their inner surface could also be considered as impermeable. It is as important to consider the degree of exposure and the local climate, as it is the standard impermeability of the material. Saturated materials may then behave impermeably. This is likely to occur when the rate of water absorption into the material is less than the rate of water arriving at its surface. In this way 'permeable' materials can act in an impermeable way.

Impermeable zones   The roofing material is usually designed to be impermeable. The materials traditionally chosen are capable of rapid soaking at their surface, thus allowing water to flow into the roof drainage system. The impermeabilty of certain materials is so good that they can act as a vapour barrier. This will trap moisture within the building, causing vapour pressures to be developed from inside and outside the facade. Water ponding will be more likely, which will stress joints. Large walls can be other zones of impermeability. The curtain walls and large-scale metal claddings have similar properties to roof coverings. Indeed, the difference between wall and roof cannot be justified for many buildings.

Impermeable areas   Areas of impermeabilty include glazed areas. These can occur in isolation surrounded by a mainly permeable envelope, or as part of a total zone. The historical growth in the provision of natural light into buildings has demanded an increased emphasis on impermeabilty of the envelope. This has been modified recently as the compromise between lighting and heating becomes loaded towards energy conservation.

Since most buildings contain some glazing, the defects associated with the impermeable envelope can exist in many buildings. The relative areas of glazed to unglazed have changed historically. The availability of large sizes of glass enabled window sizes to grow. These impermeable areas can then concentrate water runoff to the lower levels, placing an increasing load on the surrounding permeable areas.

Treatment of junctions Where non-homogeneous envelopes are adopted, an increased emphasis is placed on junctions and joints. These can be the weak link since they are the points at which the non-homogeneity of the envelope is broken. The need for the joint to accommodate movement whilst retaining a weatherproof integrity appear to be mutually contradictory. Many defects in impermeable envelopes can be traced to deficiencies in the joint.

The concentration of runoff water will add to the loading of the joints. Whilst it may be possible to keep the joints away from the main runoff paths in roofs, this is more difficult in walls.

The lightweight impermeable skin The emergence of the lightweight impermeable skin has produced a distinct change in the approach to achieving performance criteria. The structural criticality remains, but imposed over this is the need to allow for physically significant and frequently differential movement within the envelope. The well-publicised howling failures in the United States are still referred to: the cracking or loss of panels as frame and infill moved differentially with inadequate allowance. The advancements in the technology have been significant and the cladding systems emerge from being very large windows through the early single-glazed and front-sealed envelopes to dynamically responsive skins. Nevertheless there have been reports of many failures with glass, because of strength and inclusion in the facade (Green 1989).

The relationship with the frame The shrinkage of the concrete-framed structure can still produce problems. See Fig. 7.1. Where the lower floors shrink or deflect, the loading transferred to the curtain walling once the tolerance for movement in the joints is exhausted cracks the glass panels. Alternatively, at the roof level the relative rigidity of the parapet wall will limit deflection of the roof slab only. The cumulative effect of the floors sagging below can cause falling out of the glass panels (Hunton and Martin 1989).

Rainscreen cladding Recently there has been the widespread emergence of the rainscreen cladding systems which address the problem of water exclusion at the outer weatherline pragmatically by introducing the concept of controlled entry of water and planned discharge. The systems now mostly operate on a pressure-equalisation concept where voids in the external skin allow rapid variations in the air pressure behind the outer skin to match very closely those of the gusting wind driven rain. This minimises or eliminates the differential pressures across the skin which are instrumental in conventional systems for forcing water through joints. These systems have been used also in the advanced frame-and-panel systems of curtain walling.

The joint   Joints and jointing technology has been the central issue over failure in the last twenty years in modern claddings. Vast improvements have been made. The traditional mortar joint as used for the brickwork cladding panels and as connections on the construction of stone-clad structures has an average life expectancy of probably about 20 years (Gilder 1989). The emergence of the front-sealed systems a quarter of a century ago has produced serious problems (Campbell 1989). Their function was at best optimistic, and water ingress was severely problematic as the sealants deteriorated and produced capillary joints which encouraged water penetration (Josey 1986). Pumping of wind-driven rain during gusting conditions was also problematic. A schematic representation of the possible failure mechanisms is shown in Fig. 7.5.

Structural silicone   This has been significantly improved by the adoption of structural silicone, although this is still a relatively new technique. There have been very few failures reported so far. In Australia there is experience of the use of structural silicone over ten years, to 50 storeys (Hunton and Martin 1989). Green (1989) reports a failure rate in the USA of below 0.1 per cent, with no instances of glass falling out and all failures occurring within the first three years of life. The few problems are reported to have been traced to the adhesion between the silicone tensile bead and the metal substrate (mullion or transom) rather than between the glass and the silicone. Reviewing curtain wall engineering, Hunton and Martin (1989) make the following sobering comment about the Australian market: 'There is strong evidence that the concept of structural silicone glazing continues to be marketed without sufficient engineering review.'

Cleanliness during assembly is of paramount importance, and will be particularly difficult to achieve on site (Hunton and Martin 1989). The mechanism of deterioration of the joint is likely to be related to the response of the silicone to deflection of the joint under alternate positive and negative wind-loading pressures (Wilson 1989).

Obviously, wherever a component of a building is critically dependent on a single line of support, planned inspection is a must. In these cases the quality of assembly, and the adhesion and durability of the silicone are critical (Hunton and Martin 1989). Many of the systems, however, have a

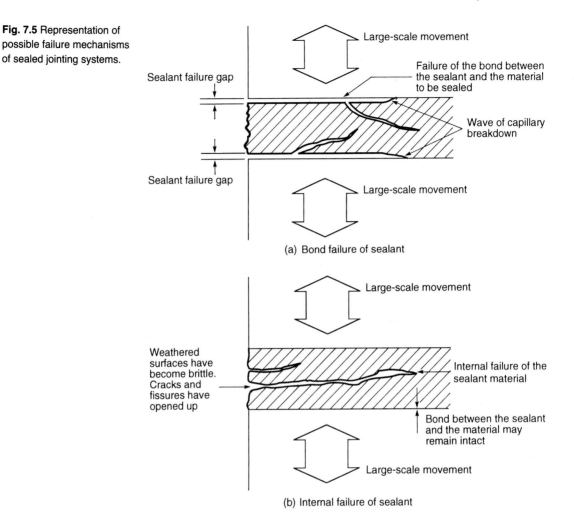

Fig. 7.5 Representation of possible failure mechanisms of sealed jointing systems.

Large-scale movement

Failure of the bond between the sealant and the material to be sealed

Sealant failure gap

Wave of capillary breakdown

Sealant failure gap

Large-scale movement

(a) Bond failure of sealant

Large-scale movement

Weathered surfaces have become brittle. Cracks and fissures have opened up

Internal failure of the sealant material

Bond between the sealant and the material may remain intact

Large-scale movement

(b) Internal failure of sealant

secondary support in the form of concealed or disguised corner cramps. It is worth noting that an initial contraction in the silicone of about 2–3 per cent has been reported (Wilson 1989). Any voiding in the silicone or damage to the surface of the exposed jointing should be avoided.

Sealants and gaskets

As with any building materials the durability in service of the material will appear only with time. In the UK gaskets followed the early gunned sealants, but had characteristic faults of springing out of the joint or being pulled out under cleaning operations (Gilder 1989).

A lot of the problems arise from the desire for maintenance-free buildings. This also is changing and expert clients are now able to determine and integrate the maintenance costs and practicality into their appraisal of designs offered to them. This in itself should produce a major advancement.

Faults There is a degree of similarity in the types of shortcomings occurring with the variety of claddings. Most are related to the environment. In particular variability in materials used and design input can create uncertainty and complexity in design image. The real problems occur where the cladding interfaces with the framing (White 1989) or the surrounding building.

The significance of faults in the stability of cladding lies firstly with the difficulty of identification because of the concealed nature of fixings, and the impracticality often of inspection. So, secondly, with the likelihood that they will emerge catastrophically. The obvious implications of falling cladding panels are sufficient reason alone for concern and attention to the avoidance and identification of deficiency. This has not gone unnoticed by the institutions and the Standing Committee on Structural Safety has reported on the problems of cladding safety (CIRIA 1989, Dore 1989). Sub-judiciary rules or agreements barring release of information identifiable with particular building failures form recurrent barriers to learning from constructional failure (Gilder 1989). In the USA information about failures is frequently more readily available in the public domain and the benefits of hindsight can be taken (Green 1989).

Structural stability The majority of the failures in claddings fortunately fall short of affecting the structural stability. This is not unknown, however. The frequent problems lie with serviceability in the envelope, a primary function with the non-loadbearing claddings. The effects on the everyday use of the building and its internal climate are more obvious, and the enormous costs associated with the modification or replacement of the cladding are a key reason for the importance of good serviceability, or fitness for purpose.

Testing of panels This distinction is clear in the standard tests and much effort is concentrated on producing water- and wind-tight claddings. The difficulty in producing this repeatably and reliably on site is twofold. Firstly, there is the translation of theoretical design into practical reality, secondly the problems of site workmanship and its supervision. There is ample evidence from the 1960s and the problems with this are sufficient to encourage testing facilities to be used, usually prior to use or completion of the building. It is not unknown for testing to occur just prior to practical completion (Green 1989).

It must be borne in mind that the adequacy of testing and the performance of the materials in practice may not directly relate. The tests are short term and do not necessarily reflect the lifetime performance (Battrick 1989). Tests on *in-situ* work only examine the potential that can be achieved (White 1989). This need not be through any avoidable shortcomings in the test, rather the testing of an *in-situ* panel leaves variability on site an unknown factor. There may also be a disparity between the mock-up and full-scale applications.

In the United States the criteria for satisfactory performance may be more extensive than for the UK, particularly in the area of seismic activity. There the testing of panels frequently requires racking loads to be imposed on the tested panelling followed by re-testing to their normal criteria. Note that for impact testing a cracking of a glass panel may not be classed as a failure (Muschenheim and Burns 1989).

Ingress of water
This involves not only moisture but also frost and snow. The wind and water permeability of the structure is an important criterion of performance, and tests make no distinction between air ingress around the whole frame or at individual locations. Damp control extends to rainscreen cladding, where the provision of ventilation into the building must avoid directly crossing the external cavity, otherwise damp is pulled into the building and introduces the risk of condensation (Hugentobler 1989).

Materials and the response to climate
Care also has to be taken with the mixing of timber and aluminium as an external shield to the framing. This can easily provide degradation and damage to the wood as the high conductivity of the aluminium lowers the temperature within the wood profile locally. Where the aluminium covers the wood completely there is a more uniform drop in temperature and the wood components seems to respond better (Hugentobler 1989).

Connections of the framing of the cladding structure to the substrate must be made with care. If the connections between the frame (subjected to the warmth and dampness of the interior environment) and the substrate are made on the cold side of the insulation layer there is a temperature drop within the connector. In effect the connection forms a cold bridge (Hugentobler 1989), and in the absence of an effective vapour barrier condensation can occur.

Differential movement
This has been the principal problem (Gilder 1989), both in the context of relative movement between the external envelope and the supporting substructure, and within multi-layered claddings. Diurnal and seasonal thermal movements within the cladding and moisture-related movements in the permeable claddings compound the dynamic relationship between the supporting structure and the envelope.

Thermal movement
Thermal movement is discussed elsewhere in more detail. It is worth noting here that the extent of movement occurring within lightweight thermally insulated claddings can be extreme, especially with dark facings which can absorb a great deal of heat. See Fig. 7.6. Where there are long runs of dark

Fig. 7.6 Schematic representations of thermal expansion problems in thin panel and insulated claddings.

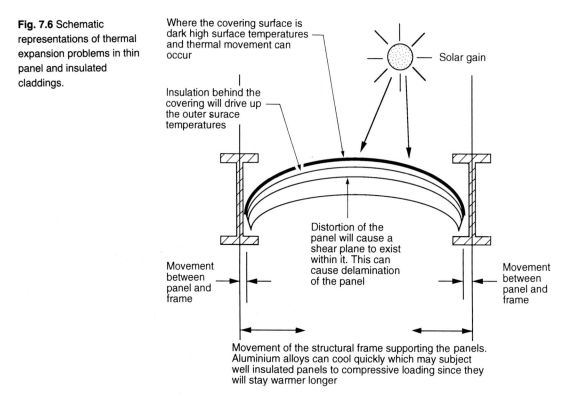

Where the covering surface is dark high surface temperatures and thermal movement can occur

Solar gain

Insulation behind the covering will drive up the outer surace temperatures

Distortion of the panel will cause a shear plane to exist within it. This can cause delamination of the panel

Movement between panel and frame

Movement between panel and frame

Movement of the structural frame supporting the panels. Aluminium alloys can cool quickly which may subject well insulated panels to compressive loading since they will stay warmer longer

material with little provision for expansion, the corner details may fail as the expansion of the coverings meets and stresses conflict at a change in direction. The movement of such panelling is usually relatively unrestricted compared with the tying-in effects occurring from self-weight in conventional or heavyweight claddings.

Workmanship   Frequently the manufacturer and site assembly staff are blamed for faults in the workmanship of the systems. Indeed, there can be severe environmental problems on-site against pressure for rapid completion. There is a factor of buildability of the design also. Poor detailing that makes correct assembly on site or in the factory impractical is also a workmanship fault (Gilder 1989).

Tolerance   Tolerances have been a problem in the past, particularly with the concrete systems of the 1960s (BBC 1984). Control and coordination of tolerance within a closed system of cladding should be readily attainable. However, the coordination with the building frame, particularly *in-situ* concrete frames is more difficult. The manufacturing tolerances achieved in the high-

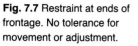

Fig. 7.7 Restraint at ends of frontage. No tolerance for movement or adjustment.

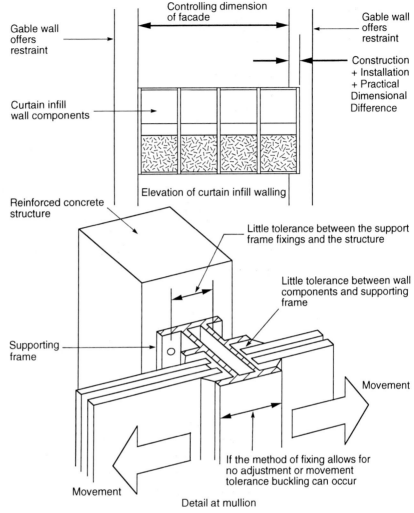

Gable wall offers restraint

Controlling dimension of facade

Gable wall offers restraint

Construction
+ Installation
+ Practical
Dimensional
Difference

Curtain infill wall components

Elevation of curtain infill walling

Reinforced concrete structure

Little tolerance between the support frame fixings and the structure

Little tolerance between wall components and supporting frame

Supporting frame

Movement

Movement

If the method of fixing allows for no adjustment or movement tolerance buckling can occur

Detail at mullion

quality modern curtain walling systems can outstrip those attainable in very high quality *in-situ* concreting, simply because of the nature of the materials involved. Instances have been recorded where ignorance of the tolerance requirements of the early infill curtain walling led to ad-hoc accommodation at the gable walls. See Fig. 7.7.

There is also a particular need for high-quality supervision where the detailing is unconventional, since genuine misunderstandings or apparently superficial changes to the detail on site can cause failure. See Fig. 7.8.

Thermal breaks   The absence of thermal breaks on the early curtain walled systems led to severe problems with surface condensation on the inside of the structure.

Fig. 7.8 Poor-quality detailing of joint in framed glazing system. Mismatch between transom and mullion is made up with excessive sealant.

Developments have not been without their problems either (Hugentobler 1989).

The design wrangle   There have been problems, because of fee structure, in getting structural engineers involved with the design of cladding for structures. Further, the legal position over responsibility is unclear and there appears to be confusion over implicit responsibility through general involvement with the project (Campbell 1989). Meanwhile there are calls for increased involvement to produce an improved product (Hunton and Martin 1989). Because of the frequently unclear lines of responsibility for design and checking there is a risk of defects going unseen at the design stage.

Testing may be unpopular because of the delays to the contract; for example, testing for salt crystallisation of stonework can take a considerable amount of time, and there may be pressures of expediency. Obviously with *fast-track* construction methods there will be acute pressure on testing or design changes during construction, where there may be as little as 10-day turn-around on design work.

Materials and life expectancy

The composite nature of modern cladding systems means that there may be a need for differential lifetime assumptions for certain components in the cladding. This may mean that although a life expectancy of 25 years may be considered a maximum for a commercial building in the South East, the maintenance or replacement of sub-components will require to be considered over a shorter time period. It is worthy of note that the life expectancy of many claddings in the USA is shorter than in the UK (Burns 1989). The situation in Europe may be variable also.

Planning maintenance

Difficulties or impracticalities in carrying out planned maintenance during the interim may shorten the life of the cladding. It should be noted also that the life expectancy of 25 years in the capital may not be achieved in the northern cities where the cost of re-cladding is less justifiable against land value and building appreciation. This relatively rapid turn-around can also relieve some of the pressure of retaining the older types of cladding system.

Moisture control

The position and reliability of vapour barriers is critical. Many specifications will be calling for sheet metal vapour barriers rather than polythene, for durability's sake. Moisture control is one of the most critical factors for serviceability and durability of the cladding structure. It is frequently associated with the premature failure of conventionally attached claddings.

There have also been problems with the staining and deterioration in polluted environments of anodised aluminium coatings (Campbell 1989).

On-site conditions are a potential cause of failure. In addition to the chronic citing of supervision as the cause, there are also possible problems with the weather and general environment in the use of sensitive sealants. Factory assembly is to be preferred (Campbell 1989, Green 1989).

8 Instability in materials

Material-induced instability

Stability-related faults
Buildings may develop stability-related faults because of weaknesses in the chemical, physical, or dimensional stability of their constituent materials. These types of instability may not be immediately significant for the soundness of the structure, but they may initiate more consequential secondary effects. Realistically, the useful life of a building will depend fundamentally on the durability of its components. This will vary according to the material type, the quality of the original design, the nature of exposure, and the attention to maintenance.

Premature deterioration
Premature deterioration is the acceleration of the inherent degradation all materials suffer. This is caused by externalities such as the environment and surrounding materials, perhaps exacerbated by design or construction inadequacies. Often the distinction between inherent deterioration and the influence of externalities is unclear. Frequently material failures turn out to be the result of incompatibility.

Identification
The defects analyst must be able to identify the causes, and their possible combinations, that have produced a defect in the building material. This is often necessary to attribute blame, and is essential if the chosen solution is to be right for the problem.

Principal mechanisms
The principal mechanisms involved are physical or chemical instability. Decay of the materials used for construction is often a natural and foreseeable event. However, the thought in the design and the nature of the environment play an important role in decay.

Chemical or physical incompatibility between materials can cause problems with reaction and degradation, or movement-related defects. Thermal movement and moisture movement incompatibility between materials is commonly one of a range of problems affecting composite structures.

Deterioration of materials

The porous and impervious contrast

The defects commonly occurring with porous building materials traditionally used as the external fabric for buildings can be contrasted with those arising since the popularisation of the impervious facade. Both have disadvantages and advantages, not least in the area of durability. The mechanism of weathering differs between the two and the defects analyst should be aware of the differences, especially if called to review a building with a combination facade. The effects are dominated by design quality. Poor design detailing is revealed as a distinctive and generally misplaced pattern staining. Only where design has been thoroughly planned can the staining effects of any deterioration be used to aesthetic advantage or the general deterioration of the facade minimised.

Crystallisation of soluble salts in porous materials

Weathering damage

The crystallisation of soluble salts in porous materials is probably the biggest single cause of weathering damage to porous building materials in the UK. The effect of salt crystallisation is to produce an unsightly efflorescence, or in extreme cases and depending on the chemicals involved, a delamination of the porous material. Brickwork is commonly affected and produces easily visible symptoms. See Fig. 8.1.

Surface effects

The mechanism of physical damage by salt deposition usually involves a salt solution passing into a dry porous material. It may occur as a surface effect; this is termed efflorescence. Small concentrations of soluble salts may exist within the material itself. Fired clay bricks, for example, usually contain sulphates of calcium, magnesium or sodium sulphate. Salts may be produced chemically by reaction with pollutants in the surrounding atmosphere such as airborne sulphur or road salt (Butlin 1989b). They may also arise from direct or bridging contact with groundwater or other specific sources such as a backing or jointing material (British Standard 1976a). Examples of this include ashes or rubble fill behind walls, or breeze blocks used as backing. Some cleaning materials or processes can introduce salts (British Standard 1976a). The salts based on chlorides and sulphates produce specific chemical and physical problems and are dealt with separately later.

Localised wetting and capillary forces

Since the process requires water, the take-up of salts will be localised to areas of the material that are subjected to wetting. Stone, concrete and

Fig. 8.1 Salt accumulations at a damp-proof course to a brick-roofed dwelling, forming stalactites.

brickwork are usually affected. Where parts of the material are dry and others are frequently wet (Taylor 1983a), capillary forces will carry the solution into the drier material. This produces a difficulty in assessment of the situation, since the salts can be carried considerable distances before emerging at a location that merely represents a good drying area (West 1970). As drying out of the moisture occurs the salts are either deposited in the pores or on the surface.

Efflorescence
When salts crystallise at the surface of the material an unsightly efflorescence is produced. This is fairly common with brickwork in the early stages of its life as the initial drying phase leaches out the free salts. Buildings constructed in wet periods such as the winter will tend to show efflorescence as they dry out in the spring (West 1970). These may also emerge internally in wet construction. Unless a detailing or construction fault allows the internal walling to become recurrently damp, this defect will only occur once, soon after construction.

Recurrent efflorescence in the external walls is usually symptomatic of defective detailing allowing water to repeatedly enter the building fabric. The efflorescence will occur as salts washed out of the brickwork, appearing where it runs down the structure and dries out.

Whether efflorescence arises depends on the solubility of the salts (West 1970). The risk of structural damage depends on the chemical composition of the salts. In severe cases pitting and spalling may occur (West 1970). For example, magnesium sulphates tend to crystallise just below the surface, and produce delamination, and may be found in mortar as well as bricks (West 1970).

Tests on the efflorescent composition may help identify the cause. Chemicals such as nitrates or chlorides will usually indicate a likelihood of external contamination (West 1970).

In the past, to counter the efflorescence problem and minimise the absorption of salts from mortar, the joint faces of stone were treated with bitumen. This can have a retrogressive effect, by restricting any movements of moisture and chemicals to the faces left free to breathe. Surface staining may actually be worsened, and exudation of the bitumen may produce a dark staining.

Delamination Deposition within the structure of a material occurs progressively with repeated wetting and drying cycles of the material. The pores gradually become blocked completely by the salt deposits. As the salts crystallise they increase in volume and create a back pressure in a zone of the porous material which corresponds to the depth of wetting. Deposition may also occur as the salt solution cools, and the pressure on the pores and mechanism of failure is similar to that of frost action, discussed later. Indeed, because many of the salts are hygroscopic, their attraction of water can render the materials more susceptible to frost attack (Taylor 1983a).

The expansive force may be directional, and will exceed the tensile strength of the material. Delamination occurs, and the porosity of the exposed surface allows renewed depositions. This progressive contour scaling may also involve shear stresses across the deposition plane induced as the material undergoes differential moisture movement. Some materials resist salt crystallisation better than others, probably because of subtle differences in pore structure and permeability (Taylor 1983a).

Chloride attack

Chloride and corrosion Chloride ions have been identified as a major cause of corrosion problems in reinforced concrete (BRS 1965b). The risk of chloride attack appears to occur when more than 0.4 per cent chloride ion by weight of cement is present. Below this level the risk is categorised as low and the chloride attack potential is weakened by the normal alkalinity of the concrete (BRE 1982b). Where the proportion of ion exceeds 1 per cent by weight the risk is classified as severe, and corrosion of the steel reinforcement can occur regardless of any high alkalinity conditions (BRE 1982a). If carbonation of the concrete has destroyed the alkalinity in the reinforcement zone the attack will be more extreme, allowing even relatively low ion concentrations to have a severe effect (BRE 1982b). See Fig. 8.2.

Environment and additives

Marine environments can produce a constant exposure to chloride attack at distances in excess of 2 km inland depending on the weather (Hudson 1988). Addition of calcium chloride as an admixture for accelerating the rate of hardening of concrete was very common in the 1970s until about 1977 (Staveley and Glover 1983). Where the chloride was added during manufacture, such as in the production of extra rapid hardening cement (ERHC), the damage was minimised because of the uniform distribution and controlled concentration of the chloride (Williams 1982).

Stuctural concrete with high concentrations of calcium chloride

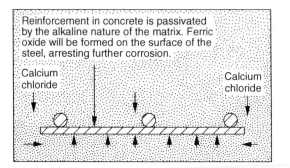

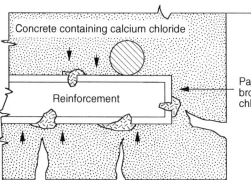

Fig. 8.2. Schematic representation of calcium chloride attack on concrete.

The role of on-site mixing
The real problems with introducing chloride into the concrete arose with on-site mixing. Inappropriate or incorrect use of admixtures has been an enduring problem, and the addition of calcium chloride on site frequently produced irregular distribution in the mix. High concentrations of calcium chloride occurred, and where these were in contact with the reinforcing steel, corrosion occurred. The practice was widespread, the problem rife, and consequently the use of calcium chloride in concrete is now severely restricted (BRE 1987a).

Additional sources of chloride
Additional sources of introduced chloride ions were hydrochloric acids used as an etching medium for the surface treatment of concrete cladding panels. If the chloride residue was left unwashed it was absorbed into the surface of the panel, producing differential concentrations across its thickness. There have also been instances where floor finishings, such as magnesite oxychloride, have been responsible for the deterioration of reinforced concrete floors (Williams 1982). There have been instances where unwashed marine aggregates have been used in concrete with disastrous consequences.

Chloride and shrinkage
In unreinforced concrete there are also shrinkage problems associated with the addition of significant amounts of calcium chloride. A 50 per cent increase in drying shrinkage can be produced by the addition of 0.5 to 2.0 per cent by weight, and this increases greatly with higher concentrations (Honeybourne 1971).

Attack mechanism
The attack mechanism operates by the chloride ion increasing the electrical conductivity of the pore water. The electrochemical process of dissolving the iron in the steel accelerates (BRE 1982a), and where there are differential concentrations of chloride, a corrosive current is developed (BRS 1965b). This produces localised and severe pitting corrosion of the steel. The corrosion occurs regardless of whether the alkalinity of the concrete is reduced by carbonation. The use of calcium chloride, however, exacerbates the effects on steel of carbonation, by promoting further oxidation of the steel reinforcement. The result is severe notching of the steel, which renders it relatively brittle. Failure may be sudden (BRE 1982b). The expansion of the corroding steel produces bulging or cracking of the concrete (BRE 1982b). Rust staining may be apparent at the surface. The localised nature means that usually the cracking will be isolated and wide. This can be contrasted with cracks produced by carbonation, which are usually less extreme but run along the length of the component. The two mechanisms can be positively identified and distinguished by chemical tests.

Sulphate attack

Volume expansion

Sulphate attack is a chemical process involving cementitious compounds and sulphated solutions. The by-products occupy a greater volume than the original gel, and physically disrupt the cement matrix by destroying its integrity and strength.

Attack circumstances

The common circumstances of sulphate attack are where unprotected concrete contains, or is in contact with, sulphate-based materials or is immersed in sulphated groundwater. Sulphate attack also occurs in brickwork; commonly the mortar is affected by sulphates washed out of the bricks.

Vulnerability of cementitious compounds

The basic chemical mechanism of sulphate attack is similar in concrete or mortar, although the accompanying increases in volume produce distinctive symptoms. Cementitious compounds are particularly vulnerable to solutions of magnesium or sodium sulphate (Williams 1982). These are present in many clay soils and emerge in acidic groundwaters which can attack the cement in ordinary concrete. Submerged concrete components such as pipes, foundations and piles are therefore highly vulnerable (Williams 1982). See Fig. 8.3. The cement in mortar can be damaged by sulphates present in the fine aggregate, or more commonly in the clays that comprise the bricks.

The aggregate effect

Aggregates containing excessive amounts of clay, coal, organic materials or sulphates pose a general threat to concrete (BRE 1987a). Sulphides in aggregates such as the clinker breeze (Staveley and Glover 1983) used in concrete during the 1920s oxidise over long periods and develop a high sulphate content (BRE 1987a).

Permeability and soils

Sulphated groundwater attack depends on the permeability and form of the concrete, the amount and nature of sulphate present, the water-table level, and variations in the soil porosity. The problematic clay soils include London and Oxford Clay, Lower Lias and Keuper Marls (Ransom 1981). The important salts are gypsum (calcium sulphate), Epsom salt (magnesium sulphate), and Glauber's salt (sodium sulphate).

Tips and marshes

Marshy ground or colliery tips may produce sulphur-based acids, also the colliery shale used extensively in the 1920s as fill beneath solid ground floors. Brick rubble with adhering (gypsum) plaster, ashes and any mining

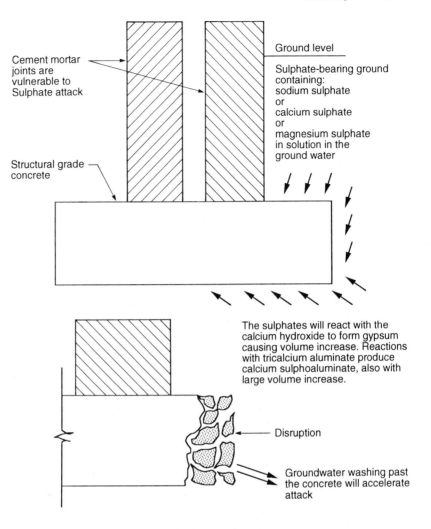

Fig. 8.3 Sulphate attack of concrete.

waste are other potential sources responsible for sulphate attack. The range of waste products commonly used as hardcore is wide. Burnt colliery waste, residual oil shale, pulverised fuel ash (PFA), and blast furnace slag can all be problematic. For instance, the high concentrations of sulphates in burnt colliery waste have been a frequent cause of concrete damage (Ransom 1981), although they are safer in the context of hardcore expansion than under-burnt shale (BRE 1973). The protection of the concrete will depend critically on the quality of waterproof barrier between them. This is often partial or absent.

Chemical reactions

The chemical process hinges on the absorption of sulphated groundwater by the concrete. The chemical reaction between the external sulphates and calcium aluminate hydrate (Ca_4AlH_3) in the hydrated cement gel produces a volume expansion of 227 per cent (Neville 1973) and disrupts the concrete. An expansive reaction occurs between tricalcium aluminate ($3CaO.Al_2O_3$) and the sulphates (Neville 1973), and calcium sulphoaluminate is formed. There is also a base reaction between calcium hydroxide and sulphates which produces a lesser expansion (124 per cent) (Neville 1973).

Calcium hydroxide

Sulphate attack can also arise in circumstances where the free lime (calcium hydroxide) in the concrete dissolves in the water held within the pores and is redistributed throughout areas of the concrete containing sulphates such as calcium or magnesium sulphate. In additon to the disruption and possible spalling of the concrete, there is a reduction in the alkalinity of the concrete, making the steel more prone to corrosion (BRS 1969a).

The chemical process depends on a continuing supply of (sulphated) water. If the substructure is totally immersed the attack will be slower than if water can evaporate, such as may occur with a basement structure or a floor (BRE 1981d). Pressure differences that can occur across floors and basement walls can also accelerate the attack. Lightweight or thin sections tend to be attacked more rapidly (BRE 1981d).

Spalling and scaling

The range of symptoms cover surface spalling and scaling (Ransom 1981), but may be more serious. In concrete floors there is likely to be cracking and lifting of the floor slab. This may extend to expansion damage of the surrounding structure at the DPC level. Doors may become difficult to open and close (Ransom 1981).

Sulphates and bricks

The presence of soluble sulphates in brickwork originates from the clay used for the brick, also sometimes in the fine aggregate for the mortar. It is usual to find sodium, potassium, or magnesium sulphates in clay bricks. The concentrations of certain salts can be modified by hard firing which decomposes or expels them.

The efflorescent link

Emergence of these salts may produce a relatively harmless efflorescence on the surface, which is removed by hard weather or brushing. Calcium sulphate is very common but of limited solubility and is considered the least harmful because of this (Taylor 1983a). It should be noted that cement or hydraulic lime mortars release efflorescent salts and some efflorescence on the mortar and brickwork from this source is particularly likely (BRE 1975a). The emergence of the sulphates from the brickwork is most apparent where the exposure is greatest, such as at the head of walls with little protection from rainfall.

Wetting bricks and mortar If there is a high concentration of salts, and there is the presence of water to carry them into the mortar, sulphate attack is likely to occur. The mortar requires to be wet for reasonably long periods if the reaction is to progress effectively (BRE 1975a), and the attack is usually limited to locations on external walls which are subject to this, such as parapets and chimneys. Alternatively, a leaking downpipe or high driving rain index may be responsible (Eldridge 1976). Obviously design detailing will make an important contribution to the incidence and severity of attack.

The tricalcium silicate compound in the cement mortar reacts with the sulphated solutions to produce calcium sulphoaluminate. This causes swelling of the brickwork as a whole, and the bricks themselves may burst. The source of the sulphate salts is usually the clay in the bricks, excepting in below-DPC circumstances where groundwater may also be involved.

Early symptoms Early symptoms include horizontal cracking in the joints between courses. As the mortar expands under the chemical attack of the sulphates, so the joints between the bricks will expand. This can produce an oversailing to the wall at the corners of the building, and also an increase in the height of the wall as net expansion occurs vertically. An expansion of 0.2 per cent is common, although significantly greater expansion has been recorded. Increases may even amount to 50 mm in height for a two-storey house (Eldridge 1976). The direct effect is usually confined to the outside leaf because of the need for water for the reaction. There may be internal cracking as the expansion of the outer leaf imposes stresses on the inner connected leaf, however (Eldridge 1976). This may be accompanied by a series of cracks in the lightly loaded parts of the wall, such as at the eaves (Eldridge 1976, BRE 1971a). Alternatively there may be differential movement between the two leafs of the brickwork producing a bowing effect (BRE 1971a).

Progressive attack Progressive sulphate attack produces whitening of the mortar, as the salts are left behind on evaporation of water. The joints become friable and may develop fine horizontal cracks (Eldridge 1976), eventually spalling. Note that a similar effect of a friable grainy mortar may also be the result of too weak a mix, or an excessive water/cement ratio. As the outer leaf expands it will carry a greater proportion of the total loads imposed on the wall. These will be concentrated near the outer surface of the brickwork, with the greatest expansion in the exposed mortar. Spalling of the brickwork may occur as an indirect effect of the sulphate attack (BRE 1971a).

Rendered brickwork Where the brickwork is rendered, the rendering may spall off as the substrate expands. Cracked rendering can allow the brickwork to become saturated for long periods, and this worsens the sulphate attack (Eldridge 1976). It is common to find inferior quality bricks used behind rendering, and if the water is not excluded the susceptibility to sulphate attack can be high (BRE 1975a). The symptomatic cracking of the render prior to detachment usually runs vertically and (predominantly) horizontally (BRS 1964a, BRE 1971a). The cement in the render may also be attacked by the sulphates, and this further reduces the bond (BRE 1975a). Detachment problems are also encountered with the use of mosaic tiling over sulphated brickwork (Ransom 1981).

Exposure and runoff

Since sulphate attack is a waterborne effect it is to be expected that the attack will be concentrated on those parts of the building exposed to rainwater, either through runoff or where it collects. Similarly to frost attack the disruption is a progressive effect depending on the porosity of the element, the degree of exposure and, particularly, the cyclic effect of drying/wetting (Neville 1973). A particularly likely location in the brickwork to suffer sulphate attack is parapet walling. This is relatively severely exposed, and the wetting/drying cycles produce rapid disintegration accompanied by vertical cracking (Mika and Desch 1988).

Chimneys

A marked expansion may be seen in chimneys built prior to the requirement in the building regulations for lining. Here the sulphur oxide gases in the smoke are absorbed into the brickwork forming the unlined flues. Slow-burning appliances are most damaging. The restraint provided by self-weight of the chimney is minimal, and preferential expansion occurs on the side most exposed to the weather, where the sulphur gases are absorbed into the mortar. The distortion on rendered stacks is likely to be worse as initial expansion allows water entry and retention (Ransom 1981). Sulphate attack of a brick chimney is shown diagrammatically in Fig. 13.37.

Time and exposure

Sulphate attack usually takes at least two years to emerge. Any oversailing of the brickwork at DPC level can be distinguished between that occurring from moisture expansion, since moisture expansion occurs in the first few months following construction.

The sulphate attack of brickwork is largely dependent on the degree of exposure. General comments regarding the influence of design and vulnerability are covered later.

Test for susceptibility

It is fairly straightforward to establish the susceptibility of a sample of the material to sulphate attack by immersing a portion of it in an acidic solution and monitoring its integrity. Methods of analysing soil water are discussed in BRE (1981d).

Environmental pollution

Sulphur dioxide

The sulphur dioxide discharged by the burning of fossil fuels is a very powerful agent of decay when in aqueous solution. The sulphur emissions dissolve in rainwater to produce sulphurous acid, which is further oxidised to sulphuric acid (Hudson 1988). A combination of these acids is deposited in the rain, and where this falls on porous materials it soaks into the surface. The calcite bonding material found in calcareous sandstones is dissolved. Where the calcites are the only bonding material the stone very rapidly becomes friable. The reaction of sulphur dioxide and sulphurous acids with dolomites and calcareous stones produces calcium and magnesium sulphates (Butlin 1989). These redissolve and become washed into the stone surface or adjacent stonework. Spalling occurs.

Emission reduction

There has been a significant reduction in emissions over the last 25 years, particularly near industrial sites (Hudson 1988). Since the 1950s measured SO_2 general levels have dropped dramatically (Butlin 1989), although values can fluctuate rapidly. Despite the emissions now being much lower, much damage has already been done to existing masonry or concrete buildings (Ransom 1981). Sulphur remains a major aggressive pollutant (Hudson 1988). Problems with acidity in the atmosphere are not restricted to sulphur. Carbon dioxide dissolved in rainwater has a pH of about 5.6. Acid rain has a pH less than 5,* but mists are more acidic and pervasive. There also appears to be a contribution of nitrogen oxides produced by vehicles to the acidity of the atmosphere, increasing the levels in urban areas (Hudson 1988, Butlin 1989). It is unclear whether it is the pH or the volume of solutions that is the more important.

Carbon dioxide and acid rain

Limestones washed by acidic rain will dissolve at the surface and produce calcium sulphate solutions. Stones that are occasionally wetted may form a skin on the surface, which will become dirty quite rapidly in smoky areas (Ragsdale and Raynham 1972). On some stones this leads to unsightly blistering, probably by some mechanism of pore blockage. This may take decades to develop, however.

* The pH scale runs from 1 to 14. 7 is neutral and represents pure water. Figures below 7 represent increasing acidity; those above 7, increasing alkalinity.

Fig. 8.4 Seismographic pattern-staining effect to stone facade in city centre.

Bleaching patterns

The bleaching effect of limestone washed by acid rain will occur only to exposed parts of the building – this produces a patterned effect depending on the design and orientation of the building (Ragsdale and Raynham 1972), and the proximity of adjacent structures. Protected sections will tend to be dirtier (BRS 1964a). It is to be expected that this effect will occur on the faces of the building most exposed to rainfall, generally the south and south-west (BRS 1964a).

Calcium sulphate absorption

The calcium sulphates washed off limestone may be absorbed by other porous materials. It is common to find limestone buildings with sandstone features to the base of the walls (BRS 1964a, Ragsdale and Raynham 1972). The absorption of the limestone wash produces staining, sometimes referred to as seismographic staining (BRS 1964a). See Fig. 8.4. It also encourages efflorescence or delamination of the plinth as the salts recrystallise below the surface of the stone. Similar defects can be found in brickwork below limestone or cast stone (British Standard 1976a). Magnesium limestone sulphates are particularly destructive (British Standard 1976a).

Durability factors: the role of pores

The durability of stone generally will depend on its chemical composition and porosity. Most sandstones are chemically durable unless they have magnesium or calcium in their structure, for example the dolomitic sandstones (Butlin 1989). These compounds react with sulphur dioxide (SO_2). Small-diameter pores are more important to weathering than large pores. Generally, the more microporous the stone, the less durable it is. In limestones the degree of microporosity is critical and has direct consequences for frost resistance and salt crystallisation (Butlin 1989). In contrast, most granites have low porosity (Preston 1989, British Standard 1976a). Problems arising from their use as plinths below limestone are not so serious or frequent as with sandstone (British Standard 1976a), but are not unknown (Ragsdale and Raynham 1972). Colour changes from grey to yellow or brown can occur, and they are no more immune than other stones to staining produced by industrial pollutants (Ragsdale and Raynham 1972).

Aggregate problems

Shrinkage and cracking

Aggregate shrinkage produces cracking to concrete components. Shrinkable fine aggregate will tend to cause map cracking (BRS 1965b) in cement paste, exhibited at the surface where drying occurs. The extent of movement is much greater than normal shrinkage cracking. In extreme cases with poor mix design it is possible for the concrete to disintegrate completely (Honeybourne 1971). This direct effect is exacerbated by frost attack.

Movement generally

The movement of coarse aggregate is usually negligible (Honeybourne 1971). The use of shrinkable coarse aggregate has largely been eliminated now through experience of the common problem aggregates. These are some of the basaltic and doleritic types, mostly from Scotland, and have a large coefficient of moisture movement. These types of aggregate can have a significant effect structurally by producing a moisture movement coefficient in the order of 0.1 per cent in the concrete. In sections which are assymetrical, there is a possibility that the differential stresses of shrinkage in the aggregate will cause a bending of the section and associated cracking on the tensile side. The consequential risk of corrosion depends also on the degree of exposure of the cracked cover to reinforcing steel (BRS 1965b). This type of effect may also be produced by variations in resistance to shrinkage caused by the distribution of reinforcement. The heavily reinforced sections resist shrinkage stresses. Relief of the stresses will produce cracking at the junction between the changes in section (BRS 1965b).

Alkali aggregate reaction

The alkali aggregate reaction is a chemical process. Aggregates containing reactive silicate respond to the presence of alkaline sodium or potassium oxides produced by the hydration of ordinary Portland cement. In this country the only aggregates identified as being susceptible are of the siliceous type, such as opal (BRE 1988a, Taylor 1983a). The mechanism requires a high moisture level in the concrete, and wetting/drying cycles accelerate the process. It is usually the fine aggregate that reacts, producing a calcium alkali silicate gel at its surface (BRE 1982b, BRE 1988a). This absorbs moisture and swells, and the concrete cracks. The effect is uncommon in the UK (BRE 1982b), but not unknown (BRE 1987a). There is currently no recognised test for the aggregates, but a British Standard is being developed. The symptoms in unrestrained concrete are a random network of fine cracks bounded by fewer larger cracks. This is generally referred to as map cracking (BRE 1988a), and can usually be brushed up to make it more visible. In restrained concrete the cracking will tend to be more ordered, and will follow the lines of restraint (BRE 1988a).

Development symptoms

The defect develops slowly, taking about five years for any symptoms to emerge (Williams 1982, BRE 1988a). These can have a similar appearance to shrinkage or frost-related cracking, but may precede any serious structural implications (BRE 1988a). Care must be exercised when identifying an alkali silicate reaction. Generally, drying shrinkage occurs early in the life of the component (within the first year or so of casting). Frost attack can be differentiated where there is a surface spalling of the concrete. In serious cases of alkali silica reaction the colourless jelly may be visible, and confirmation may be made using microscopic analysis. The effect on the component is a loss of structural strength caused by the disruption of its integrity. The detachment of the surface concrete exposes any reinforcement to corrosion, and the concrete to frost attack (Williams 1982, Honeybourne 1971).

Loading and deformation

Columns shorten over a number of years, and beams deflect. This is often unimportant, but there may be problems for pre-stressed components and secondary effects on claddings and especially brick infill (Williams 1982).

Creep allowance

If there is no open joint between the beam and the head of cladding to allow for creep, then the compressive loads will be transferred to the cladding in a similar way to the drying shrinkage problems discussed and illustrated later. Bulging and failure of the panels will occur.

Frost attack

Porosity and risk

This is not the most important weathering agent in the UK, and affects porous materials only. For frost failure to occur in materials they usually require to be frequently frozen whilst very wet. This means that the occurrence is usually restricted to features of the building that hold water, such as window sills, or those in direct or bridging contact with the ground. Parapet copings are particularly vulnerable because of their higher degree of exposure, as are boundary walls which usually have no DPC, no elaborate copings and are subjected to extreme conditions on both faces. Winters with temperatures oscillating around freezing, and freezing rain will therefore be more destructive than prolonged severe frosts (West 1970).

Unrestrained or restrained forces

It is common to consider that the force damaging the porous structure of materials is the hydrostatic pressure of the water expanding as it freezes. Unrestrained, this produces about 9 per cent expansion in volume, and confined it can produce theoretical forces greater than any porous building material can withstand in tension. A theoretical concept and testing procedures to analyse the percentage of voids, and therefore the ability of the porous material to accommodate the expansion at freezing, were developed. In practice though, very little correlation has been found between this 'saturation coefficient' and the frost resistance of porous materials. The concept has been all but abandoned (Taylor 1983b).

Ice formation: temperature and pore size

Whilst the hydrostatic pressure is an important factor in the development of stress in the material, it does not appear to be the predominant force. In addition to the hydrostatic pressure developed during the phase change, current theory holds that ice formation occurs at different temperatures in different sized pores. The larger pores freeze first, whilst the water in the smaller pores remains liquid. Below freezing point, water has a higher vapour pressure than ice, consequently the fine pores 'feed' water to the growing ice front rather than themselves freezing. This produces very large growths in the large pores, and the resistance of the sides of the pores to the growth is translated into tensile stress in the material (Taylor 1983b). With each of these contributory mechanisms the pore diameter, type and distribution are important features, so too is the proportion of cul-de-sac pores. Bricks with a large proportion of fine pores have been found to have a poor frost resistance, probably because of the 'feeding' phenomenon (Taylor 1983b). Hard-firing is a stipulation in BS 3921 (British Standard 1974a) for bricks satisfying frost-resistant applications. This appears to reduce the proportion of fine pores.

Resistance of large pores

For example, Otterham Stock bricks with around 40 per cent by volume of large pores are resistant to frost (Taylor 1983b). The fine-grained Fletton brick with a larger proportion of small diameter pores is easy to cut and dimensionally stable, but much less frost resistant (Taylor 1983b). The misapplication of bricks with poor frost resistance does not inevitably mean failure, because of the variability in exposure of individual buildings and details, but it may be a good indication during inspection.

Surface effects

Frost attack occurs near the surface of materials for a number of obvious reasons. Moisture gains easy access, and in the case of hydrated materials the pores near the surface are largest because of the nature of drying-out. Further, the temperature drop in the material during cold weather is most marked and rapid near the surface. See Fig. 8.5.

Material composition factors

Since the degree of pores and their size is important to the mechanism of physical attack, the water/cement ratio of concrete can be expected to play an important role in determining frost resistance. The significance of the water/cement ratio is non-linear; increasing the ratio above about 0.4 has a marked effect on durability.

With bricks there appears to be little quantifiable link between compressive strength and frost resistance (West 1970), except in the most extreme and obvious cases. This is to be expected, since the mechanism of failure relies on tensile strength. Some bricks appear to be more resistant than others, and BS 3921 (British Standard 1974a) makes a distinction on the basis of grading for use and exposure.

Flaking and friability

Frost attack produces surface flaking and a friability of the porous structure. The material fails in tension under the build-up of stresses. Where this produces spalling, the extent of water penetration may be indicated by staining from pollutants absorbed in solution. In brickwork, both frost attack and sulphate attack produce a friability of the brick and/or the mortar binding, but may be identified by testing. Bricks that are poorly burnt will have a reduced durability to frost. Note that underburning is associated with high salt contents and hence problems with salt attack also. The

Fig. 8.5 Frost attack to brickwork. The attack is progressive and causes delamination of the bricks. They become friable, and the more advanced the attack the deeper the damage. In this instance a leaking guttering on a dilapidated property ensures that the brickwork is thoroughly and frequently saturated. There is also some organic growth on the surface of the affected bricks.

friability symptomatic of frost attack is accompanied by expansion of the joints and possibly the bricks also. Bad samples or batches may appear obvious in the brickwork. The symptoms are most obvious in the winter or early spring (Eldridge 1976).

Longitudinal and vertical expansion

A longitudinal and vertical expansion of the walling occurs. This may produce an oversailing effect or, where restrained, a bulging of the wall.

Stability of the structure

Where the mortar has been affected by frost action this may have occurred whilst it was immature, or where the exposure of the building has not been properly taken into account at the mix design stage (Eldridge 1976). The presence of clay minerals as aggregate in concrete may be responsible for frost susceptibility and excessive movement (BRE 1987a).

Exposure and liability

Certain areas on the building which have a high exposure rating will be more liable to frost attack. The comments about design attention are particularly valid and should be examined to identify the cause. The risk is exacerbated by leaking downpipes and overflowing hoppers. The high-risk areas for stonework are copings and cornices (British Standard 1976a). These are best shielded with lead flashings and a DPC placed under them to protect adjoining walls.

Slaking and spalling

Frequently, minor handling or placing damage to bricks will only appear after weathering deterioration (Ragsdale and Raynham 1972), and although symptomatic of frost attack, it would be unfortunate to attribute all the blame falsely to the material. Spalling of brick surfaces can also arise from lime blowing. This is a regional problem, caused by the inclusion of fossil constituents in the clay during manufacture. As the lime slakes it expands, spalling the face of the brickwork and revealing a whitened lump. That the failure is caused by a point source will probably be obvious compared with frost effects occurring across an area.

Carbonation

Shrinkage and exposure

Carbonation shrinkage is the result of exposure to the environment (Neville 1973), and the problem is only frequent with external concrete. The carbonation mechanism involves a solution of carbonic acid, formed when the carbon dioxide (CO_2) present in the air dissolves in the free moisture lying in the pores just below the surface of the concrete. This acidic solution dissolves hydrated cement minerals, in particular calcium hydroxide (free lime). As the alkaline calcium hydroxide ($Ca(OH)_2$) decomposes it produces calcium carbonate ($CaCO_3$) (limestone). The result is that the compressive stresses induced by drying shrinkage are relieved by the dissolving calcium hydroxide. This allows a volume reduction in the concrete. In this sense, although shrinkage of concrete associated with drying and carbonation may coexist, they remain distinct mechanisms (Neville 1973). This process operates very slowly in good quality concrete (Honeybourne 1971). See Fig. 8.6.

Surface layer mechanism   The carbonation mechanism means that the process occurs mostly in the surface layer of the concrete component. Hence the speed of progression is highly dependent on the density and permeability of the concrete, and the depth of attack progresses relatively slowly (Neville 1973). See Fig. 8.7.

Lightweight aggregates   Concretes based on lightweight aggregates are particularly vulnerable to carbonation and are critically dependent on good compaction to minimise permeability. The carbonation rate has been measured as double that in mixes using normal aggregates (Neville 1973), and in such circumstances the depth affected is likely to be greater. This will obviously make the depth of cover over reinforcement more critical, and the requirements of CP114 were amended for lightweight aggregate concrete to reflect this (BRS 1965b). The use of air-entraining or workability admixtures can produce similar problems because of increased permeability caused by reduced density (BRS 1965b).

Surface strength and permeability   Generally, as carbonation proceeds the compaction effect produces a marginal increase in the surface strength of most concretes (Neville 1973, Taylor 1983a), and a reduction in permeability (Neville 1973). To some extent the process is self-limiting, and in good quality concretes it is unlikely to extend more than 10 mm below the surface (Taylor 1983a).

The role of moisture   The carbonation process requires moisture, making it dependent on humidity. Opinions vary about the most favourable humidity range, either between 25 and 50 per cent (Taylor 1983a) or 50 and 75 per cent RH (BRE 1982a). There is agreement that at humidities above the upper end of the range (Taylor 1983a) or below 25 per cent (Neville 1973) there is either too much or too little moisture, and the rate of carbonation is restricted. Within the intermediate range of humidities the rate increases. The rate of progression is also dependent on the variations in concentration of carbon dioxide such as occur between built-up and rural areas. It continues even at very low concentrations.

Cracking: the time factor   The carbonation mechanism produces surface cracking distinguishable from that caused by drying shrinkage because of the much greater time before drying shrinkage cracks develop (Honeybourne 1971). Carbonation affects reinforcement steel generally, producing a uniform corrosion. The expansion as the steel corrodes causes a hair-line crack along the line of reinforcement and probably along its whole length, to give a useful visual indication (BRE 1982b). It is possible for cracks caused by carbonation shrinkage to appear

Fig. 8.6 (a) Carbonation damage revealed to a concrete system-built house. The reinforcement has corroded in the acidic environment. **(b)** The depth of penetration is considerable, through the rendered concrete. This close-up view shows the extent of the damage to the reinforcement. (Reproduced by permission of Ian Chandler 1988).

after a few months of exposure (Honeybourne 1971). The limestone produced by the carbonation process will tend to fill up the cracks and make them less obvious (Honeybourne 1971).

Testing The extent and depth of carbonation can be tested by using either manganese hydroxide or, more commonly, a phenolphthalein solution. Phenolphthalein reacts with free calcium hydroxide ($Ca(OH)_2$) present in

Fig. 8.6 (b)

uncarbonated cement to produce a pink indicator. The carbonated cement (with no residual calcium hydroxide) produces no reaction, and so remains clear.

Alkaline to acidic environment

The important consequence of carbonation is the loss of the alkaline environment in the concrete (Neville 1973), which protects mild steel reinforcement from corrosion (Ransom 1981). It is worth noting that

Fig. 8.7(a) Schematic representation of carbonation attack on reinforced concrete. Exposed components are particularly vulnerable. **(b)** Emphasis on testing of samples using phenolphthalein solution to establish depth of carbonation attack.

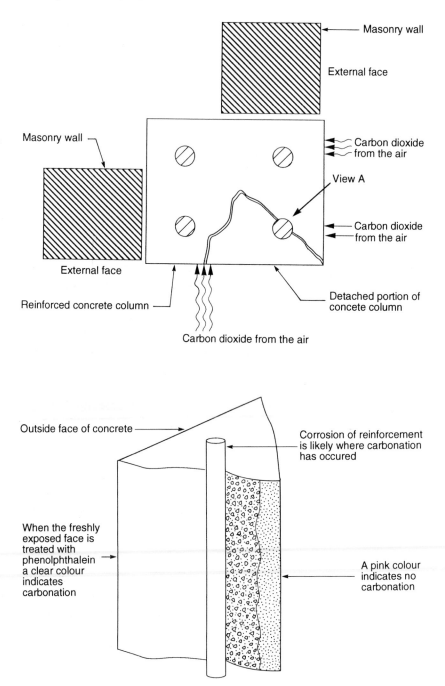

Masonry wall

External face

Masonry wall

Carbon dioxide from the air

View A

Carbon dioxide from the air

External face

Reinforced concrete column

Detached portion of concete column

Carbon dioxide from the air

Outside face of concrete

Corrosion of reinforcement is likely where carbonation has occured

When the freshly exposed face is treated with phenolphthalein a clear colour indicates carbonation

A pink colour indicates no carbonation

exposure to acidic environments, such as those produced by chloride attack, increases the depth and rate of carbonation (BRE 1982a). With dense, relatively impermeable concrete and adequate cover to the steel, the risk of corrosion has conventionally been considered unlikely (BRS 1965b). Where carbonation reaches the reinforcement zone, however, rusting of the steel produces an expansive disruption to the concrete. The reduction in cross-sectional area produces strength and fatigue problems. Surface cracking mirrors the reinforcement pattern, and spalling exposes fresh free lime to continue the carbonation process. The high frequency of failure in practice, because of insufficient cover, or inadequate density, is evidence of a widespread problem. It is frequently possible to find depths of cover in the order of 8 mm, where it should be at least 20 mm for even the most sheltered of external applications (British Standard 1983a), and more commonly 40 or 50 mm.

Steelwork placed flush with the surface is not unknown. Manufacturing tolerances in precast work were investigated by the Building Research Station, and results indicated a significant variability in placement tolerance of the steelwork in precast work similar to that of *in-situ* work (BRS 1965b). There has been much criticism of the quality of manufacture of systems components, and this appears to be a contributory factor in many of the long-term durability problems.

Significance of reinforcement cover

The significance of cover is confirmed by the range of minimum covers specified according to the exposure rating by BS 882 (British Standard 1983a). This may be cover to the main bars or only the stirrups or binding wire – the effect is still significant. The probable carbonation depth of 10 mm produces an obvious risk of corrosion. In cases where there is a suspicion of inadequate cover, use a cover meter to establish the reinforcement depth across any undamaged parts of the concrete.

Thin and exposed sections

A large number of carbonation problems and rust-induced spalling can occur on thin and highly exposed sections such as window mullions or sills. There is a likelihood that the reinforcing steel or its accessories may have insufficient cover.

Corrosion

Reinforcement corrosion

The corrosion of reinforcement steel is an electrochemical process. Areas where corrosion can and cannot occur form the anode and cathode respectively, and may arise between steel reinforcement and the concrete

Fig. 8.8 Details of possible corrosion of steel reinforcement.

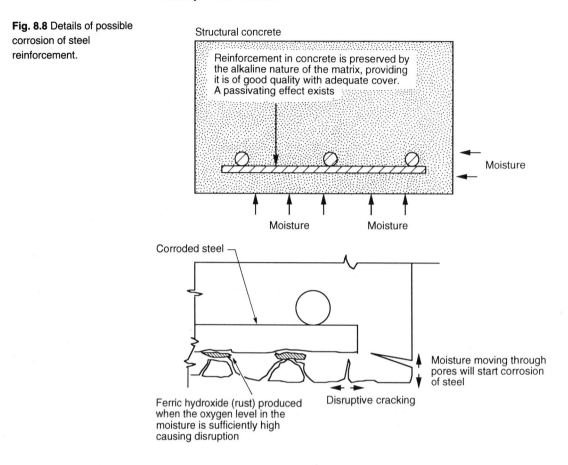

Structural concrete

Reinforcement in concrete is preserved by the alkaline nature of the matrix, providing it is of good quality with adequate cover. A passivating effect exists

Moisture

Moisture   Moisture

Corroded steel

Ferric hydroxide (rust) produced when the oxygen level in the moisture is sufficiently high causing disruption

Disruptive cracking

Moisture moving through pores will start corrosion of steel

surrounding it. Microgalvanic corrosion cells are set up by differential concentrations of oxygen held in the pore water or by anionic solutions of carbonates, chlorides or sulphates (BRS 1965b, BRS 1969a). These break down the thin protective oxide film formed in alkaline environments on the surface of steel, and corrosion occurs (BRS 1969a). See Fig. 8.8.

Preferential corrosion

Galvanic cells can also be set up when different metals are in contact directly, or through a conductive medium. A typical example in the case of steel reinforcement to concrete can occur with composite cladding panels. It has been common practice to use silicon bronze for the ties between the inner and outer skins of sandwich cladding panels (BRS 1965b). Where these ties are allowed to come into contact with the mild steel reinforcement, a galvanic cell is formed which leads to preferential corrosion of the anodic steel. Other problems may occur with concrete cladding panels because of the fine tolerances of cover to reinforcement to limit the weight by minimising the cross-sectional thickness.

Depth of cover   In the early stages of corrosion, there may be patterned staining of the concrete surface caused by the leaching out of the iron to the surface. After testing with a cover meter it may be possible to distinguish between corrosion caused or exacerbated by inadequate cover to the reinforcement and that produced by carbonation and/or chloride attack. Any spalling or cracking may help the analysis and confirm cover depth; a phenolphthalein test may confirm further. In cases where the iron staining to the surface doesn't appear to follow a pattern, the iron content in the aggregates may be responsible.

Loss of protective coating   The corrosion of zinc-coated steelwork (galvanised) will occur when the zinc has been dissolved – this can take a considerable time and depends largely on the amount of sulphur in the air and the humidity. In aggressive environments it may be worth considering this possibility. Note that the possibility of perforation or variability in thickness of older galvanising coatings is considerable (de Vekey 1979). Chemical or metallurgical testing may be necessary to assess the residual life expectancy of galvanised steel not showing obvious signs of corrosion, but considered prone to it.

With the system-built properties of the 1960s, the combination of marginal design and poor control over fabrication and construction has now been revealed by a high incidence of rusting to reinforcement.

Symptoms   Corrosion of the reinforcement in concrete is possible in the full range of elements constructed of reinforced concrete. The quality of the concrete and the depth of cover over the reinforcement are vital factors affecting the degree of protection given to the concrete.

The symptoms of this defect emerge firstly as directional cracks in the concrete surface. This is followed by some displacement of sections culminating in spalling. The mechanism is expansion of the reinforcement as it rusts. This pushes the cover off, the minor cracks that are produced allow water ingress, and the problem is self perpetuating. The spalled concrete reveals rusted steel, which will be poorly bonded to the existing concrete. Porous concretes such as those produced using a high water/cement ratio are more susceptible. Indeed, those with a w/c ratio greater than 0.7 are likely to remain permanently porous. The effect is exacerbated by chlorides added as an accelerator in the concrete.

Corrosion problems may arise where ferrous metal connectors are used in construction, such as the connectors between stone blocks or copings (Eldridge 1976), and where metal railings are housed in stonework.

History of wall ties   There are a number of references to the first use of cavity walling and wall ties. It is frequently reported that the use of cavity walling was introduced experimentally around 1923, but it is quite possible to find properties older than this with tied cavities. Indeed, these are frequently the source of reported failure (Williams 1986). Earlier, the Public Health Acts of 1875 influenced the adoption of the cavity wall, but did not produce the first recorded use. Infrequent examples have been cited back as far as 1866, and even 1805, using cast-iron ties. These have proved to be quite durable and the cast iron may help account for fewer failures in ties in older properties.

Alternatives to cast iron were wrought iron after about 1850, and these appear to have also been quite durable (de Vekey 1979). Earlier than this simple header bricks were used to span the cavity. It was soon realised that their porosity conducted moisture, and by the turn of the century salt-glazed bricks were used, sometimes with an iron core. In fact, the situation appears not to have been general, and the dates at which cavity walled construction started appear to be regional. The method was slowly accepted, but only became countrywide around the turn of the century (Williams 1986). However, it is possible to find large numbers of houses built in London in the 1930s with solid walls. Occurrences into the 1950s have been recorded in sheltered areas (de Vekey 1979, Hatchwell 1985). Note also that early cavity wall construction was sometimes disguised, such as by the adoption of a mock Flemish bond using snapped headers (Sealey 1985).

Failures

Large-scale failure of wall ties has not yet been recorded, although the incidence of occurrence has been noted since the 1950s and is increasing regionally. The early galvanised steel ties that succeeded the bitumen-coated metal ties were still prone to rusting. Their life expectancy is directly proportional to the thickness of the coating, and wall-tie failures have been attributed to coatings that are incomplete or of inferior thickness. Where this is suspected it may be checked using chemical or metallurgical analysis.

The Building Research Establishment has reported indications that the expected life of the bitumen-coated and galvanised ties may be less than the design life assumed for the walls as a whole (de Vekey 1979). Following the discovery that the early zinc-coated ties were inadequately resistant to corrosion (Seeley 1987), the minimum thicknesses were revised and increased in 1982.

Walls cracking

The expansive failure of ties within the leaves of the wall produces characteristic cracking symptoms and sometimes a loss of stability in the walling. The cracking in the leafs will be differential because the corrosion rate in the wet outside leaf will advance more quickly. To some extent exposure must be important to the occurrence of failure, and should be borne in mind when evaluating continuing weather resistance of the wall.

Walls leaning   The wire twist ties tend to fail more rapidly than the thicker ties (Staveley and Glover 1983). With the thicker ties, however, the outer leaf expansion may be considerable and readily measurable. As the outer leaf expands vertically it will begin to carry a proportion of the roof load. This will exacerbate any buckling caused by the expansion of the ties and, depending on whether there is continuity of the ties, the inner leaf may buckle outwards also (de Vekey 1979). Both of these features will require checking if measurable expansion has occurred. This may be done using an endoscope to inspect the cavity. The physical inspection of embedded ties will require the removal of selected bricks.

Chlorides and sulphates   A contributory cause of wall-tie failure may be the use of black-ash mortar, which was regional and reacts in contact with ferrous wall ties, including galvanised ties (Eldridge 1976). A similar effect to that caused by the sulphates in black-ash mortar can be produced by the chlorided aggregates or the presence of excessively permeable or perhaps the chloride-based mortars. These both allow easy moisture penetration and retention around the wall tie (Seeley 1987), and permeable mortars allow early carbonation of the mortar (de Vekey 1979). Williams (1986) also confirmed the link between chlorides and deterioration of galvanised coatings and hence ties, and that high concentrations of chloride ion can render galvanising of little use.

Constructional faults which keep the wall particularly wet, such as cracked render, will accelerate the corrosion pace. There may also be a significant increase in risk where non-absorbent cavity-fill insulation has been installed (Hillel 1984, Addleson 1982).

Structural stability   The ties in cavity walls are important for the overall stability of the building, and usually require to accommodate some differential thermal movement between the leafs. Any failure in their strength commits the leaf structures to act as separate thin walls, and bowing under thermal expansion forces and induced roof loading may occur. Corrosion between courses forces the brickwork apart, producing an accompanying but undefined loss of integrity of the cavity wall. Horizontal cracks occur between the courses at intervals corresponding to the location of the wall ties (Addleson 1982), and may be accompanied by vertical or diagonal cracking. This will usually be every four courses or so, if the ties were installed properly, and is most likely to occur in lightly loaded zones of walling, such as below the eaves. Repointed, thicker joints may be indicative of previous damage.

Where corrosion is extreme there may be no remaining link between the two skins of the walls. This is probably quite likely in very old properties, but not exclusively. It is possible for the ties to corrode completely where they are exposed within the cavity, but give no external symptoms in the walls (Atkinson 1985). Speculative internal inspection will be necessary where there is the likelihood of this having occurred.

The significance of the tie failure will depend on the inherent stiffness of the structure, which in turn will be related to its form (Staveley and Glover 1983). Large masonry elevations such as gables are particularly likely to need repair. In cases where there is disruptive expansion, some remedial action will be obviously necessary. The loss of integrity renders the wall prone to collapse under high wind loads, particularly the suction forces created over closely separated large gable walls (de Vekey 1979, Williams 1986).

Remedies

Remedial work involves the removal of the old ties to prevent further disruptive expansion caused by the corrosion, and replacement with new ties, either stainless steel or plastic. Proprietary kits and processes are available based on epoxy-bonding ties in drilled holes. Note that the corrosion risk is not confined to conventional cavity-wall ties. Fixings used in the formation of the system-built panels are also susceptible to similar corrosion problems.

Corrosion of steel

Unprotected steel will corrode rapidly in wet oxygenated conditions (Hudson 1988). The UK climate is particularly damp. Rust is hydrated ferric oxide ($Fe_2O_3.H_2O$), produced as a result of a complex set of reactions. Water and oxygen must be present together (Hudson 1988). On exposed steel the detailing can make a significant difference to the durability. For instance, stiffeners which create catchment areas for water are problematic. The collection of acidic dirt and water can be avoided by the insertion of drainage holes in horizontal details, and the bases of columns should be designed to avoid water and dirt collection (Hudson 1988).

Steel piles, with minimal oxygen levels in the subsoil surroundings, corrode very slowly (0.03 mm per year) regardless of the nature of the soil (Hudson 1988). The rate will be different above ground and at the interface. With exposed steels such as claddings the durability will depend on the environment as well as the protection method (Taylor 1983a). Internal environments which are warm and wet, perhaps with pollutants, will encourage steel to corrode more quickly than externally. Examples of highly corrosive atmospheres include foundries, paper manufacture, chemical works and swimming pools (Taylor 1983a).

Fixings

Non-ferrous fixings in stone present few problems (British Standard 1976a). Ironwork is a major problem because of the large coefficients of expansion and oxidation (British Standard 1976a). The delamination of wrought iron as it corrodes is very disruptive.

Corrosion of reinforcement

Cracking usually follows the direction of reinforcement (BRE 1982b), and this distinguishes the symptom from crack formations that are random and unrelated to the reinforcement location (BRE 1982b). There may also be rust staining indications. Note that any ferrous sulphide inclusions in the aggregate will produce rust stains, with a soft grey material at the origin. Check whether the rusting is main reinforcement or merely ties (BRE 1982b) – there is unlikely to be cracking where the ties alone are staining.

A number of protective measures can be taken to prevent steel becoming exposed to the atmosphere and rusting. Concrete is but one. The pH of alkaline concrete is about 12.6–13.5 (BRE 1982a). General steel reinforcement corrosion will reduce the tensile capacity of the steel in direct relationship to the cross-sectional area, and make it more prone to fatigue (BRE 1982b).

Depth of cover

Measure the depth of cover with a cover meter. Where the steel, or the disruption its expansion causes in the concrete, makes a component unfit to support the stresses imposed on it, the component serviceability will degrade. The stiffness of components will reduce and deflections or bowing may occur.

Prestressed reinforcement corrosion

Where cracking occurs within a hollow component, failure may be the first indication (BRE 1982b). Corrosion in the tendons of post-tensioned components can be very serious, especially if the grouting is damaged by corrosion expansion. Where prestressed tendons are bonded this is less serious.

Elemental considerations

Elemental considerations should be incorporated into an analysis of the whole building structure, and reviewed in the light of their possible significance to the structure as a whole (BRE 1982b).

Metals generally

Oxidation

A major factor affecting the longevity of exposed metals is their behaviour when exposed to the atmosphere. Ionisation of the oxygen in the air and a balancing reaction with the metal occurs at the interface of the two. Electrons are drawn from the metallic bond and the process termed oxidation occurs. The pressure of oxygen vapour will attack the surface, causing oxidation, but the core of the metal will remain unaffected. Metals may be alloyed to enhance the oxidation of the parent metal, or to improve their mechanical characteristics (Taylor 1983a, Everett 1975).

The oxides formed from certain metal ions occupy less space than the original metal. This can result in a porous oxide layer, allowing oxidation to continue into the bulk metal. Where the oxide is of similar size, or larger than, the parent metal ion, increasingly impervious layers are formed, and the surface may become distorted. The common metals used for external applications are of the latter sort. Oxide densities may change across the surface of sheets and sections. The obvious disadvantage of increased access for water and other agents of change is linked to degree of oxide porosity.

The stability of the oxide layer also varies with the type of metal and the surrounding oxygen concentration. The greater the oxygen vapour pressure, the more stable the oxide. External metal applications generally experience only minor variations in atmospheric oxygen on the exposed surfaces compared with the reverse side. See Fig. 8.9.

Creep

Creep can be termed a time-dependent strain that can occur over extended periods. The loading necessary to develop creep can come from the self-weight of the metal. Sheets may be pulled apart, which will open up joints and cause stress concentrations at junctions and changes of direction. Laps and vertical upstands in metal roofs are a particular risk. The creep performance of metals is related to their elasticity, with lead generally more likely to creep than copper. Where metals are of large dimensions and discontinuously supported the risk of creep is increased. Alloying materials to provide increased durability may also provide increased creep resistance, e.g., zinc/copper/titanium is generally better than zinc/lead.

Patination

The production of patination on external metals is seen as part of the graceful ageing of certain buildings. The colours are related to the spectral reflectance characteristics of the oxide layer, with copper becoming the characteristic green. The warm greys of Terne-coated stainless steel should appear in 18–24 months. Patination of zinc can be complete within 9 months, and produces a greyish appearance. Where water can wash over the metal it takes with it the patina, which can be deposited on new surfaces causing staining, or set up new electrolytic corrosion cells. The constant erosion and reformation of the patination can etch the surface of the metal. The patination colour of aluminium will approximately indicate the pollution of the ambient environment: light greys in rural areas and black in urban areas (*Architects' Journal* 1987a). This can be a useful guide when considering other defects associated with environmental pollution, e.g., acid deposition.

Abrasion

Trafficking over metals can cause severe abrasion and impact damage risk.

Fig. 8.9 Accumulation of rust around metal window frame has produced buckling.

Any surface deposits may be ground into the metal. The use of timber duckboards may remove this threat but could create damage from projecting fixings. There could also be chemical incompatibilty between the metal and the timber and/or its preservative. Wear on moving metal parts will abrade surfaces and may increase joint widths. This can reduce weather resistance in windows and doors, leading to increased closing loads by users, causing impact damage.

Hygro-erosion The natural dust deposits from the atmosphere contain a range of compounds, some of which can attack the surfaces of metals. They also contribute to the external appearance of the building, and when washed away by precipitations may cause pattern staining. Where water flows over a metal surface it will wash away the ambient detritus and small amounts of the metal. The greater the flow, the greater the washing effect. Areas of flow concentrations adjacent to gutters and beneath discharging rainwater downpipes are particularly vulnerable. Water may lie in stagnant pockets, and can contain concentrations of aggressive materials. The washing effect can therefore be beneficial and extend the useful life of the metals (*Architects' Journal* 1987a).

Physical wear Where physical wear on metals exists, the arrises and irregular shapes are particularly vulnerable. The stress concentration may have produced a work-hardened region which will behave in a brittle way compared with the bulk metal. Cracking can result, as in the case of angular rolls in metal roofing. Pedestrian traffic can bend over-standing seams, and metal-clad timber rolls can suffer physical wear. Access and maintenance traffic provision can result in mechanical impact damage, high point and moving loads.

Dissimilar metals The mechanism of electrolytic corrosion has been commented upon earlier. This type of corrosion will be particularly likely where dissimilar metals come into contact either physically or electrolytically. The relative position on the electrochemical series for two metals will determine their potential reactivity when linked through an electrolyte. It will also enable the identification of the corroding (anodic) metal. The process of electrolytic corrosion is shown in Fig. 8.10.

The concentration of noble cathodic metals in runoff water may cause corrosion cells to develop some way away from the parent metal. Gutters and cesspools are particularly susceptible. The flow of water through the corrosion cell will encourage the depletion of the anode, and this is common in roof and guttering applications. Where different metals are used for fixings these can corrode rapidly, due to electrolytic or differential oxygen concentrations (BRE 1985d). Dissimilar metals are commonly used for fixings because of their superior physical properties. Where water can gain access to the interface, severe corrosion can occur. Fixings may link together several metals, different backgrounds, and produce corrosion mechanisms. Copper fixings may support lead sheets and be fixed to timber rolls. Joints and seams offer harbourage for small amounts of water and debris. These are sites of potential differential oxygenation corrosion. The corrosion process will generally increase with the ambient temperature of the cell,

Fig. 8.10 Diagrammatic representation of electrolytic corrosion of a metal.

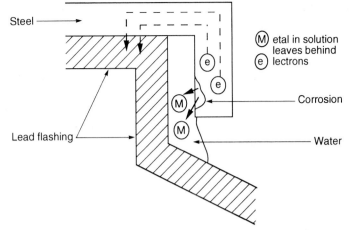

Steel ⟶

Ⓜ etal in solution leaves behind
ⓔ lectrons

——— Corrosion

——— Water

Lead flashing

This describes the process which is driven by the fact that at the atomic scale metals are made up of large centres (the nucleus) around which are orbiting electrons. These are small and move quickly. When in contact with water metal atoms can dissolve into it, although different metals do this at different rates. They all tend to leave behind some electrons. These can flow through the metal to other metals when they are in contact. This will enable the original metal to continue to dissolve, depleting it and causing a loss of strength. It is possible for corrosion sites (anodic) and generally non-corrosion sites (cathodic) to exist within the same metal. This will enable electrolytic corrosion to progress without influence from other metals, since differences in the relative electrical potential of metals will either encourage or discourage a flow of electrons. When the metal in solution comes in contact with amounts of oxygen it can oxidise, these oxides tend to be heavy and are commonly formed on the surface of metals, e.g., rust. These deposits are commonly close to the areas of depletion and are the visual indication of corrosion.

warm and damp conditions, as may occur in leaking claddings, and warm roofs are at risk.

Movement   In structural terms the movement of metals is related to their elasticity. The measured strain values for ductile metals can be considerable, although brittle metals exhibit little deformation. Where the structural movement of metals is excessive, the nature of the loading and condition of the metal should be examined. Abrupt and sudden failures are possible. The amount of movement associated with the non-loadbearing external use of bulk metals is related to their coefficient of expansion. This can be significant, since typical values are 17.1×10^{-6} per °C for copper and 29.7×10^{-6} per °C for lead. Failure to allow for this at joints and changes of direction will produce buckling, distortion and cracking. The creep of metals will claim further dimensional allowance from the joint. The exposure of matt metal surfaces to direct solar gain from a clear sky can produce very high surface temperatures (*Architects' Journal* 1987b), and will be increased by the placing of insulation behind the metal. Surface temperatures will increase rapidly and be maintained. This will provide better conditions for electrolytic corrosion. The shading effect of projections may produce some cool regions, but the generally high conductivity of metals will assist in maintaining steady temperature.

Flexure of the background to the metal will impose additional bending stresses upon it. These can produce deformations including tensile cracking. Decking and supporting materials may possess different moisture- and temperature-induced movement to the supported metal, forcing dimensional change upon it. When this movement mechanism is considered in relation to the background and enclosing structure, the interface between the two becomes a key movement area (Hinks and Cook 1989c).

Stress-induced corrosion

Stress-induced corrosion can occur around regions in the metal which have been cold-worked. Cracks can develop. The changes induced in the microstructure of the metal will influence the occurrence of anodic and cathodic regions. This can also occur following forming and welding. Where this raises the temperature of relatively small sections of an alloy, the temperature gradient to other sections at ambient temperature may pass through critical regions of its phase diagram. The cooling process may not conform to that adopted during manufacture, and internal structural changes may occur. Cold-worked fixings may also provide crevices for differential oxygenation effects to develop.

Fatigue

Metals can suffer from fatigue when subjected to a loading and unloading cycle. It will depend on the level of loading applied and can emanate from imposed loads, self-weight, thermal movement, and live loads. The total loading in a detailed situation will also need to take account of the orientation of supporting surfaces, and the performance of surrounding materials. This cyclic movement of a metal can induce work hardening, and will transform part of a ductile metal into a brittle one. The physical performance characteristics are likely to be different, and this will exacerbate the problem. The oxide layers commonly formed on metals can be brittle, and will assist in the general embrittlement of the metal.

Aggressive environments

The internal and external environments around buildings can be aggresssive towards its fabric and structure. Metals are traditionally selected because of their ability to resist attack. Recently this selection has been based on the results of several on-site investigations, although there is a general lack of information concerning current corrosion rates (Butlin 1989b). The macro-scale involves the disposition of sulphurous acids on claddings and roofs. When placed into aqueous solutions they can form the electrolyte for

Fig. 8.11 Diagrammatic representation of acidic corrosion of a metal.

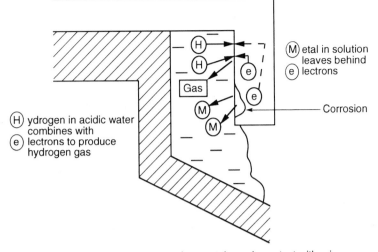

This type of corrosion can occur when metals are in contact with rainwater since this can be a weak acid due to dissolution of compounds from the atmosphere. The basic process of electrolytic corrosion is driven by the flow of electrons from the anodic site combining with hydrogen in the water to form hydrogen gas molecules. Where the gas forms a layer over the surface of the metal, further use of electrons can be arrested. This may slow down corrosion, although where an electrical contact exists to a dissimilar metal then electrolytic corrosion may continue. Where water can move over the surface of the metal the gas layer can be continually swept away, encouraging acidic corrosion.

corrosion cells. This is shown diagrammatically in Fig. 8.11.

Industrial processes inside buildings can deposit a range of aggressive compounds onto metals. Damp conditions will assist their attack of the metals. The micro-scale environment can include the corrosion of aluminium and zinc nails in damp timber treated with copper-chrome-arsenic preservative. This can occur after the initial application of the preservative (BRE 1985d). The cavities within constructed elements can develop their own aggressive micro-climates. Water vapour may be deposited on surfaces or interstitially to trigger corrosion. The disposition of aggressive compounds on a metal surface in a dry environment is no guarantee that they will not develop electrolytically since they, or their corrosion products, may be hygroscopic. Moisture may be drawn out of the air at temperatures above its dew point (Butlin 1989b) to fuel this process.

Organic damage

The growth of vegetation on the external faces of buildings will produce concentrations of organic material when washed by rainwater, or when rotting. Mosses and lichens are considered to add to the general appearance of buildings. Their deposits can accumulate on poorly weathered metal surfaces, in joints and rainwater goods. Stagnant pools caused by lack of falls may become increasingly concentrated as more and more debris accumulates. Timber which contains quantities of organic acids, e.g., oak and western red cedar, may attack zinc when in direct or aqueous contact. The organic compounds in some bitumen products used as backing layers for sheet metals may attack some metals, e.g., copper (*Architects' Journal* 1987a).

Alkali attack   The presence of strong alkalis in the ambient air around buildings is rare. The source is commonly from materials of construction. The alkaline nature of cementitious materials containing lime assists in the preservation of buried ferrous metals. Where these can leach out onto metal surfaces they may attack them. Aluminium is particularly vulnerable to this type of attack, and anodising of the surface provides little protection. Lead can also be vulnerable to attack from water washed from cementitious materials.

Where concrete structures support metal sheets and components, a risk of condensation corrosion exists. This risk is enhanced where vapour barriers are omitted, inadequately positioned, or jointed.

Friction   Moving metallic parts are sources of frictional wear. This is a localised and severe form of abrasion, and leads to poor alignment of parts. This can have implications for weather resistance in the case of windows and doors. The surfaces of the metal may be repeatedly stripped of any protective oxide film, and subjected to work hardening. Sheets or sections sliding relative to one another may also induce frictional wear.

Anodising   The artificial production of an oxide film on aluminium offers the additional benefit of colouring the metal's surface. The general protection offered will be better than for plain metal, although the relative purity of the aluminium will influence its durability. The higher the purity, the better the resistance. Anodised aluminium will require much protection against heavily polluted atmospheres, even with its inherent resistance to chemical attack because of the oxide coating (Preston 1989).

Heat treatment   A considerable range of heat-treatment processes are applied to construction materials. The process of hot-dip galvanising of mild steel is one of the earliest forms of sacrificial protection. This method is shown in Fig. 8.12.

The applied thickness will be dependent upon the immersion time and the complexity of the sections. Where thin coatings and cooling cracks occur, the resulting corrosion cells should not deplete the bulk metal. Where water gains entry, the bonding between the zinc and steel can be broken

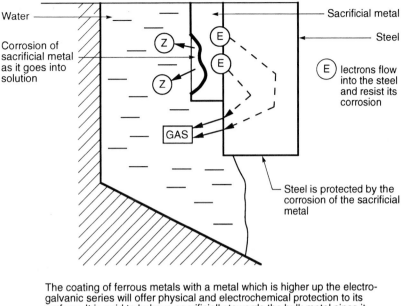

Fig. 8.12 Diagrammatic representation of sacrificial protection of a metal.

The coating of ferrous metals with a metal which is higher up the electro-galvanic series will offer physical and electrochemical protection to its surface. It is said to behave sacrificially towards the bulk metal since it will go into solution before it. This process will continue until all of the sacrificial metal has entered solution. At this point the rate of corrosion of the bulk metal will increase.

Thin layers of metal, e.g., zinc, offer such protection, however where the layer is detached and cracks exist, corrosion may proceed beneath the layer due to the differential concentration rates of oxygen at the surface of the crack and deep inside it.

down by differential oxygenation corrosion. The spraying of anodic metal coatings onto bulk metals may not achieve the same coverage or thickness as an immersion process.

Heat treatments which alter the physical properties of the metal are more complex. Site-based changes involving the heating of metal can have implications on strength and corrosion resistance. These changes may not be immediately apparent. Alloys of metals are particularly vulnerable to these changes.

Condensation corrosion

The incidence of condensation corrosion of metals appears to be low. This appears to be related to the general application of metals in construction, since the incidence of general condensation in construction is high. An area of concern is the metal coverings to flat roofs. The sealing of the traditional 'Cold' flat roof ventilation path seems to suggest that the incidence of condensation will increase at, or under, the covering (Building 1987). Corrosion may then occur. Where large fluctuations in outside temperatures occur, the isolation of the covering from the main building can result in water being drawn through joints. This is due to the reduction in pressure associated with differences between inside and outside temperature. Once the corrosion area has become established, it will penetrate the covering. This process is driven by a variety of mechanisms, which may include low CO_2 concentrations, chemicals within the timber (e.g. preservatives and flame retardants) and variable oxygen levels.

Pitting   The pitting of a metal's surface caused by localised corrosion can rapidly penetrate. This will result in local failure of the material, allowing weather to enter, or the load-carrying capacity to be reduced. It can occur rapidly since the edge of the pit will become generally cathodic relative to the inner point of the pit, as shown in Fig. 8.13.

The surface of aluminium becomes roughened by surface pitting, which will affect its appearance. In polluted atmospheres isolated pits can reach depths 100 times greater than the average corrosion depth of bulk metal (Butlin 1989b). Pitting can also be a symptom of calcium chloride attack of buried reinforcement, since a more general corrosion is usually associated

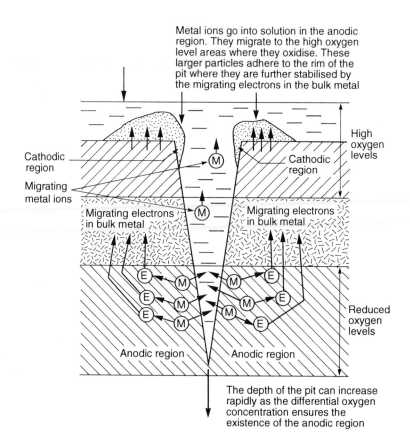

Fig. 8.13 The pitting corrosion of a metal.

with carbonation-induced corrosion. Where small amounts of cathodic metal come in contact with a bulk anodic metal, very small corrosion cells can develop. These can corrode into the general surface, creating a small reservoir of water to drive the process onwards.

Buckling   The growth of corrosion products from the parent metal is commonly associated with an increase in their volume. This is seen when mild steel reinforcement in concrete corrodes. This production of ferric oxide can cause deformation of steel components, and cause protective films to become detached. Glass in metal windows can be subjected to compressive loads, causing random cracking. In certain cases entire window frames can buckle and distort. See Fig. 8.9.

Where sheet metals are restrained at their edges, temperature-induced movements may cause them to bow. This may place further strain on the fixings so that when the panel contracts it may be inadequately restrained. The applied loadings on the edges may lead to separate compression failures.

Delamination of steel   The hot rolling of structural sections and the forging of other steels can impart an intrinsic layered structure to the metal. The interfaces between layers may have different properties compared with the inside of a layer, and these differences can be exploited by corrosion. They may become planes of differential oxidation attack, and may also be regions of reduced primary bonding. This will produce regions vulnerable to crack propagation, which will further assist in the corrosion process. Badly corroded surfaces may become completely detached, being held in place by the assistance of the protective covering.

Tearing   The differential movement problems common when metals and other materials meet are exacerbated when rigid fixings have been provided. The junction of calcium silicate brickwork and metal roof sheeting may need to accommodate both materials moving in opposite directions at the same time, since bricks can shrink and metals expand under drying conditions. Where sufficient differential movement can occur the shearing stresses may tear the metal, this can be assisted by the cyclic nature of the defect work-hardening the metal. Metals are not susceptible to moisture movements. When they are used in conjunction with moisture-sensitive materials differential movements can cause problems and strain the metals, causing tearing.

9 Dimensional instability: moisture movement

Dimensional instability of materials

Material movement requirements
One of the major sources of building failure is inadequate attention during design to the material-movement requirements of a structure (Rainger 1983). Material movements can be physically substantial and involve large forces. If these movements are restrained in any way by the method of incorporation into the building, the build-up of stresses in addition to the service loading may exceed the strength of the material and cause failure in the form of cracking or buckling (Everett 1975, Addleson 1982). Consequently, it is the stresses set up by restrained movement that are more important than the extent of free movement that would occur in unrestrained conditions (BRE 1975a). Frequently, errors in design or construction lead to provisions for movement being excluded from the construction, a possibility which requires to be verified during investigation of suspected movement faults. A generalised view of the dimensional instability of materials is shown in Fig. 9.1.

Modern movement problems
The developments in modern construction technologies, new materials, and their combination have been accompanied by greater movement problems. Rainger (1983) has produced an excellent analysis of the movement problems of building fabric and the materials used for construction. In this he suggests that the main causes of the increase in movement-associated failures are:

Thinner sections
(1) The trends towards thinner sections with reduced thermal capacity. The thermally lightweight components respond rapidly and effectively to small or frequent changes in temperature, and their thermal movements are consequently more extreme.

Larger sub-components
(2) The use of larger sub-components each having more movement spread over fewer joints. Large continuous components involve large physical movement and may develop considerable movement-induced stresses. Accommodation of the physical movement at the joints is particularly critical with these systems and their failures more dramatic.

Fig. 9.1 A generalised view of dimensional instability of materials and their relationship with the built form.

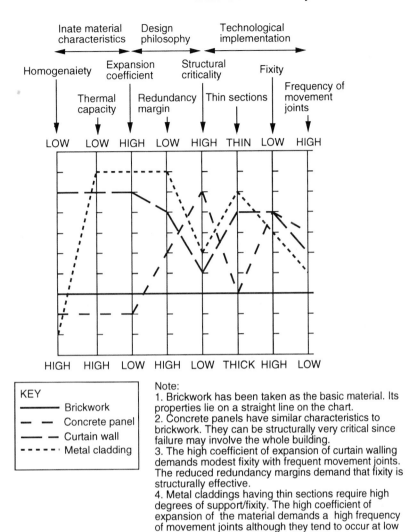

KEY
—————— Brickwork
— — Concrete panel
— · — Curtain wall
- - - - · Metal cladding

Note:
1. Brickwork has been taken as the basic material. Its properties lie on a straight line on the chart.
2. Concrete panels have similar characteristics to brickwork. They can be structurally very critical since failure may involve the whole building.
3. The high coefficient of expansion of curtain walling demands modest fixity with frequent movement joints. The reduced redundancy margins demand that fixity is structurally effective.
4. Metal claddings having thin sections require high degrees of support/fixity. The high coefficient of expansion of the material demands a high frequency of movement joints although they tend to occur at low frequency.

High thermal movement coefficients

(3) Incorporation of materials with high thermal movement coefficients. This ensures there will be extreme physical movement response to temperature changes.

Dry-jointed assemblies

(4) Widespread adoption of dry-jointed assemblies. Movement accommodation problems arise from the methods of construction and the coordination of numerous different materials with various dimensional stabilities within a dimensionally coordinated grid.

Slender structures

(5) More slender, highly stressed structures. There is a reduced margin for thermal or moisture-induced stresses, and the buildings generally have less structural redundancy to cope with traditionally minor defects.

Extrinsic and intrinsic forces

In the context of movement of the building fabric, it is possible to make a distinction between extrinsic forces, which are imposed on the building, such as solar heat gain, and intrinsic forces, which are the result of internal physical alterations to the materials, for example irreversible moisture changes. These are significantly different causes of movement. Material movement caused by climatic conditions will be strongly dependent on the nature of the exposure (Wilcockson 1980). Similar buildings in dissimilar circumstances will be affected differently. The intrinsic forces may be common to any building; however. Refer to Table 3.1 for the principal causes of intrinsic and extrinsic causes of movements and their effects.

It is most likely that rapid or extreme variations in humidity or temperature required to produce significant reversible moisture or thermal movement will occur predominantly to the exterior of the building. Since the function of the exterior is to act as a buffer against climatic extremes, it is to be expected that in normal circumstances the interior will be in a state of long-term equilibrium.

Moisture-related movements

Moisture movement

Moisture movement is the term used to describe the dimensional variation of materials as they respond to changes in the moisture content of their surroundings (Everett 1975). In general, the greater the permeability of a material, the more rapidly it will respond to fluctuations in the ambient moisture content.

Porosity and movement

Most traditional building materials are porous and subject to moisture movement (Everett 1975, Addleson 1982), but the degree of deformation this induces varies widely. There are some notable exceptions, including the highly impermeable materials recently in vogue as building enclosures. The plastic, glass and metal sheets will discharge water with little or no absorption. They experience no conventional moisture movement, and consequently the quality and long-term performance of sealing to joints is critical to the overall watertightness of the fabric. The severe problems that may arise with such situations and their other distinctive troubles are discussed elsewhere.

Moisture movement and problems

Uncontrolled moisture movement may create a nuisance, such as a deterioration in appearance, without being structurally critical (Everett 1975). It is unusual to find components that are constructionally isolated, however. Commonly the incidence of moisture movement itself will produce greater problems, perhaps only indirectly. External cracking produced by moisture movement may allow water ingress to initiate corrosive, sulphate and/or efflorescent chemical attacks, and will encourage physical frost attack (Humphreys 1988). Tracing back through the possible combinations of these defects is an unenviable but necessary exercise.

The role of climate

Climate and structure

The climate is an important factor in the occurrence of moisture movement in the external fabric of buildings, and the appearance of its symptoms. It has been shown (Parker 1966), for example, that clay brickwork may require three weeks to dry out fully after exposure to only one day's driving rain. This leads to a distinct possibility that the outer portion or skin of unrendered brickwork can remain almost totally and continually saturated during prolonged wet weather (BRE 1975a). Such periods can extend into months in the winter season, particularly in exposed locations such as the west of Scotland (Parker 1966). Conversely, in the summer, any prolonged dry periods can produce the opposite extreme conditions.

The materials respond dynamically and reversibly to extremes of climate with extremes of moisture movement or stress development. The implications of this for the long-term performance of materials during the life of the building is only one feature of the moisture-movement problem. There is also an influence on the drying-out period that follows the traditional wet-trade construction. This is when the majority of cracking occurs, produced by irreversible moisture movement (BRE 1975a). The climate affects moisture adjustment of porous materials generally. For instance, a building constructed in late autumn may not dry out properly until late into the following spring (Parker 1966), or even before the end of the summer (Seeley 1987). Until then it will not experience its full drying shrinkage, thereby delaying the maturity of any associated faults. Obviously the timing of any assessment of the building is important, and a full knowledge of the local climate, the nature of exposure of the specific building and (where relevant) the date of construction is critical.

Moisture and thermal movement frequently act in conjunction. The effects can be similar (Addleson 1982) and when designing for movement accommodation they require to be considered collectively. They are not necessarily additive, however (Rainger 1983), and where climatic conditions such as hot, dry weather combine to produce conflicting movements in a material, one force will generally predominate. For timber this force is moisture movement; for concrete it will be shrinkage.

Moisture, shrinkage, and restraint

Problems can emerge directly as stresses arising from restrained movement lead to failure of materials, producing a loss of structural strength and/or watertightness. Movement restraint in one direction only may lead to greater deformation in a direction normal to it. This can produce bulging or bowing of panels or elements. Effective restraint in more than one direction is likely to cause more severe failure. Restraint by attached buildings is a frequent occurrence. In such instances, or where a building has sufficiently large sections to warrant subdivision, the incorporation of movement joints would avoid the problems. Check for their presence.

Irreversible and reversible moisture movement

There are two principal categories of moisture movement that porous materials exhibit: irreversible and reversible movement. Buildings respond to moisture movement by producing a number of characteristic defect symptoms (BRE 1975a). Irreversible moisture movement occurs early in the life of some construction materials, and hence the building. The nature of the movement is characteristic of the particular material, producing either a physical expansion or contraction. This defect occurs only once, and in the absence of further stresses will not move subsequently (Melville and Gordon 1979). Reversible moisture movement is generally less extensive than irreversible movement, but it occurs repeatedly throughout the life of the building as a dynamic response to variations in relative humidity.

Serious faults can arise if the design accommodation of these short- and long-term effects is neglected or underestimated. Although concrete, bricks and timber are not the only materials to suffer moisture movement, they tend to produce the most serious problems for buildings because they are commonly used as structural materials.

Symptoms generally

Appearance of moisture movement symptoms

The occurrence and appearance of moisture movement symptoms are related to the type of structure by the nature of its reaction to intrinsic stresses. Where the resultant forces are unrestrained the structure will move, perhaps differentially. This is often accompanied in simple masonry walling by an oversailing effect (Melville and Gordon 1979). The occurrence of constrained forces can produce much more significant problems. The defect may reveal itself as bowing of the structure. Stress concentrations at changes in direction and or section will concentrate the symptoms, although the forces may be relatively widespread within the structure itself.

Symptomatic cracking

Symptomatic cracking can often be a useful tell-tale in itself. By assessing the location and characteristics of cracks, bowing or oversailing, it may be possible to make an initial assessment about the cause(s), and devise a suitable monitoring method to confirm them. Note, however, that many of the cracking symptoms associated with moisture or thermal movement are superficially similar to those arising from ground or substructure instabilities.

It will be necessary to assess carefully all available information and criteria before arriving at a correct decision. This may require monitoring of the building for an adequate period to establish whether faults are progressive or stable. Such information can give valuable assistance in the correct analysis of the defect, and the selection of any necessary appropriate remedial works (Melville and Gordon 1979).

Vertical and horizontal movement

In conventional masonry low-rise buildings, the walls are usually relatively lightly loaded, and provide little restraint to vertical movement. The fabric can move vertically in response to the moisture-movement forces. Comparatively little movement may be tolerated horizontally though, because of the resistance to distortion derived from the constructional form (BRE 1975a). Consequently, as failure follows the line of least resistance (through the masonry), the emergent symptoms of moisture-related shrinkage will be the formation of vertical tensile cracks. The incorporation of reinforcement between courses of brickwork is sometimes used to improve the tensile capacity of the wall, but the stresses are not dissipated by this and the cracks will either emerge elsewhere, or perhaps as smaller, less unsightly openings (BRE 1975a), making assessment difficult.

Compressive bulging

Compressive bulging is indicative of moisture-related expansion, but because compressive strength is a fundamental requirement of loadbearing walling materials this is a less frequent occurrence than cracking.

Moisture movement and structural weakness

The points of least resistance in the building fabric are usually around openings, where there is minimal constructional tying in of the walls and where the loading on the structure is minimised by the arching effect of lintels. Movement faults have a tendency to appear at such weaknesses (BRE 1975a). They may also emerge at the tops of walls where the loads imposed on the brickwork, and self-weight, are at a minimum. Shrinkage-related tensile cracks in the external envelope will tend to run vertically between the sills and lintels of windows in adjacent storeys, or between a ground-floor window sill and the damp-proof course (BRE 1975a). See Fig. 9.2. It is unusual for drying-shrinkage cracks to pass through the damp-proof course, since the brickwork below usually remains relatively damp, and the DPC acts as a convenient slip plane for the movement in the wall

Fig. 9.2 Variety of simple drying-shrinkage cracks. Windows and other openings create weaknesses. Tensile cracking occurs where contraction stresses are concentrated.

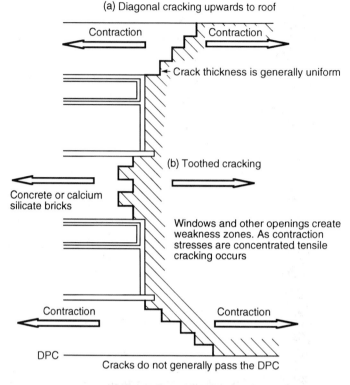

(a) Diagonal cracking upwards to roof

Contraction

Contraction

Crack thickness is generally uniform

(b) Toothed cracking

Concrete or calcium silicate bricks

Windows and other openings create weakness zones. As contraction stresses are concentrated tensile cracking occurs

Contraction

Contraction

DPC

Cracks do not generally pass the DPC

(c) Diagonal cracking downwards

above. This is particularly common with the asphalt DPCs (Melville and Gordon 1979).

Moisture movement internally

Internally, the moisture movement cracking of partitions may follow vertical or horizontal lines, or create a stepped vertical crack which follows the joints between the units. See Fig. 9.3. The style will depend on a combination of factors. These will include the type and degree of movement exhibited by the materials, which is likely to be excessive; the strength and loading of the wall, which is frequently lightweight; the degree of tying in of the wall to the floor and roof structures; and for masonry walls, the balance between the strengths of the bonding mortar and the blocks or bricks.

The larger masonry units tend to produce stepped cracks, and in long runs the cracking may subdivide the wall into a number of sections of similar length as the shrinkage stresses are relieved at intervals.

The influence of constructional form

The nature of the constructional form will obviously influence the style of response of the fabric to moisture movement. For instance, cross-wall

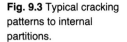

Fig. 9.3 Typical cracking patterns to internal partitions.

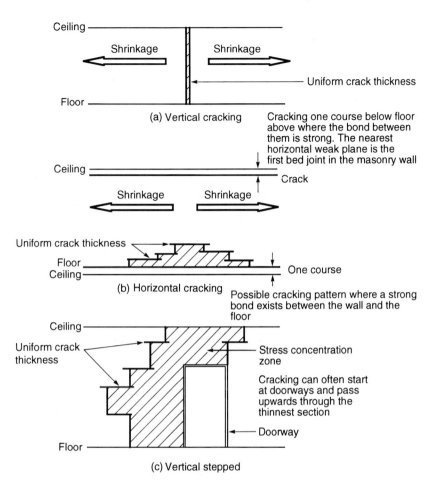

Ceiling

Shrinkage   Shrinkage

Uniform crack thickness

Floor

(a) Vertical cracking

Cracking one course below floor above where the bond between them is strong. The nearest horizontal weak plane is the first bed joint in the masonry wall

Ceiling

Crack

Shrinkage   Shrinkage

Uniform crack thickness

Floor
Ceiling

One course

(b) Horizontal cracking

Possible cracking pattern where a strong bond exists between the wall and the floor

Ceiling

Uniform crack thickness

Stress concentration zone

Cracking can often start at doorways and pass upwards through the thinnest section

Doorway

Floor

(c) Vertical stepped

construction will buttress and subdivide long spans of walling. This will strengthen the wall against buckling tendencies, but may limit expansion movement accommodation. It will be necessary to analyse fully the structural character of the building, which itself is not always a straightforward task.

Irreversible movement defects

Adjustment of moisture content

Irreversible movement occurs with manufactured or processed materials as they adjust their moisture content to the ambient humidity after production. The exact nature of irreversible movement depends on the material substance and the nature of the manufacturing process. The physical extent of movement will depend on the difference in moisture content between the material immediately after production and its *in-situ* equilibrium, and the coefficient of moisture movement for the particular material. Obviously, the degree of restraint will also be important. Although measurements of unrestrained movement have limited relevance to restraint conditions, it is at least possible to identify and categorise the materials that are the likely sources of problems, and also the degree of movement that should have been accommodated (BRE 1975a).

Restraint of irreversible moisture movement

Moisture movement can be restrained either at the edges of materials, for example at returns and enclosures to walling, or by the fixing of the materials in the case of laminated sheets or layered panel constructions. This must be accounted for, since the nature of the fault symptoms will differ according to the homogeneity and fixing of the panels.

Induced stress

In the context of claddings, the stresses on the fixings may produce concealed defects, particularly if the fixings are of a poor grade and provide a weak link. The jointing of panels should be adequate to accommodate the movement requirements of thermal and moisture response of the panels (Seeley 1987). Examples abound from the multi-storey systems era of inadequate design or constructional accommodation of all types of movement, and defects produced by movement stresses continue to occur widely. A common symptom of movement problems is the extrusion of sealant at the joints. These faults may develop further to produce cracking or spalling of the panel edges (Seeley 1987).

Internal and external moisture levels

There will normally be a difference in moisture content equilibria between internal and external locations, which may influence the time it takes for irreversible shrinkage or expansion to occur. The total physical movement will also be determined by the relative difference in moisture content of the material at installation and the equilibrium position. Materials that undergo drying-shrinkage will shrink more in internal locations than externally. Materials that irreversibly expand will do so more in external locations. The moisture content equilibria will vary seasonally, and it should be appreciated that the emergence of faults due to irreversible movement may be seasonal also, rather than occurring at a set period after construction. Rapidly drying out the structure using the heating installation to save construction time will almost certainly accentuate the problems (BRE 1975a).

Irreversible expansion

Ceramic products

Ceramic products, including clay bricks and flooring tiles, undergo an irreversible expansion as they adjust from the drying-out that occurs during firing to the ambient-moisture conditions (Eldridge 1976). The amount of irreversible expansion varies with the type of clay brick; values start at around 0.1 per cent and range up to 0.2 per cent. This represents a greater coefficient than those for either reversible moisture movement (about 0.02 per cent) or thermal movement (5.5 to 8 × 10^{-6} per °C). Note that brickwork irreversibly expands only a little more than half as much as the same unbonded bricks. This is because of the restraint imposed by the bonding.

Early irreversible moisture movement

About half of the total irreversible expansion of clay bricks takes place within the first week following production (Ransom 1981), and the process of moisture adjustment is mostly complete within the first three months (Rainger 1983), (Eldridge 1976, Melville and Gordon 1979). The incorporation of immature bricks into structures within this period, such as may happen in construction booms, is very likely to lead to problems of movement accommodation in the fabric soon after construction.

Whilst the symptoms of irreversible expansion may emerge within this three-month time period (Melville and Gordon 1979), they may be held over until an appropriate seasonal change, or there may be a lengthy delay. It has been suggested that five years is not an overestimate in some cases (Eldridge 1976). In such protracted incidents, a separate long-term expansion of fired-clay products has been recorded as producing failures (BRE 1975a).

Irreversible expansion defects

The irreversible expansion of fired-clay products will produce failure at the location in the structure which offers least resistance. In clay brickwork this may produce oversailing of long sections of walling (see Fig. 9.4), as the DPC provides a slip plane, or as the horizontal bond between courses fails. Thermal expansion may also be contributory and exacerbate the physical movement, again oversailing is symptomatic.

Expansion in masonry walling

The point weakest in resistance to expansions in a conventional masonry wall will be at the corner, where two forces converge, and the only resistance to movement is the tensile strength of the materials. Vertical tensile cracks may appear in external walls here. Incorporating effective movement joints into the wall should be sufficient to eliminate this problem (Eldridge 1976).

Fig. 9.4 Oversailing of walling at corners caused by expansion of long runs of brickwork.

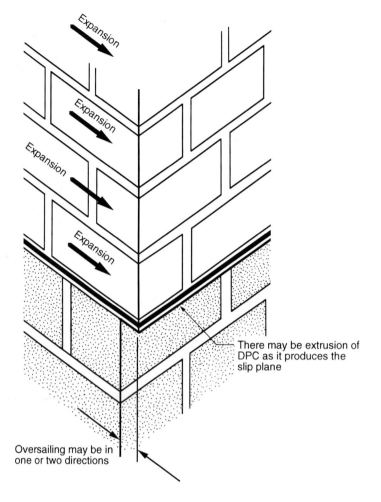

Expansion

Expansion

Expansion

Expansion

There may be extrusion of DPC as it produces the slip plane

Oversailing may be in one or two directions

Vertical cracking   An alternative form of vertical crack in a conventional cavity external wall also occurs at the quoin, about a half brick from the shorter return. This acquired defect (Richardson 1985b) is a rotational effect caused by irreversible expansion (Eldridge 1976). The distance of a half brick from the quoin corresponds to the thickness of the return wall. There will usually be two tensile cracks but the rotational nature will hide the cavity side crack from view. See Figs. 9.5 and 9.6.

Crack stability   Once formed, the crack movement is likely to be reasonably stable unless subject to excessive thermal movement, and need not be structurally significant (Eldridge 1976). Bricks with a particularly high coefficient of expansion are most likely to produce the problem, also short returns that are less than about 700 mm in length (Eldridge 1976, Melville and Gordon 1979). Returns greater than this can usually accommodate the rotational

Fig. 9.5 Rotational cracking in brickwork at short returns **(a)** in a long wall restrained at the far end, and **(b)** in stepped terrace construction.

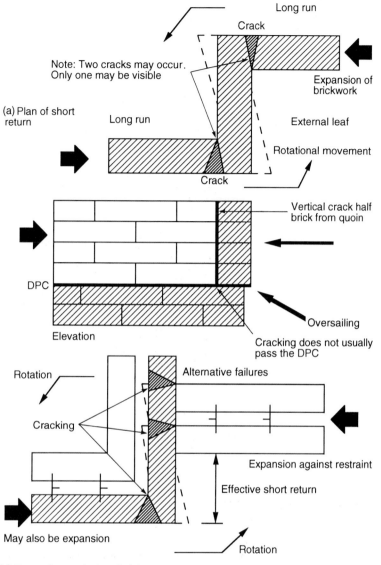

(a) Plan of short return

Long run

Crack

Long run

Note: Two cracks may occur. Only one may be visible

Expansion of brickwork

External leaf

Rotational movement

Crack

Vertical crack half brick from quoin

DPC

Elevation

Oversailing

Cracking does not usually pass the DPC

Rotation

Alternative failures

Cracking

Expansion against restraint

Effective short return

May also be expansion

Rotation

(b) Expansion against restraint

movement without cracking at the quoin. The crack is unlikely to extend below the DPC unless it is approximately a metre above ground level (Eldridge 1976).

Other symptoms   A further type of expansion-related crack may appear diagonally above DPC level. This may run in a single direction or may double back on itself to produce a toothed-arrow effect. The crack width is likely to be fairly uniform

Fig. 9.6 Example of cracking at the end of a long wall with a short return. The short return is unable to accommodate the expansion movement by rotation so a vertical stress fracture occurs, usually half a brick from the quoin in half brick thick walling.

and, depending on the bond strength of the mortar, may pass through the bricks or around them. The crack occurs as part of the clay brickwork around a weakness in the wall (such as an opening) moves under irreversible expansion. The problem is made worse if accompanied by thermal expansion. The remaining brickwork if restrained separates at this point of relative weakness, and a crack is formed and the end of the unrestrained brickwork oversails at the quoin.

Irreversible shrinkage

Cement-based products

In contrast to fired-clay products, most cement-based products undergo an irreversible shrinkage as the water used in the initial mix migrates. After the initial evaporation of surface moisture, the remaining free water trapped within the concrete also evaporates. This escapes through the porous structure at a much slower rate and produces the significant irreversible drying-shrinkage as the product adjusts from being virtually saturated to its new relatively dry equilibrium.

The influence of free water

Remaining free water is held within very fine pores and is lost over a period extending into years (Addleson 1982). This itself is susceptible to changes in ambient moisture content. With thick concrete forms the surface layers may dry more quickly than the main body, and this can lead to fine surface cracking. This need not be problematic (BRE 1973a, Reid 1973), and more serious than this is any differential shrinkage within the body of a casting. It tends to occur if the thickness of the casting is variable. Smaller cross-sections will dry out more quickly and the tension at the joint as the shrinkage takes place may causes cracking. Similar effects may arise with the jointing of different mixes (BRE 1973).

Obviously for concrete the water content of the mix will be important. The convenience of placing a wet conventional mix will be outweighed by the consequentially high degree of drying shrinkage. This tends to produce a more porous, weak concrete, which will be liable to increased reversible movement.

Drying shrinkage

Drying shrinkage in cement-based products depends also on the hydration of the cement paste to form the gel structure; indeed this is a primary cause (BRE 1975a). The greater the cement content of the mix, the more marked the drying shrinkage. It has been commented elsewhere that undue attention is placed on the strength of concrete, when in many instances moisture movement is the more important of the two factors (BRE 1975a). Problems can be expected.

Aggregates and shrinkage

Aggregate type can also affect the total amount of movement significantly (BRE 1975a). Rigid aggregates will withstand the drying-shrinkage forces and give some dimensional stability to the concrete. Softer aggregates, or those that are themselves highly responsive to moisture content changes, will allow and even contribute to the instability of the mix. Lightweight concretes have been found to be particularly susceptible to drying shrinkage because of the nature of the aggregate (BRE 1975a). Glass-fibre reinforced cement is also known to undergo drying shrinkage on exposure to low humidity and/or high temperature conditions (Seeley 1987).

Variability in moisture movement

Consequently there is a wide variability in the irreversible shrinkage of cement-based products, particularly concretes. In instances where the defects analyst has little background information, it may be difficult to be conclusive.

Accommodation of movement

Unless there is adequate provision for the irreversible movement, as the material completes its dimensional adjustment *in situ* it produces cracks or other disruptions in the building fabric. The amount of movement that requires to be accommodated is usually significantly greater than the reversible movements that follow (BRE 1975a). Products such as concrete blocks that undergo drying shrinkage, yet are allowed to become wet before incorporation because of neglectful materials storage, will suffer greater moisture movement. The irreversible shrinkage of *in-situ* concrete can produce cracking in buildings over a long period. Indeed, cracks produced this way may occur and increase in size up to 5 years after construction (Eldridge 1976). There is some general encouragement, however. *Principles of Building* (Vol.1) (BRE 1975a) reported that: 'Generally, shrinkage cracking is important only because of its effect on appearance. It does not materially affect structural stability nor, in cavity-wall construction, does it affect the exclusion of rain.' (See also Melville and Gordon 1979.)

Precast products

Precast products can be dried out before they are incorporated into the building (Reid 1973), and if this is done properly the problems of large irreversible moisture movements are eliminated. Residual moisture movement problems will be of the reversible, dynamic form.

Common symptoms

The common symptoms of irreversible shrinkage are cracks between weaknesses in the structure. In walls, for example, cracks may appear between openings. The effect is unlikely to bridge the DPC, since products in contact with the ground will not necessarily dry out as completely. The DPC acts as a slip plane and allows the differential movement to release the stresses (Eldridge 1976).

Problems with rendering shrinkage

Renderings applied to a stable background will have a tendency to crack as they shrink. This also increases with the cement content. The quality and uniformity of bond across the interface will be important, and determines whether any cracking is fine and evenly distributed, or infrequent and more substantial. Where the background is insufficiently sound to withstand the force of the render shrinking, spalling of the finish may occur by delamination of the background (BRE 1975a). This has been known to happen with poor brickwork. In instances where the background is sufficiently strong to withstand the movement stresses of the foreground, the overall dimensional stability may be maintained (Reid 1973).

The importance of background stability

A similar problem with poor background stability can arise with wood-wool slabbing used as a decking for roofs and skimmed with a cement/sand mortar (BRE 1973). The drying shrinkage of the mortar is usually sufficient to compress the unstable wood-wool. This causes an opening up of the joints between the slabs and can cause a patterned stress damage to the bonded roof covering.

The calcium silicates

In addition, these types of moisture movement also apply to calcium silicate, and concrete bricks and blocks, produced using a wet or semi-wet method.

Other symptoms

The common symptom of irreversible shrinkage during the initial stages of their life is a clean, vertical, tensile crack. It will either pass through alternate brick courses, if the mortar is strong enough to allow the brickwork to move en masse, or produce a toothed effect if the mortar is weak, and the bricks can move independently (Eldridge 1976). Evidently, as with expansion of clay bricks, the relationship between the strength and rigidity of the walling units and the mortar binding them is important. Either type of crack should be of a reasonably uniform width (Melville and Gordon 1979) and the edges appear suitable to fit together again. In changeable weather the crack size and appearance may alter as the bricks expand and contract under reversible moisture movement (Eldridge 1976). Obviously the toothed crack will be less obvious and less aesthetically objectionable than the vertical crack that passes through alternate bricks. See Fig. 9.7.

Fine cracking in the jointing may be relatively unobtrusive. An example of cracking in calcium silicate brickwork is shown in Fig. 9.8.

There is also a possibility of diagonal or V-shaped cracks occurring in free-standing walls, where the initial irreversible shrinkage followed by reversible moisture movement causes the wall to move. A similar effect to that with clay brickwork can occur (see earlier) with the ends of the wall oversailing. The defect will occur above the DPC only, indicating a fault within the main structure rather than below the ground.

Fig. 9.7 Tensile shrinkage
cracks in calcium silicate or
concrete bricks.

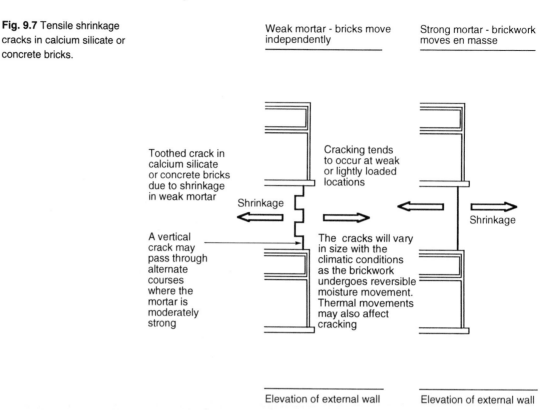

Weak mortar - bricks move
independently

Strong mortar - brickwork
moves en masse

Toothed crack in
calcium silicate
or concrete bricks
due to shrinkage
in weak mortar

Shrinkage

Cracking tends
to occur at weak
or lightly loaded
locations

Shrinkage

A vertical
crack may
pass through
alternate
courses
where the
mortar is
moderately
strong

The cracks will vary
in size with the
climatic conditions
as the brickwork
undergoes reversible
moisture movement.
Thermal movements
may also affect
cracking

Elevation of external wall

Elevation of external wall

Cracking and weakness in the structure

The cracks are most likely to appear at weak points in the structure of the wall, and where there is little load on the bricks to tie them in, and so limit their free movement. A common location is between windows or at door openings (Melville and Gordon 1979), where lintels divert the loads around the opening. In long runs, the cracks commonly appear at intervals of about 1.8 m (Melville and Gordon 1979).

Internal contractions

Where calcium silicate or concrete bricks are used for lightweight internal partition construction, the irreversible shrinkage problems can occur within the partition or at the juncture between the internal partitions and the external envelope. These movements usually produce vertical cracks, but sometimes they may be horizontal if the head or base of the wall is tied firmly to the bounding floor, or roof structures, see Fig. 9.3. With large-scale components such as blockwork, stepped openings can be produced. The factors affecting the shrinkage of the blocks depend on its composition and history (Eldridge 1976), and the weight of the wall. A single vertical crack in the wall will usually mean that the movement has not been accommodated within the joints because the mortar bond is very strong.

Fig. 9.8 Cracking to calcium silicate brickwork.

Conflicts of expansion and contraction

Conflicting irreversible stresses

New construction that integrates cement-based and ceramic products can produce problems as the stresses of irreversible dimensional changes conflict. The strength of many building materials is greater in compression than tension (Reid 1973), and cracking will occur relatively easily during restraint to contraction (BRE 1979b, 1979c, 1979d). A common example of the problem where such forces are restrained occurs with clay brickwork infill construction to an *in-situ* concrete frame (see Fig. 9.9).

Fig. 9.9 Contraction of *in-situ* concrete frame and expansion of brickwork panels.

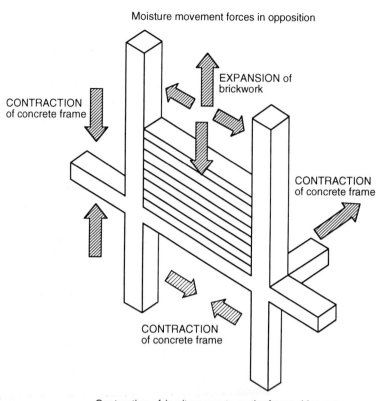

Moisture movement forces in opposition

EXPANSION of brickwork

CONTRACTION of concrete frame

CONTRACTION of concrete frame

CONTRACTION of concrete frame

Contraction of *in-situ* concrete as the frame dries out.
Expansion of rigid brickwork as moisture is absorbed

As the concrete frame dries out, it irreversibly shrinks and compresses the enclosed brickwork. If the brickwork expands, the moisture movement forces are in opposition. The situation is exacerbated by any thermally induced forces. The larger the size of the bay, the greater the physical movement restrained.

Tensile cracking   Tensile cracking of the frame, and perhaps also the floor slab, will occur soon after construction. The nature of cracking will depend on the shape of the building, and openings in the fabric will appear at the weakest point(s).

Resistance to internal tensions   Whether a concrete frame can withstand the stresses imposed on it by drying shrinkage depends largely on its resistance to internal tension. Reinforcement may protect the concrete against its inherent tensile weakness, and of course this is a primary function in conventional structural design. This cannot be relied upon to withstand drying-shrinkage stresses without careful design, so tensile cracking remains a distinct possibility even with reinforced concrete structures (BRE 1975a).

Bond failure The mechanism of bond failure between steel and concrete involves the steel taking the compressive load of the drying shrinkage in the concrete. This requires a continuous and effective bond between the steel and concrete, which is usual. The concrete is left to take the tensile force of the contraction, and cracks. In such instances it has been suggested that the bond stress of the composite will not be discontinuous (BRE 1973). This is not to say that it will continue to support the same load, or that the steel reinforcement will have effective cover. This is an analytical task for the structural engineer.

Concentrations of reinforcement can accentuate moisture movement stresses at their boundary with less highly reinforced concrete, for example edge beams (BRE 1973a). Shrinkage cracking will demarcate the relatively rigid and relatively flexible zones, unless some additional provision is made during the design. In contrast, the numerous small reinforcing bars such as are found in shell constructions may produce numerous small cracks in the concrete rather than eliminating them (BRE 1973, see also Reid 1973). The patterned cracking will help with identification of the cause.

Conflicting stresses An example of failure in an *in-situ* concrete frame is provided by Hodgkinson (1983). If conflicting stresses from clay brickwork expansion and concrete shrinkage are imposed on the external corner of a regular grid concrete frame, the weakness caused by a change in direction gives an increased degree of freedom for movement in the fabric. The resultant force of the conflicting stresses pushes the corner out from the building. As the frame fails in tension its movement is punctuated by diagonal cracking to the floor slab.

In some instances reinforced concrete floor slabs may lift at the corners, and this alone may produce cracking in the frame and supporting walls (BRE 1973). The degree of anchorage of the slabs will be an important factor in restraining this movement tendency (BRE 1973).

Restrained expansion Where the expansion of brickwork is restrained this can cause problems for the restraints. The use of concrete support nibs may accommodate normal loading, but can fail in shear, due to the expansive loading of the brickwork (Hodgkinson 1983). This is a fairly common fault, caused by the omission of a simple preventative measure. Incorporation of a suitable movement joint around the brick infill panels could have accommodated such movement.

Defects in floors Defects can also arise early in the life of carelessly laid fresh clay flooring tiles, particularly if this is done on a concrete floor, normally within a year (Seeley 1987). The clay tiles will expand irreversibly, and a large proportion of this movement takes place in the first two weeks after firing (Ransom 1981). If there is no movement accommodation, the surrounding structure constrains the movement of the floor finish. The problem is exacerbated by laying onto a relatively green concrete floor slab, since this itself will still be undergoing irreversible drying shrinkage. The two movements conflict, and the compressive force on the floor finish debonds the tiling cleanly (the weak link in this case), causing it to bulge. The thinner the tiles, the more easily this occurs (Seeley 1987). The failure is worsened if the tiling is rigidly bonded, and if there is omission of a separating layer to act as a slip membrane (Ransom 1981).

Vertical shrinkage

With cast-*in-situ* frameworks vertical shrinkage of concrete can have serious secondary effects (Eldridge 1976). This is distinct from creep, but the effects are superficially similar. The shrinkage of the framework may put non-loadbearing cladding into compressive loading it was not designed for. This pressure may be vertical or horizontal, and causes a bulging of the panel and possible failure (Melville and Gordon 1979). Any weakness in the fixing of the panel will be tested. The same effect can be produced with infill brickwork to an *in-situ* concrete frame. Since the shrinkage continues for five years or so after casting, albeit at a decreasing rate, the problem may not emerge completely immediately.

Reversible movement defects

Reversible movements

Similar types of movement symptom to those produced by irreversible movement can arise from reversible movements alone. Reversible shrinkage and expansion occur with porous materials, usually on a smaller scale than the irreversible movements (Everett 1975, BRE 1975a). Whilst some of the irreversible movements are avoidable or can be mitigated by careful product management, the reversible movements are unavoidable. Their emergence depends largely on design accommodation and forethought. This type of moisture movement is sometimes approximately rather than exactly reversible (Ransom 1981), and is a dynamic response to a dynamic equilibrium. Whilst minor faults arising from irreversible movements are usually stable and would therefore require no remedial works, the reversible movements will continue indefinitely, and may produce recurrent instability.

Unrestrained movement

Where the moisture movement tendency of a component is unrestrained, the movement will occur simply. This is straightforward for a free-standing component. Where one or more of the dimensions of the component are restrained, then subject to the inbuilt stresses not exceeding the strength of restraint, any deformation under movement is in another unrestrained or less restrained direction. The deformation takes the line of least resistance. This may produce, in the case of expansive moisture movement, a buckling of the restrained element (see the example on brick infill panels earlier). Cracking may accompany this in the subject element, if the strength of the element is insufficient to withstand the internal forces brought about by expansion under restraint.

Contraction

In the case of a contraction under moisture movement, the element is likely to crack as the elongation required by the edge restraint is stronger than the internal strength of the material. Buckling is unlikely to occur here.

Inadequate restraint

Where the restraining component is inadequate for the purpose intended or imposed upon it, it will yield instead of the element producing the load. An expansion of an enclosed element produces a buckling and/or sliding shear in the wall. In the case of a contraction, the contracting element is pulled out of the restraint, or the restraining element is distorted, buckling or bowing. With a reasonably straightforward structure, it is usually relatively simple to assess whether the buckling in a wall has been caused by expansion or contraction of the cross-wall construction. Expansion produces a convex buckling; contraction produces a concave buckling, but requires an effective connection between the elements greater than the strength of the external wall itself. An alternative mode of failure is a horizontal crack and movement as the restraining element fails under shear.

Free movement

Table 9.1 compares the unrestrained moisture movements for a range of common building materials.

The variation in moisture expansion coefficients between different materials is a marked if generalised statement. Note that the irreversible and reversible movement of materials does not occur exclusively; if the circumstances are suitable they will coincide during the early life of the structure.

Differential moisture movement

Materials may be subjected to differential moisture movement across their thickness (see also Chapter 10) if the environment on both sides is different. In addition, non-homogeneous composites may combine materials with differential moisture movement properties. In either instance, the consequence is that the composite tends to deform and disintegrate as changing moisture conditions create internally opposed forces. Where the two

Table 9.1 Typical moisture and thermal movement characteristics of a range of common building materials

Material	Moisture movement (%)		Thermal movement (Coefficient of thermal expansion 10^{-6} per degree C)
	(R)	(I)	
Brick (clay)	±0.02	+0.070	5.5 to 8.0
Brick (calcium silicate)	±0.03	−0.025	9.0 to 15.0
Concrete (dense)	±0.038	−0.05	11.0 to 13.0
Concrete (lightweight)	±0.03	Neg.	8.0*
Glass (soda lime)	Neg.	Neg.	5.6 to 9.0
Aluminium	Neg.	Neg.	24.0
Steel (structural)	Neg.	Neg.	12.0
Steel (stainless)	Neg.	Neg.	17.0
Timber†	±3.20 to ±4.7	Cont.	3.0 to 6.0
Plastic (polythene)	Neg.	Neg.	143 (HD) to 198 (LD)
Plaster (PVC)	Neg.	Neg.	65
Polyesters	0.02‡	Neg.	17–24

Key

(R)eversible and (I)rreversible movement characteristics and the coefficients of linear expansion of materials are highly variable, and the values contained here are taken from standard texts and the authors experimental work. They should be recognised as representative values which may well vary in specific instances.

* Many forms of lightweight concrete exist and this will cause this typical value to vary.

Neg.: Amounts are generally so low that effects are ignored.

Cont.: Timber can be considered to be constantly moving in order to remain in equilibrium with the relative humidity of its surroundings. A constant movement can be considered to have occurred when the timber is originally seasoned. This lowering of its moisture content to a working level will then vary with the moisture present in its surroundings.

† The anisotropic nature of the material will mean that its major axis will exhibit differences in movement characteristics.

HD = High density; LD = Low density
‡ Typical value for glass fibre reinforced polyester.

instances are coincidental the movement is maximised. This deformation itself is a form of restraint. Disruption may occur across the juncture of the opposing forces, leading to delamination.

Reversible moisture movement in brickwork

Clay brickwork may crack due to reversible moisture movement as expansion causes bulging. This is particularly problematic if the wall is restrained. In low-rise construction the crack is often greater at the base, and will start at the DPC. As the brickwork expands and is constrained, so the surface bulges, depending on the amount of restraint from wall ties. The crack appears at the apex of the bulge. Moisture movement problems are exacerbated by using bricks of high thermal or moisture expansion coefficients, and by long runs of brickwork without proper allowance for movement.

Moisture movements in timber

Timber is worth highlighting because it is anisotropic, responds noticeably to changes in moisture content, and is widely used for construction. The transverse and radial moisture movements are significant but differ markedly, and in some species of timber the longitudinal movements may be virtually disregarded (Everett 1975). The class of timber varies according to the potential movement, in turn dependent on the species. Averaged movements are shown in Table 9.1. Timber hygroscopicity means that it will adjust its moisture content to approach equilibrium with the surroundings, producing seasonal variations. Correspondingly, it will absorb moisture if in contact with a damp surface (Everett 1975).

Movement in timber and structural damage

The forces produced by the expansion of timber can be considerable and sufficient to cause structural damage. The common areas of problems are floors and roofs. Structurally less significant are defects with door and window assemblies. These are exposed to climatic extremes and often differential conditions.

Usually the joints between floorboarding are designed to allow for expansion without distortion or loss of appearance. Where the edge gaps to boarded floors have been omitted or are inadequate for the size of room, however, the expansion across the boarding can be sufficient to push out the walls (BRE 1975a). Similar effects are also possible with restrained timber joists, and flooding of the under-floor void to suspended timber ground floors can produce a damp enough environment for maximum movement. The upper floors are less susceptible to moisture movement, since they are unlikely to become or remain damp for long periods or as frequently. The move to using fabricated timber boarding in large modular sections produces greater physical movement to be accommodated. The extreme sensitivity of some of these materials, and the absence of internal structure, can produce dramatic moisture movement responses to differential conditions. Indirect problems can arise with other elements of the structure not easily coordinated with the extreme and anisotropic moisture movement of timber.

Expansion of timber flooring

The expansion and bulging of parquet flooring has been a frequent problem in the past, although it is less common now because of changes in construction technologies. The failures illustrate well the problems that arise with any material that exhibits moisture movement and is laid in large areas with little accommodation for that movement. Indeed, with either moisture or thermally related movements there is a general criticality of size of component (Reid 1973). Where large expanses of moisture-sensitive material exist the net physical movement that can be produced, even from materials with a relatively modest response coefficient, is enough to produce obvious disturbance.

Other cellulosic products

The move to artificial sheeting of roofs (Scotland) and other timber-based products such as wood-wool slabbing incorporates susceptibility to extreme moisture movement. These components are also frequently used for decking of flat roofs, which in turn are renowned for their poor waterproofing durability.

Climatic influence

The physical amount of reversible moisture movement in materials will vary with the climatic conditions (Everett 1975). The problems for the building can be expected to be at a maximum in extremes of climate. In the UK the climatic variability affects the extremes experienced by the buildings, and emphasises the dynamic nature of the problem. Problems can be expected with buildings constructed or clad with moisture-sensitive products and located in exposed surroundings.

Constructional detailing

Location of detail

Location of the detail is also important to the penetration of water, and so to the degree of moisture movement that will be experienced. It is important to appreciate that if the exposure of a building or a single element such as a simple wall is not uniform, then the moisture movement will not be uniform either.

The external envelope   A conventional cavity wall during a prolonged rainstorm may well become wet on the inside face of the external leaf. This can happen in two ways: either from openings in the pointing allowing water to bypass the outer leaf or, less frequently (Parker 1966), by a porosity characteristic. In this latter case the whole of the brick approaches saturation, and this will produce a maximum moisture movement load on the wall (in such cases the frost susceptibility of the brickwork may be at a maximum). With fine-tolerance construction these dimensional changes need to be considered.

Waterproofing the structure   Problems may arise with attempts to waterproof the structure. This approach will reduce the water absorption of the surface, but will not eliminate it. What it does do is impede the evaporation of the water at the surface and so the material remains wetter (Eldridge 1976).

Severe exposure   Building locations that constitute a high exposure rating will be more prone to the climatically induced defects than those which attract a moderate exposure rating, or are sheltered. The divide between these criteria is broad and some buildings or parts of buildings may be unusually protected or exposed. This demands some intuition in the analysis of the possible causes contributing to defects.

Good and bad practice   The design detailing of the building can be instrumental also, as this will influence the impact of the environment on the structure. The acceptance of design practice as being good or bad should not be solely based on the rigid acceptance of codes and standards, since these can only be general. Details should be worked up for the specific location and building type, taking into account the material and component characteristics. The role of feedback in this design process is important, although it will sharpen the differences in selection criteria between traditional/established materials and new types.

10 Dimensional instability: thermal movement

Temperature-related movements

Criteria for thermal movement

Thermal movement differs from moisture movement importantly in that all building materials exhibit thermal movement. Some are more unstable than others, however. Thermal movement derives from seasonal trends and rapid diurnal fluctuations, and may coincide with moisture movement.

Temperature difference and colour

The amount of thermal movement that an unrestrained element will exhibit will depend on the temperature difference it experiences, the coefficient of expansion, and the length of the relevant dimension (BRE 1975a, Rainger 1983). The movement is commonly linear.

The temperature difference that the material experiences will depend on the temperature range, the degree of exposure and the colour (Rainger 1983). Also, it will depend on the extent and location of any thermal insulation. In the UK the common temperature ranges experienced by building components vary with the type of external material and its colour.

The role of climate

Although the macro-climate is generally well documented, the localised effects of the micro-climate around buildings is more difficult to determine. Local shading caused by the building or those in close proximity will strongly influence surface temperatures, solar blocking and wind channelling.

Symptoms generally

Direct effects

The direct effects of thermally induced stresses are cracking, buckling of elements or finishes, or detachment of components. These direct faults may give rise to secondary effects such as the ingress of water (Addleson 1982), a breakdown of structural integrity (Addleson 1982), or at least a change in appearance (Addleson 1982).

The climate   There are a number of climatic thermal influences that will produce relatively rapid and considerable changes in thermal conditions. These include most obviously solar radiation, and also the wind, rain and snow.

Stress concentration   The build-up of stresses occurs in structures that are restrained. As with moisture movement, the real problems are with restraint or the sheer physical movement associated with large components or elements.

Stress fluctuation: the day/night cycle   Stresses that fluctuate slowly will often produce fewer problems than if imposed rapidly. The material and the structure has time to creep and redistribute the stresses through joints (BRE 1973). Consequently, the 50 °C (Addleson 1982) seasonal range likely throughout the year in the UK is unlikely to produce drastic thermal movement problems (BRE 1973), whilst a rapid change of 15–20 °C (Addleson 1982) during a summer day and night cycle may produce significant stresses in the structure, leading to physical damage. Differential movement can occur with rapid temperature changes. This is a frequent movement-related problem, which tends to produce significant disruption to the building fabric.

Elasticity   The amount of stress induced in a structure by restrained thermal movement is dependent on the modulus of elasticity of the materials rather than the physical movement that would occur in unrestrained conditions. For example, compare strong and medium-strength bricks, both laid in cement mortar. The modulus of elasticity of the strong brickwork may be three times that of the medium-strength brickwork. Although they exhibit similar thermal expansion coefficients, the stress set up by restrained movement of the strong brickwork will be three times that in the medium-strength brickwork (BRE 1975a). The significance of this is that the stress set up in the strong brickwork may represent a far greater proportion of its total compressive strength.

Envelope and frame   Thermal movement affects not just the enveloping fabric, but also the frame components. The envelope materials are usually relatively thin sections. When they undergo thermal extension they may also buckle since they are not inherently stiff. The method of fixing claddings varies widely, but there will be a degree of restraint at the edges of the external panels which will encourage buckling failure. Stresses may occur at junctions between elements of different thermal mass. Similarly to moisture movement, the thermally heavyweight element will resist movement stresses induced in the lightweight component. Depending on whether these stresses are tensile or compressive, either a crack may form at the joint between the two zones or deformation of the thinner section may occur.

Wall reaction   The reaction of the walls depends on how well they are built. A thoroughly built wall with no slack in the joints to allow for expansion movement to be accommodated will exhibit maximum movement at the ends. The DPC will act as a slip membrane, and it is unlikely that the cracking will extend below the DPC. Where combinations of constructions exist, such as solid traditional construction and cavity walls, the weaker section of the two will provide the least resistance to thermal stresses and will fail. Long elements of structure will be exposed to great stresses and if unrestrained or unsupported may move significantly. Parapet walls are a common example. Extension at the corner point caused by expansive movement in two directions against very little restraint, produces a movement fault and oversailing of the wall at the corner.

Roof reaction   A similar effect can occur with the roof slab. A large proportion of flat roofs in the UK are reinforced concrete, and the expansion of a monolithic roof slab can produce significant dimensional change (BRE 1973). Post–Second World War terraced housing was commonly built using long roof expanses without expansion joint provision.

Ineffective restraint   Movement of the roof slab occurs comparatively easily because the loads on the walls at roof height are at a minimum, so there is little effective restraint. The normal rigidity of the roof structure will assist in the destruction of the surrounding walls and any adjacent buildings may be damaged (BRE 1973). The damage will most likely be restricted to the top storey. In buildings with conventional masonry walls, those walls in line with the expansion movement will crack in a jagged manner (BRE 1973). The jagged shape represents the bulging effect that tends to occur in the centre of the roof as this reaches maximum temperature and spreads. This buckling within the slab itself may be significant (BRE 1973). Those walls normal to the expansion movement will exhibit a simple horizontal crack between courses. This will occur, for example, in the external gable walls at the extremes of the roof movement (Melville and Gordon 1979). Damage will extend to include plasterwork at junctions between walls and the roof slab and across the face of damaged walls.

The integrity of framed buildings   Framed buildings respond differently. The integrity of the structural frame means that a movement of the roof slab will distort columns and beams

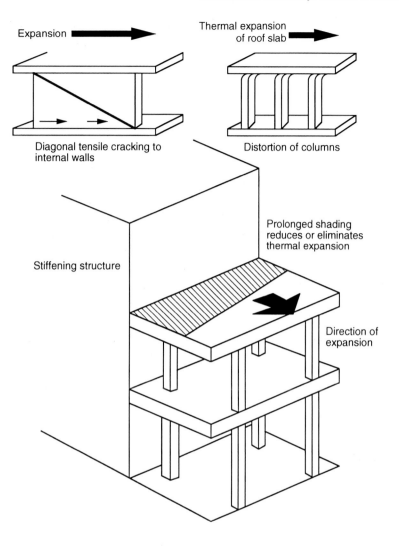

Fig. 10.1 Diagrammatic representation of the effects of thermal expansion of a concrete roof in a framed building.

Expansion ➡

Thermal expansion of roof slab ➡

Diagonal tensile cracking to internal walls

Distortion of columns

Prolonged shading reduces or eliminates thermal expansion

Stiffening structure

Direction of expansion

attached to it. In the same direction as roof movement, beams will tend to bulge and column lines running parallel to the movement will exhibit tensile cracks as they try to match the movement. Walls parallel to the movement will tend to crack diagonally as the movement stresses create tension and produce a lozenging effect. See Fig. 10.1.

As the distance from the centre of the movement increases, so will the physical movement that must be accommodated. The cracking will usually become worse towards the perimeter of the building. This may be uniform in the case of a regular structure with no restraint from adjacent buildings. Where there is resistance provided by the stiffening effect of adjacent structures, the cracking patterns will be distorted in the direction of easiest movement (BRE 1973).

Stability of the structure

Shear cores and shafts

Any shear cores such as lift shafts or other stiff zones within the structure will act as a stabilising restraint for the frame and roof. Thermal movement and any cracking patterns will emanate from such stiff zones, but generally be limited in their immediate vicinity. In cases of suspected expansion problems, it will be necessary to check whether there are expansion joints present and if so, whether they are at suitable intervals.

Effective restraint is also derived from well shaded portions of the building or component concerned. In such cases movement will be focused in the direction providing least resistance to movement as if there was conventional restraint involved (Addleson 1982, BRE 1973) (see Fig. 10.1).

Rigidity and flexibility

Material response

It is worth distinguishing here between rigid and flexible materials. The rigid materials will make a marked impact on their surroundings as they develop thermal stresses. However, flexible materials, such as bitumen roof coverings, whilst being highly responsive to thermal changes, and generally dimensionally unstable, can exert little force when they expand, because they are flexible. They may produce a significant force on contraction, and it is possible to find instances where the force of contraction of a bituminised roof covering is sufficient to rip the skirting out of the upstand at abutments.

Mitigating factors

Blistering and loss of adhesion

Other faults arising with the extreme range of temperatures experienced by roof coverings include blistering and loss of adhesion. The reversible thermal movement will produce fatigue in the material, particularly at points of concentrated stress (for example, fillets, projections which provide restraint, etc.). These direct faults are associated with serious secondary defects. Here in essence we are concerned about the overall reversible thermal movement patterns of the material, rather than its expansion alone.

Material properties

The obvious cause in some situations is the material properties themselves. Dark roofs, large spans in exposed locations, rigid materials with inflexible or inadequately designed joints placed at too infrequent intervals, all combine to produce thermal movement. These are not infrequent occurrences singularly, or indeed in combination.

Partial restraint   With partial restraint of elements of structure, the extension will occur at the weakest point. The cracks will indicate a tensile failure, and in the case of a roof restrained more effectively at one end than the other, cracking faults will appear at the end most free to move.

Restrictions on movement   In cases where movement is restricted by effective restraint at each end of a wall, the stress build-up is greater than the capacity of the wall. Bulging or buckling will occur. In the case of walls this may often produce an arching of the copings which experience the least loading of the wall and have greatest degree of movement freedom.

Differential thermal movement   Differential thermal movement can be problematic with composite constructions. Curtain walling, for example, integrates materials with high thermal movement responses into a rigid structure with low thermal capacity. It undergoes extensive thermal movement, and the allowance for differential movement in the jointing of the components is critical (Seeley 1987). Frequently the importance of this was not appreciated with the early systems, and there were numerous dramatic failures. Problems also arise when the fixing of the curtain walling framework to the structural framework of the building is inadequate to allow for relative movement under thermal or any other forces.

The orientation effect   Distinct from the differential response of different materials is the fact that the building will not necessarily be subject to the same thermal conditions on all faces. On walls predominantly facing the sun, a higher extreme temperature can be reached. The north wall, for example, may receive little or no effect from the sun's heating rays. This also has an implication for moisture movement, since the sun's rays will affect the building envelope differentially, setting up differential stresses in the contraction that may follow. Walls with different thermal capacity because of their thickness, attachments or coverings will absorb heat differently and accordingly will move differently. It is this differential movement that is most likely to cause problems for the structure, and provide a useful set of clues for the analyst.

For instance, it is usual for the roof to get hotter than the walls because it is more exposed to the sun's rays and receives them at a more efficient angle of incidence (BRE 1975a). It must be borne in mind that a flat concrete roof which is expanding under thermal expansion will be abutting against relatively thin walls with little weight on them and no restraint other than the bond strength of the brickwork. The roof, on the other hand, responds to the resistance of the walls in a compressive manner and is more capable of sustaining this stress build-up than the walls. The walls will effectively be suffering shear stress across the horizontal jointing.

Internal stresses

The consequences of movement of the roof will not be restricted to the external walls. The internal walls will also experience shear stresses, and may fail as the roof moves. This is the interface between the exterior-responsive structure and the internal-buffered structure. The symptoms will be horizontal cracks running at or just below the ceiling level. For a uniformly restrained roof slab, the greater the distance from the centre, the greater the stress and effective movement, so the more likely and more severe will be the damage to the internal partition junction (BRE 1975a).

Constructional detailing

Surface temperature and colour

Obviously where materials are heated to different temperatures and also at a different rate, there is maximum scope for damage (BRE 1973). Panels exposed to the sunlight may undergo extreme thermal movement. Not only will they respond to normal changes in temperature, but the incidence of sunlight will produce a direct heating effect on the panels. Components of different colour will absorb heat energy and warm up differently. Addleson suggests that materials of a dark colour with backing insulation may rise in temperature by as much as 140 °C, whereas dark coloured masonry may rise 85 °C (Addleson 1982). Note that the lower ranges of temperatures are consistent for materials of different colour: it is the upper range that is affected significantly by colour. Also note that the assumed temperature for building design of 10°C (Rainger 1983) is rarely near the centre of the temperature range. Consequently most materials will exhibit more expansion than contraction (Rainger 1983), and the associated problems will be greater and more frequent because of this. This is a radiative process. It may produce differential expansion across the face of a wall with different colours.

A lack of conduction

The darker colours are significantly more problematic. In addition to the significance of the radiative process of heat gain and heat loss (which affects cladding and roofing materials by their colour), the conductive nature of the elements is a significant aspect. The cause of many defects in the building envelope is related to excessive heat gain on the surface and the absence of means to dissipate the heat.

High insulation and extreme temperatures

The relatively recent development in highly insulated building envelopes has led to claddings with highly insulated backgrounds. Apart from radiative heat transfer the other predominant method of heat loss from materials is by conduction. The inclusion of thermal insulation allows extreme temperatures to be attained within the surface layers of composites or thin claddings, which makes the thermal distortion and movement much worse. The move towards a warm roof detail to overcome the condensation problems of the cold roof detail produced an insulated covering that reaches much more extreme temperatures than before. This can accelerate the deterioration of the covering, but also accentuate the physical movement associated with thermal gain (Addleson 1982).

The further consequence of this is that the extreme thermal range of the surface is accompanied by movement or stress build-up at the boundary between the surface layer and the insulation. Differential movement stresses occur and shear stresses across the interface destroy the unity of the composite. The alternative of having insulation on the inside of the envelope leads to extreme temperature fluctuations for the envelope and part of the frame (Addleson 1982). The logical development has been the inverted roof.

Induced temperature gradients

Problems with direct solar gain are not restricted to the sheer physical movement – there is a temperature gradient and this produces differential movement across the thickness of the panel itself. This can produce a bowing effect in panels. Where the panel is homogeneous there may be fewer problems than for a non-homogeneous composite. A rigid substrate may restrain the surface movement, or failure may occur at the interface. A flexible substrate will allow relatively free movement at the surface and distortion and buckling of the panel will be evident.

Dimensional instability and accommodation

The problems of thermal movement are exhibited in dimensional instability of materials used for the structure – roof coverings and claddings are exposed to the solar-gain effect and this maximises the problem. The problem is internal in that the components and the materials themselves have to accommodate their own movement without disruption, but also where they are integrated with other elements of structure with characteristic movement characteristics – for example, the edge details of curtain walling abutting the rigid gable ends of a building. Problems of accommodating the sheer scale of physical movement occur with curtain-walled buildings at the corners where maximum movement occurs from two directions.

Part II Hygrothermal deficiency

11 Introduction

Relationship between building and environment

The hygrothermal performance of a building can be considered as the dynamic interrelationship between the internal and external environments. Some degree of control is essential, but it would be a fallacy to produce an envelope totally impervious to moisture or without any mechanism for limiting the internal humidity. The constructional impracticalities are only the start: it is simply not desirable or healthy to produce buildings that do not breathe.

The environmental difference

An environmental difference occurs across the entire external envelope. Traditional porous materials tend to possess sufficient tolerance for moisture absorption and re-emission to allow the structure to act in harmony with the external environment. It will only fail where the actual exposure is excessive, or the detailing is inadequate.

Unambitious expectations

Traditionally, the primary function of the envelope was to keep most of the weather out. This was a coarse method of environmental control, but expectations were also unambitious. The solid wall was a standard construction solution. Its popularity appears to be increasing, particularly as an alteration to an existing, possible defective cavity wall.

Modern demands

The modern demands placed on the external envelope are different. This is due to a range of social, economic and political forces acting upon the nature of design, construction and the use of buildings. A diagrammatic comparison between the hygrothermal performance of traditional and modern forms of construction is shown in Fig. 11.1.

Dynamic instablity

The move from traditionally built, dynamically stable, highly ventilated, constantly occupied buildings, of high thermal inertia, towards modern types has brought with it a range of new problems. Many of these manifest themselves as defects. The adoption of isolated improvements, may have

Fig. 11.1 A comparison of
the hygrothermal
performance of traditional
and modern forms of
construction.

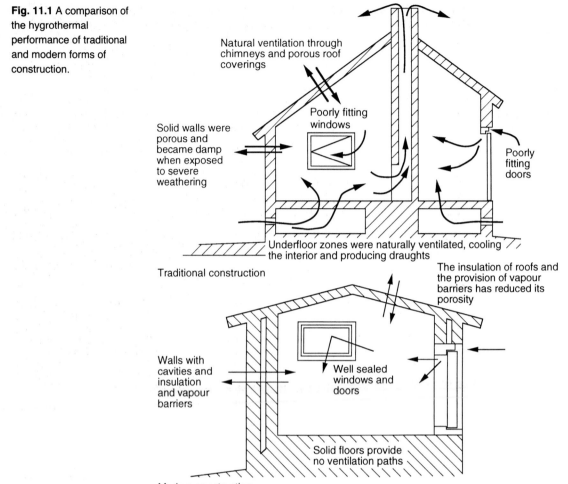

Fig. 11.1 A comparison of the hygrothermal performance of traditional and modern forms of construction.

Natural ventilation through chimneys and porous roof coverings

Poorly fitting windows

Solid walls were porous and became damp when exposed to severe weathering

Poorly fitting doors

Underfloor zones were naturally ventilated, cooling the interior and producing draughts

Traditional construction

The insulation of roofs and the provision of vapour barriers has reduced its porosity

Walls with cavities and insulation and vapour barriers

Well sealed windows and doors

Solid floors provide no ventilation paths

Modern construction

adverse effects on the internal environment of the building. The more common pattern of intermittent building occupancy, where the cooling and heating cycles may become dynamically unstable, can influence the probability of condensation.

Design processes appear to be geared to produce standard/traditional solutions. They seem unresponsive to the new conditions, when compared with the forces driving the changes. It should be recognised that the variable nature of the UK climate contributes to the complexity of any responsive solution to these problems.

Heat and moisture   Where heat- and moisture-producing machines/processes have been subsequently incorporated into the building, this has totally changed the

Fig. 11.2 A variety of different weathering techniques are shown as hygrothermal solutions to similar occupancies.

internal environment and the hygrothermal relationship across the external envelope. Many building structures are now operating under environmental regimes for which they were not designed.

The general move towards creating buildings which are watertight and airtight has implications for their exteriors. Impervious materials tend to demand close tolerances. This places additional criticality on the performance of the joints.

Absorption and runoff The rate of water runoff from facing materials is inversely related to their absorption. Where traditional run-off rates have been applied to the new impervious materials, design defects may have been incorporated. The external envelope will usually comprise mixtures of traditional and new materials. Their disparate moisture and thermal performance characteristics re-emphasise the need to consider the provision of joints and jointing methods. This is needed not just to ensure waterproofing, but also to accommodate their movement characteristics.

Fig. 11.3 Ceramic tiles delaminating to expose underlying materials of reduced durability.

Quality and effective performance

The overall probability of failure of an element is likely to increase with its frequency. Therefore in a multi-jointed envelope failure would appear to be more likely. Behind the facade the backing structure may only be designed for the reduced durability levels expected, with an emphasis upon small sections and reduced weight. See Fig. 11.3. For this to provide effective performance, good quality construction must be achieved, since the implications of its failure are greater.

Layers and pathways

The provision of an envelope which contains various layers of material will mean that several vertical interfaces between different materials exist. These may become hygrothermal barriers against any equilibrium effect, and also add to the total number of joints. They may be linked to provide complex hygrothermal pathways through which water can move. Small-sized pathways will permit little hygrothermal equalisation to occur, whereas larger pathways may add to the thermal inertia of the envelope by requiring large amounts of energy to produce hygrothermal equalisation. When the above factors are considered, it appears that the tolerance level of the modern external envelope has been reduced.

12 Penetration of the envelope by solids

Wind and air tightness

Porosity of buildings

Most buildings can be considered as porous, since there is a facility for air to move through them. This may be as part of the general provision of natural ventilation, or as the result of poorly jointed material, components and elements. The individual materials may be porous, although the interconnection of pores is needed if they are to allow air into the building. For penetration, the pressure exerted by the trapped air pocket must first be overcome. Attempts have been made to measure the porosity of windows and walls, but these have failed to simulate accurately the action of the wind in real situations (Marsh 1977). Air-porous buildings may contain passageways which can be exploited by water and water vapour entry. The size and shape of pathways may change with the wind pattern, as in the case of chattering slates, and flapping felt in pitched roofs.

Ventilation or draughts

The movement of air can be considered either as ventilation or draughts. The latter could be coarsely defined as air movement which causes discomfort or physical deficiency in the fabric or structure. This general deficiency can be classified as either poor fit of the external envelope, or use of air-porous materials. A substantive amount of air entry is through joints and gaps, since varying degrees of impermeability are characteristics of the materials used for the external envelope. See Fig. 12.1. The flow rates of air through joints will vary with the applied air pressure and the movement characteristics of the materials bounding the joint. This in turn will be related to the temperature and relative humidity of the materials. The type of construction will also influence the leakage rates. Some evidence exists to suggest that multi-jointed claddings give high leakage rates (Thorogood 1979), although a major leakage of one joint can have a considerable effect. The sealing of joints between different materials demands movement accommodation.

Provision of joints

Joints in construction are required at changes of material and to

Fig. 12.1 Advanced rot in timber mullions has exposed the edges of panes of glass and allows direct entry of wind and dust.

accommodate movement. The downward-facing open joints common with precast concrete panel systems should allow water to drain from them. Where this is not achieved, water drains into the fabric (Addleson 1982). The open nature of these joints will allow wind to blow into them, as shown in Fig. 12.2. Poor or inadequate sealing will allow wind behind the primary defence. Where a route through the primary and secondary defences exists, wind and dust may enter the building. Caulking to resist the flow of air is commonly placed at the rear of panel systems (Edwards 1986); this may also act as the final barrier to water entry.

Hygrothermal deficiency

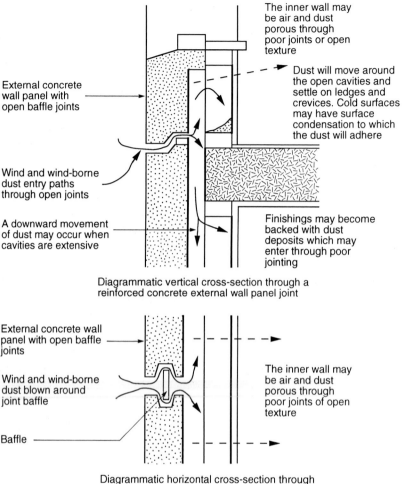

Fig. 12.2 Model of dust and wind penetration paths through the external skin of the building.

External concrete wall panel with open baffle joints

Wind and wind-borne dust entry paths through open joints

A downward movement of dust may occur when cavities are extensive

The inner wall may be air and dust porous through poor joints or open texture

Dust will move around the open cavities and settle on ledges and crevices. Cold surfaces may have surface condensation to which the dust will adhere

Finishings may become backed with dust deposits which may enter through poor jointing

Diagrammatic vertical cross-section through a reinforced concrete external wall panel joint

External concrete wall panel with open baffle joints

Wind and wind-borne dust blown around joint baffle

Baffle

The inner wall may be air and dust porous through poor joints of open texture

Diagrammatic horizontal cross-section through a reinforced concrete external wall panel joint

The size of clear opening and the thickness of the material will affect air flow rates. Air flowing through very small gaps, as in the case of distributed cracking in brickwork, will have to overcome the frictional resistance of the sides. Under a no-wind condition gaps, holes and poor sealing will also allow air warmed from within the building to escape. Where joints between dissimilar materials are filled with sealants, their adhesive bond can fail. The resulting air gap may change size in accordance with temperature and humidity, giving an uneven airflow rate.

Pressure difference across the envelope The wind will only attempt to enter the building when an internally negative pressure exists, compared with that outside. Very porous buildings, by definition, are excellent at equalising these pressures. New sealed buildings are relatively poor, since they attempt to maintain an average pressure at all times. Where their seals fail, there is an increased risk of air ingress, since the pressure differentials are potentially greater. Seals between moving elements and components can fail through wear. In the case of doors and windows this may actually make their operation easier. Materials based on foamed plastics which offer gas-tight seals may not be able to accommodate the maximum movement between the components. Compressive or tensile failure can result. At the micro-scale gaps in the primary defence will assist in equalising pressures, as in the case of open-jointed panel systems. Pressures inside inner cavities will approach wind-pressure values; this will then search out gaps in the secondary defence, causing draughts and air leakage into the building. Clean open weep holes and pipes can equalise pressures, although water has been blown through them into the building (Edwards 1986).

Effect of buildings on wind

The variable nature of the wind is complicated by the effects of surrounding buildings. This can cause very high local wind speeds, when the general wind speed is low (Cook 1976). The flow of air around buildings influences the flow of air through the building. On leeward walls the external pressures will be generally negative compared with those existing in the inside of the building. This may suck internal air out through the external envelope. High winds are more likely as the distance above the ground increases, which can make upper storeys more vulnerable to air leakage. The airflow pattern around buildings may change with the life of the building due to adjacent buildings, and this may exert new additional loads on joints and seals.

Forced venting

Recent regulations, as part of the European harmonisation process, recognise a statutory need for mechanical ventilation in high-risk areas, e.g., kitchens and bathrooms. Although a detailed examination of the influence of ventilation is outside the scope of this book, an increase in local air flow rates will place additional loads on the integrity of sealed joints. The link between ventilation and condensation avoidance is well known, although with the provision of substantial levels of insulation a comprehensive analysis is needed. See Fig. 12.3.

Fig. 12.3 An example of
user-driven venting
required to protect a
high risk area – inelegant
but effective.

Dust penetration mechanisms

**Direct entry through
gaps and joints**

Dust deposited on the external face of the building will affect appearance, as
in the case of soot (BRE 1964a, 1964b) found in urban areas from times of
widespread coal burning. It can also influence the nature of chemical and
physical attack, due to its size and general content. Where buildings are
located close to busy roads this can be a major source of dust and pollution,

particularly in dry weather. The dust from this and other sources which enters the building can bring with it a range of germs, viruses and micro-organisms. These can become a source of discomfort for the users of the building, and are particularly important in the case of clean rooms (British Standard 1989).

The major sources of dust entry are shown in Fig. 12.2. These include ill-fitting joints. These may be provided to accommodate movement, or to provide weather resistance of openings. Poorly fitting seals to moving parts of doors and windows allow dust entry. The diffuse cracks in masonry will allow wind-borne dust to enter.

Blown in by winds and pressure differences

Dust is borne along by the wind and adheres to the surfaces of the building. The amount and its distribution on the facades of buildings is dependent on the wind flow path and the relative cleanliness of the environment. The wind flow patterns around buildings are complex, and are related to the frictional resistance of the facade, its shape and the influence of surrounding buildings (Penwarden and Wise 1975).

The need for highly sealed windows and doors has produced complex frame sections. These can act as dust arrestors when partial failure of the seals occurs. Dust accumulations may then enter in bulk when the window opening light is moved. This is shown in Fig. 12.4.

Where material pores are open to the wind they may accumulate dust. These may become pockets of chemical attack. Poor or inadequate sealing of the primary defence will allow wind to deposit dust on hidden internal surfaces. Ledges and unweathered surfaces are vulnerable. During its passage through the primary defence the reduction in wind speed may allow dust to drop out into the joint, which may eventually seal it. The small cracks common in brickwork can behave in this way. The visual examination of larger cracks to determine their standard of cleanliness is a valuable part of defects appraisal (Eldridge 1976).

The wind action on lightweight coverings can cause them to move, opening up their joints. This will produce wear on the fixings and allow dust entry. In the case of small unit lapped roof coverings dust can accumulate on the felt, encouraging the growth of germs and bacteria. Suitable conditions for nesting birds are also provided (Cornwell 1979).

Positive pressure outside

The wind can exert a positive pressure on the faces of buildings. Its quantity can fluctuate as the wind gusts. This can pulse air through any interconnecting spaces linked to the outside. Internal cavities may channel air currents and eddies. Where the pressure is sufficient the air can break through into the building, creating draughts and spreading dust. In buildings which are naturally ventilated this natural porosity assists this

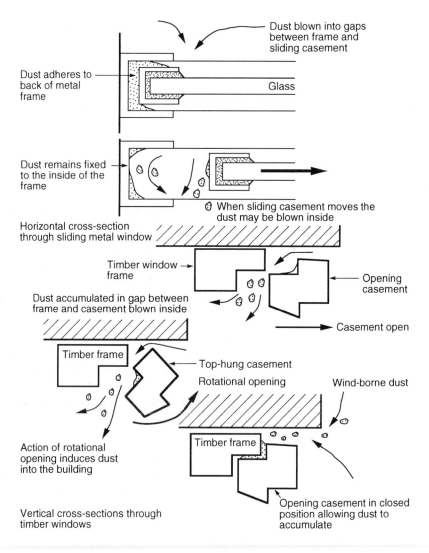

Fig. 12.4 Dust entry paths associated with opening windows.

Dust blown into gaps between frame and sliding casement

Dust adheres to back of metal frame

Glass

Dust remains fixed to the inside of the frame

When sliding casement moves the dust may be blown inside

Horizontal cross-section through sliding metal window

Timber window frame

Opening casement

Dust accumulated in gap between frame and casement blown inside

Casement open

Timber frame

Top-hung casement

Rotational opening

Wind-borne dust

Action of rotational opening induces dust into the building

Timber frame

Vertical cross-sections through timber windows

Opening casement in closed position allowing dust to accumulate

process, although the use of mechanical (exhaust) ventilation will demand a better degree of sealing of the fabric (Thorogood 1979). The high degree of sealing common with 1980s domestic buildings has meant that the pressure difference across the fabric has increased compared with earlier forms.

Negative pressure outside

The magnitude and direction of the wind can be highly variable. This can alter the pressures exerted on the building, and change positive zones into negative zones. Where this happens over a short time scale, as where wind veers about the corner of a tall building, the pressures created may increase the pulsing of air through the fabric. These changes in pressure are important for wind loading and the general movement of the building.

Oscillation and movement in the structure will increase the risk of joints and seals in the primary defence opening up. A generally negative pressure outside will permit air and dust to be sucked from the building. Where the internal air has a high dust burden, this may be deposited on outside surfaces, affecting appearance and possibly increasing the risk of chemical attack.

Leakage related to movement cycles

The joints and seals in the primary defence should be sufficiently robust to resist wind action and accommodate the movement due to moisture and temperature variations in the materials. The modern reliance on mastics has offered the opportunity of truly sealed joints. Failure mechanisms of these joints are generally associated with embrittlement or lack of adhesion to surrounding surfaces. Both failures will allow dust entry. Indeed, it is an approximate method of dating the failure. Where contraction of the materials occurs, the sealant can be ripped away from the moving surfaces, causing cracking (Addleson 1982). Where the build-up of dust is severe this may enter the joint during the contraction phase, and thereby restrict its performance in accommodating expansion movement.

The open joints common in some types of precast concrete panel construction provide a harbourage for dust. The air movement should be arrested at the rear of the panel, as in Fig. 12.2, and failure of this barrier will increase the risk of dust and draught entry.

Indirect entry through layered structures

Masonry cavity walls can permit the wind to carry dust through the outer skin, by means of cracks, weepholes and failed movement joints. Older buildings having open cavities will assist its distribution to the inner skin. Where the sealing of the inner skin is poor, dust will gain entry to the building. The built-in ends of joists which penetrate the inner leaf can have gaps around them. Poor sealing around skirtings where no plasterwork exists can be vulnerable areas (Thorogood 1979). Door and window jambs are regions where changes of material and directions assist in providing dust entry gaps.

The move towards semi-solidification of cavity walls through the use of insulation materials has restricted the possibilities for dust entry to the building. Additional layers are obstacles, however, where they are poorly placed and fixed additional dust harbourages exist. The wind acting upon a well sealed building will continue to pressurise any dust entry paths that exist.

Use of the building

The general domestic building does not place exacting demands upon the degree of dust ingress. The majority of dust enters through open windows and doors, and these are controlled by the occupants. Reasonable standards of cleanliness of general surfaces are possible, since they are generally impervious and well sealed. The influence of dust entry to these buildings is not a major factor for most people. In the case of clean rooms the demands are more exacting (British Standard 1989). The legal and commercial implications of dust entry into food processing and health laboratories, etc., are such that minor fabric defects cannot be tolerated. Special considerations also relate to the construction of nuclear shelters, where the need to keep out radioactive fall-out dust is essential (British Board of Agrément 1983).

Access to the building

Pedestrian and vehicular access to the building allows direct dust transmission to occur. The immediately adjacent external pedestrian surfaces may generate dust and debris, as in the case of busy roads. A positive pressurisation of the interior by mechanical ventilation may assist in reducing draughts and dust ingress, although doors will continue to remain critical areas. The provision of increasing standards of cleanliness on routes to 'clean rooms' can fail where the integrity of the dividing structure is incomplete.

Consequences of dust penetration

Dust as a pollutant

The ingress of dust into the building causes various degrees of internal pollution. The particles are generally small, typically a few microns in diameter, and can adhere to internal surfaces or be blown around the building. The dust particle may be a solid cocktail containing acids, organic matter, unburnt and partially burnt products of combustion, also inert material. The total amount admitted will affect the standard of internal air quality, in some cases placing this below acceptable limits. Critically clean areas are most vulnerable.

Germs and bacteria

Germs and bacteria can adhere to surfaces and internal pores of dust particles. The ingress of dust to the building will also mean the admission of these health-affecting organisms. Their concentration will in general be similar to that naturally occurring in the air, and as such they may present no large-scale risk. However, where dust concentrations can build up, so too can bacterial colonies. Cracks and crevices which are difficult to clean, or zones of the building which are virtually inaccessible, e.g., floor voids and cavities, are potential danger areas.

Discoloration The space within buildings may be used for purposes which generate dust. This may present health and danger hazards to the occupants (BRE 1978b, BRE 1984), and cause discoloration of internal surfaces. The ingress of dust from external sources, including that exhausted from the building, will add to the discoloration. The acceptable standard of cleanliness for each building will vary, but high dust burdens on surfaces will have health and aesthetic implications.

Where the conditions are suitable, it is possible for the dust layer to support mould growth (BRE 1985b). Areas where moisture can accumulate due to condensation or leaks, as well as areas which are difficult to clean, are likely sites. Their natural colours of browns and dark greys will stain decorations. This will have implications for the illuminated environment in the room, since the reflective performance of the surfaces is likely to be reduced, affecting natural and artificial light flow.

Pollens During the summer months the burden of solids in the air is increased by the addition of seeds and pollen. These organic particles can be present in sufficient densities to cause allergic reactions in certain persons, e.g., asthma and hay fever. Natural ventilation offers no filtering mechanism, and direct entry is possible. The incidence of allergic reaction is likely to be similar to that present in the general population. Where mechanical ventilation systems are operated this may direct concentrated air flows towards reactive occupants, causing some distress. This may cause an increased incidence of complaints. The underlying cause of this may be the malfunction of the air-handling plant. A full analysis of this cause is outside the scope of this book. Specialist advice should be sought.

Capillary effects With the changes in energy levels of the atmosphere around buildings, the general dust concentration of the external air will alter. At most times there is some present in sufficient quantities to allow a gradual deposition to occur on buildings. This may reduce the reflective properties of some materials, e.g., solar reflecting treatments to flat roofs. The dust may be blown into regions where it can cause distress to the building. Dust and debris washed from roofs may accumulate in gutters, due to inadequate falls or low flow rates. This is an ideal site for moss and mould growth, which can introduce organic acids into the outflow water. The gutter may be more prone to overflow, which may enable the primary defence to be breached.

Dusts may be partially hygroscopic and this will encourage the flow of water through them; they may form a thin capillary conducting water layer. A build-up at the base of cavity trays caused by dust entry through weepholes may allow water entry through any lapped joints. Where insufficient upstands are provided for flat roofs, dust can build up to sufficient levels to provide a capillary path into surrounding walls or between layered and sheeted coverings.

Sealing pores and joints

Since many materials used externally are porous and textured, they can trap dust. This accumulation may have effects on the material, due to the composition of the dust. This process can also occur internally when the dust burden is high. The cleaning process will be complicated by the general texture of the surfaces and their configuration, with heavily textured or complex pore-shaped surfaces being especially difficult. The sealing of pores by dust may affect the mechanical behaviour of the material, since the moisture and temperature movements of the dust are likely to be different to the bulk material. Generally the dust is weaker, although cementitious dust can occur. The total sealing of major joints between materials and components is less likely, although any ingress of solid material into movement joints is to be avoided.

Blockage to vapour pressure equalisation

The sealing of pores and joints may also have an influence upon the performance of the material with regard to moisture vapour. The faces of materials exposed to dust accumulation may become partially or completely sealed with a solid material, and this layer may possess a very different water vapour permeability to that of the bulk material. Vapour may move through the bulk material at one rate, and then slow down when approaching a sealed surface. Where this occurs externally this may increase the risks associated with chemical and frost attack.

Under the action of wind pressure, air can migrate into and out from various cavities within the building. This may be viewed as a form of ventilated cavity, and as such it will equalise wind and vapour pressure within the cavity. When dust accumulations are such that joints are either reduced in size, or are completely sealed, this equalisation process can no longer occur. This may cause moisture transfer problems, and increase wind loading in certain areas.

Pattern staining of the external facade

Variable weathering

The weathering of external materials will be related to their physical and chemical ability to resist the attacks from the environment. The qualitative assessment of the appearance of buildings is a subjective process, and is spiced by the opinions and preconceptions of the viewer. This text is not concerned with the passing of value judgements concerning the aesthetics of buildings, only in this instance the effects of weathering patterns on their external faces. This is partly a natural ageing process, but is still capable of producing facades of poor appearance. The differential ageing and wear of materials can cause textural differences to become exaggerated. Weaker materials can be washed away, or washed over more durable materials. Where water washes over dust accumulating surfaces this can cause the 'seismograph' (BRE 1964a) pattern. A failure to provide adequate drips and weathered surfaces will exacerbate the problem.

Water runoff patterns will also be influenced by the wind flow around the building. Concentrations into corners may produce the 'moustache' effects produced by white streaks below the ends of projecting window sills (BRE 1964b). The appearance may change as the facades are subjected to different weathering. The long-term trends will produce generally dirt-encrusted areas and cleaned areas. The position of the boundary between them may shift with time. See Fig. 12.5.

Fig. 12.5 Pattern staining of external face of building. Inadequate throw from the window sill drips dust and pollutants onto the raking plinth.

Influence of shape and texture

The relative prominence of parts of the external face of buildings will leave some permanently protected, whilst others are permanently exposed. This will heighten the colour changes which tend to occur when materials weather. The homogeneity of the original will disappear with time, and differences will become more marked. The patterns of weathering can be useful in identifying where concentrations of water flow occur. This cleaning process may also expose other defects in joints, materials and/or components. Deeply textured materials may retain collections of debris which weathering will not remove. Exposed aggregate precast concrete panels are examples of this. Water runoff may emphasise the difference between the dark recesses of the aggregate crevice, and its smooth clean outer face.

Wind patterns

The wind patterns around complex building shapes can be highly variable. Tall buildings can experience upward and downward air flow on certain facades (Penwarden and Wise 1975). Both of these processes will influence the weathering and appearance of the building. Corners can cause swirling vortices which may produce considerable washing of the facade. Clean regions adjacent to corners can be caused by this effect. The driving of rain into channels and corners may have little influence on the appearance of low-rise buildings with generally porous facades. In the case of tall buildings with generally impervious facades the total amount of water passing over the facade at low level is considerable. This will have an impact on the weathering process, and in turn the appearance of the building.

Material differences

The smooth impervious surfaces of modern facings can be made matt by the action of weathering. Exposure of the underlying layers of the material may reveal a more porous texture, where dust accumulation can occur more readily. Where differential weathering occurs on facades of the same building, dust accumulation will vary. Materials placed adjacent to one another may weather at differing rates, and these differences may affect the dust disposition upon them. Where ledges and channels accumulate quantities of debris this may be washed off in large chunks which can adhere to surfaces, or fall into new sites. The sites of substantial dust and organic matter are also suitable sites for plant growth. Gutters, cracks in walls, and chimneys are common sites, since the nutrients required for very hardy forms of plant life are minimal.

The staining associated with chemical reactions, as efflorescence and iron stains, is in many cases part of the general staining of facades. These stains may be minimal in size, although irritating in character (Fidler 1988). Various colours from the greenish stains of copper salts to the black stains of manganese are possible. The causes may well be the natural weathering of adjacent materials, or an incorrect positioning of materials which allows the washings of one material to react with another.

13 Penetration of the envelope by liquids

Liquids and the building envelope

Sources and locations of dampness

Sources and locations of dampness may be differentiated as deriving from within the building as a result of its use, from within the structure (for example, the water of construction), or from some other external source.

External sources of penetrating dampness

Water present in the ground and entering a building by direct and continuous contact with the structure and that which reaches the structure as precipitation are deficiencies that are usually distinguished. However, both represent an external source of penetration, and it is only the mechanism of immediate entry and the sources that differ fundamentally. Various and characteristic symptoms can be used as reasonably reliable aids to diagnosis; nevertheless the appraisal of the cause of the fault has to extend back from this.

The water of construction

The water of construction is considerable in traditional forms of building. This will be distributed throughout the building and will gradually dry out during the first year or so following construction. There will of course be a seasonal influence on the drying rate depending on when the water was introduced. Consequently, the emergence of symptoms will be seasonally dependent, but they will appear early in the life of the building. Obviously the water of construction will not be a continuing source of damp problems; however, it may provide the catalyst for chronic problems.

Distribution of dampness

Although the water is initially centred around wet applications, it will migrate under the action of ventilation and heating and be deposited on surfaces and interstitially where the dewpoint of the air is reached. The damp deposits may not be directly visible, and may be noticeable only by the indirect symptoms associated with the high moisture content.

Traditional and dry forms of construction

Traditional building techniques allowed sufficient time for natural ventilation to remove a significant proportion of the water, but with the drive towards short production times for well-sealed and highly insulated buildings, the natural drying process cannot provide a full treatment. The shift towards dry construction has reduced the water of construction, but in doing so it has served also to emphasise the need for sufficient drying-out periods. With incomplete buildings the entry of damp is by no means confined to the water of construction. To compound this, highly sealed structures composed of impervious materials retain water as effectively as they exclude it.

Implications of forced drying out

Forced drying out of traditional construction serves well to diversify the dampness in the building and therefore to increase the incidence of condensation. The artificial heating used to accelerate the drying out of the structure frequently only reduces the surface moisture content. The moisture content of the core structure will remain high, and after decoration this may re-emerge to disrupt the surface finish. There are also implications of dimensional instability of the components if they are dried out rapidly and differentially. Materials which are inherently porous may not allow rapid migration of water from wet to dry regions.

Dampness penetration vertically downwards

Most walls are designed to resist typical driving rain storms but the concentrations of water from downpipes and broken guttering can create chronic dampness. This can fill cavities and cause a range of internal defects once it has penetrated the building. Penetration paths can be omnidirectional, depending on the pore size of the materials and their juxtaposition.

Exposure

Assessment of the degree of exposure at the design stage is critical. This exposure may change during the life of the building due to micro-climatic changes and the configuration of adjacent buildings. Both of these factors can influence the air flow pattern around the building's exterior, and therefore its relative exposure.

Dampness penetration laterally

The response of the structure to lateral water pressure may result in the horizontal movement of water through the envelope. A frequent example is the basement structure, which is commonly subjected to a constant water pressure. This can frequently be of such severity that specialised constructions are essential if a dry interior is to be provided. The reliance on mass waterproof construction demands that movement joints receive special treatment if they are not to fail. Tanking demands a high quality of workmanship and continuity, since it provides a blanket damp-proofing membrane. Since minor ground movement may cause cracking in basements, its structural integrity must match its water resisting properties. Service connections through walls are vulnerable. Where flooding does occur it can commonly follow a seasonal variation.

The porous land drains sometimes laid around basements to reduce the lateral water pressure may become blocked with silt and other debris. This may allow the water pressure slowly to rise, perhaps over many months, and force water into the basement.

A drastic form of lateral penetration can be that emanating from flooding by sea water. The salts carried by the water can damage buried metals and walls may remain almost permanently damp, because of their induced hydroscopicity. This can be identified by salts precipitating out on surfaces of the buildings. These may leave coloured stains. Litmus tests can be used to identify the pH of the salts.

Flooding may also be due to other causes, perhaps following plumbing damage or fire-fighting. The effects on the structure can be severe due to the loading from the water, and its consequent drying out. Significant moisture movement may occur. There may also be a degree of dirt and debris collection in the floor sub-voids and other cavities, which will require cleaning. Ventilation paths may be blocked, which may generate secondary fungal attack of the structure. Cabled services may be severely affected since the relatively small ducts and trunking provision are difficult to inspect and clean. Corrosion and waterlogging of equipment can have serious operational and safety implications.

Precipitation

Inadequate weathering of surfaces which does not take into account the direction of movement of water across the building facade may provide areas for water to lie. These areas will allow a greater absorption of water into the surface of the material, which in turn may make it more likely to suffer from frost attack. Ponding will act in resisting evaporation from within the material.

Snow falls provide an additional form of dampness penetration. A major problem with snow is its loading of the structure. The amount of loading will change with its state, and where it can slide over a surface some live loading can occur. The snow also contains chemicals which may be aggressive to the external materials of the building. Deep-lying snow can bury joints and subject them to hydrostatic pressure during its melting phase. This may force water into the structure. In the case of parapet walls the depth of snow can easily extend above the DPC level and soak the wall. Upstands and the backs of chimneys are particularly vulnerable. A range of possible effects is shown in Fig. 13.1. The general depth of lying snow on roofs has increased with the amount of roof insulation.

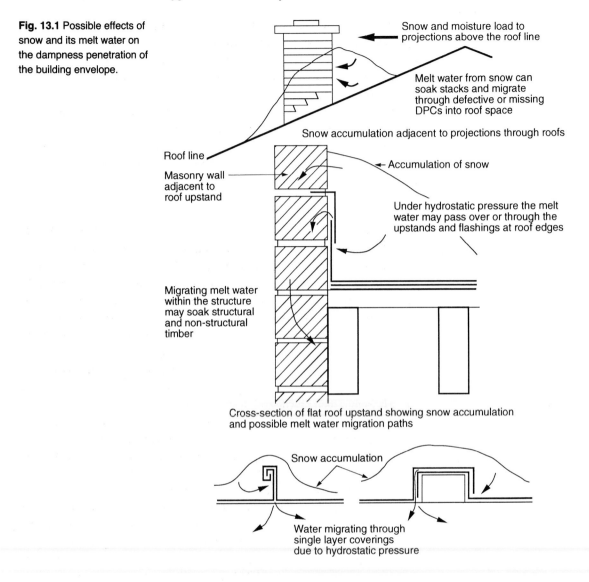

Fig. 13.1 Possible effects of snow and its melt water on the dampness penetration of the building envelope.

Snow and moisture load to projections above the roof line

Melt water from snow can soak stacks and migrate through defective or missing DPCs into roof space

Snow accumulation adjacent to projections through roofs

Roof line

Accumulation of snow

Masonry wall adjacent to roof upstand

Under hydrostatic pressure the melt water may pass over or through the upstands and flashings at roof edges

Migrating melt water within the structure may soak structural and non-structural timber

Cross-section of flat roof upstand showing snow accumulation and possible melt water migration paths

Snow accumulation

Water migrating through single layer coverings due to hydrostatic pressure

Wind-driven rain will apply a fluctuating load to the building. This can force water into crevices, joints and large pores. Where the forces are sufficient, water can be driven over the building surface in an omnidirectional manner. This may be vertical, placing new inverted demands upon the weathering of surfaces. Where the wind can generate a pressure difference between the outside and inside of a porous material, it is more likely to let water pass through it. Where the pressure is more readily equalised the net force is minimised or eliminated. This is one of the principles behind the modern rainscreen pressure-equalisation systems.

Where there is no net pressure difference across the structure of the envelope the ingress of water should theoretically only be achieved by the agencies of absorption and capillarity. With non-absorbent materials there is ingress of the water at the joints only. The joints may therefore become regions where a marked pressure difference occurs, which may allow water behind an impervious outer skin. This has implications for the loss of any water trapped in the structure. Saturation of backing materials and water accumulation may create conditions similar to, or more severe than, condensation. The unhealthy environment which this can produce may allow mould and fungi growth to develop. Furthermore, the areas of saturation associated with joints can become the sites of hygrothermal cold bridging.

Constructional variations

The problem of joints and driving rain was most obviously illustrated with the early designs of curtain walling. The incorrect positioning of components and deficiencies in mastic sealants for the joints produced capillary pathways. This allowed water to penetrate and accumulate inside the wall. The connections are also subject to water runoff, since the external skin of the building is generally non-absorbent. It is relatively simple for a lightweight building skin to become positionally out of phase with fluctuating air gusts and driving rain. This will open joints and allow water to be driven into the structure as differential pressures occur across the envelope. The amount of water ingress may therefore depend on the degree of resistance to air movement.

The pragmatic development of the rainscreen pressure-equalisation method has been incorporated into the overcladding rainscreen concept. As with conventional small unit lapped roof coverings, it is assumed that water will pass through the outer covering, but only in minimal quantities. A vertical cavity provides for drainage of the water.

Traditional porous materials will absorb water in accordance with their exposure, and degree of porosity. This will generally provide conditions suitable for mould growth and frost attack as shown in Fig. 8.5.

There have been recorded instances where plastic tubes have been incorporated within the sills of timber-framed windows to carry away inner surface condensation from a collection channel in the inner frame. This is shown in Fig. 13.2. The water is discharged below the external timber sill but over the underlying brickwork.

The condensate may freeze, causing disruption to the face of the brickwork. Inadequate projections of the sills, poor weathering and a lack of throating may add to the problems of water discharge. Where porous materials are likely to remain damp for prolonged periods the risk of mould growth and pattern staining increases.

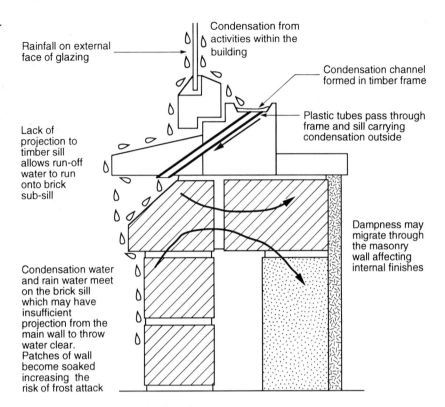

Fig. 13.2 Condensate water discharge causing a secondary effect of water penetration of the primary defence below a timber window.

Rainfall on external face of glazing

Condensation from activities within the building

Condensation channel formed in timber frame

Plastic tubes pass through frame and sill carrying condensation outside

Lack of projection to timber sill allows run-off water to run onto brick sub-sill

Dampness may migrate through the masonry wall affecting internal finishes

Condensation water and rain water meet on the brick sill which may have insufficient projection from the main wall to throw water clear. Patches of wall become soaked increasing the risk of frost attack

Thermal and vapour pressure differences

Allen has identified and described a concept of thermal pumping which occurs within the fabric of buildings (Allen 1989, 1990). The mechanism operates when cyclic temperature differences create differential vapour pressures between a cavity or moisture-absorbent layer and the interior spaces. Water vapour, and eventually moisture, travel across any internal openings such as cracks or open joints in the finishings. Whilst this may arise across any part of the building envelope, roofs or walls exposed to high solar gain are likely to be affected most. The symptom is a cyclic yet superficially paradoxical water leakage from the opening. Allen cites examples of this process discovered below a concrete (cold) roof, and under a well-sealed insulated lead roof covering. The leakage appears in dry and particularly hot weather, but not usually during wet or cold weather. It is the cycling of the external temperature and the difference between it and the internal temperature that is important, rather than the external dampness.

During cold or wet periods, which cool down the roof slab or envelope skin, there may be a lower vapour pressure in the fabric laminate than in the internal spaces. Moisture will be drawn into the cavity or absorbent layers of the fabric from the inside of the building, and retained. The vapour

Fig. 13.3 Thermal pumping mechanism (after Allen 1989, 1990). Where a low-grade vapour barrier produces a partially sealed cavity, differential pressures across the boundary alternately pump water in and out of the cavity. The symptoms may appear paradoxical set against prevailing conditions. They can also be superficially identical to a conventionally leaking roof.

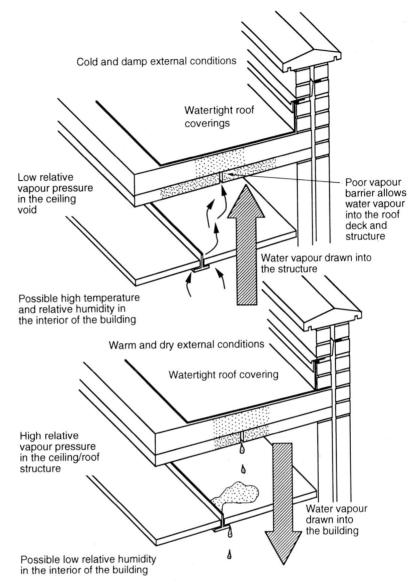

Cold and damp external conditions

Watertight roof coverings

Low relative vapour pressure in the ceiling void

Poor vapour barrier allows water vapour into the roof deck and structure

Water vapour drawn into the structure

Possible high temperature and relative humidity in the interior of the building

Warm and dry external conditions

Watertight roof covering

High relative vapour pressure in the ceiling/roof structure

Water vapour drawn into the building

Possible low relative humidity in the interior of the building

can pass through low-grade vapour barriers, or those where joint sealing is ineffective. Humid atmospheres will be a more significant risk. During warm weather the roof slab or envelope skin can increase in temperature greatly, especially if it is insulated at the rear. This increases the vapour pressure of the entrapped moisture and, if it exceeds the internal vapour pressure, moisture is forced out through the opening and into the interior of the building. See Fig. 13.3.

Consequently the moisture is forced out of the envelope at the times when conventional leakage is unlikely, and conversely not when it would

Fig. 13.4 Breaks can occur at entries where door seals and bars offer little resistance to entry.

be most probable. Evident from the pattern of occurrence and perhaps an obvious stagnancy of the moisture, the contribution of conventional leakage of the envelope is unlikely or of secondary importance only. Significant leakage of the external skin would probably prevent a differential vapour pressure occurring across internal openings to any significant degree, and a pattern of conventional water ingress would predominate. See Fig. 13.4.

Primary and secondary defence against water penetration

Buildings usually have a primary and secondary line of defence against water penetration. The primary line of defence is the structural position of the outer skin of the envelope and its quality of detailing will determine whether water is deflected or not. Consequently the commonly important details for this are the overhanging of roofs, sill details to windows and thresholds of doors and the water-shedding treatments around any other breaches in the integrity of the external envelope. This is an entirely passive aspect of the building function, as is the secondary line of defence.*

* The exception being such active approaches as electrostatic damp-proof courses.

Fig. 13.5 Representation of the primary and secondary defence provision of the vertical external envelope.

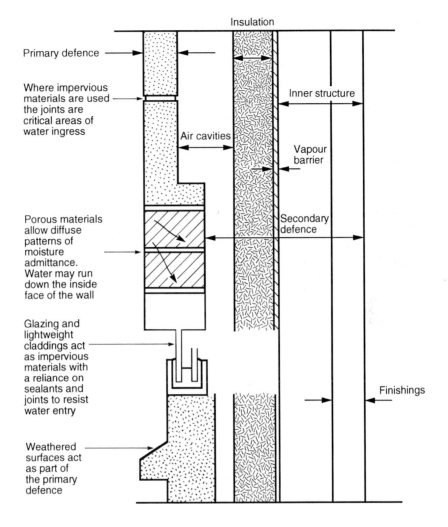

Insulation

Primary defence

Where impervious materials are used the joints are critical areas of water ingress

Inner structure

Air cavities

Vapour barrier

Porous materials allow diffuse patterns of moisture admittance. Water may run down the inside face of the wall

Secondary defence

Glazing and lightweight claddings act as impervious materials with a reliance on sealants and joints to resist water entry

Finishings

Weathered surfaces act as part of the primary defence

The secondary line of defence is drawn from the back of the primary line of defence, and includes all of the structure through to the inside of the building. The general arrangements of the primary and secondary lines of defence are shown in Figs. 13.5 and 13.6.

Primary water resistance of the envelope

The weathering and discharge of water is a primary defence mechanism. Joints should be a secondary mechanism and kept free of moisture wherever possible. Joints appear to have been the weak link in many external weathering systems. Expansive material surfaces are largely well behaved. Homogeneous materials which are cast/laid in place have inherently overcome these jointing problems, as in the case of asphalt tanking and roofing. The comparison can be drawn with the use of bitumen felt roofs

Fig. 13.6 Representation of the primary and secondary defence provision of the sloping external envelope.

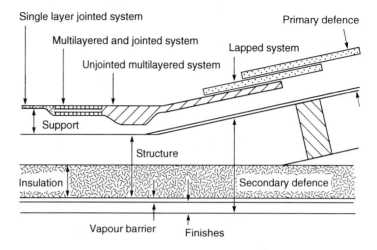

where joints are areas of failure. The inadequate falls on flat roofs and weathering surfaces will place additional stress on dry jointing techniques. Hard impervious materials absorb little moisture and therefore most is discharged. This will also impose an increased loading on jointings and water-shedding details.

Solid walls

Solid walls offer only one defence line to water ingress. The external part of the wall will behave in a predominantly primary defence role, whilst the inner region will act as a secondary defence line. A feature of its total defence mechanism will be its resistance to dampness penetration, and the time required for water to arrive at the inside surface. Laminated walls will place different resistances and interfaces across the direction of water flow.

Secondary water resistance of the envelope

Where breaches have been made in the primary envelope, the secondary defence will begin to operate. This may be localised around failed joints and details, or across large areas. The poor jointing together of dissimilar materials and components will provide ready entry points in the external envelope for water entry. The use of air cavities to stop the water flow is a traditionally effective method. This may perform satisfactorily in general but fail specifically around junctions and openings where the cavity is bridged. Where thin flexible barriers of waterproof material are used beneath the primary defence, these can fail due to impact damage or poor jointing. Where this important part of the secondary defence has been breached the remaining structure may offer little resistance to dampness.

Water penetration of the primary defence

Porous materials and water absorption

Many materials used in construction are porous. Ceramic and masonry components have various degrees of porosity depending on their composition and manufacturing process. The general durability of a material and its resistance to weather ingress are commonly described by reference to the driving rain index (BRE 1971b), and this should be considered in relation to the general exposure of the site.

The traditional methods of measurement of porosity are commonly concerned with determining the relative density and the bulk density of the materials and interpreting any differences as evidence of pores (Walker and Morgan 1975). A total reliance on this method of assessment ignores the effects of either separate or linked pores. Individual pores which fill with water may be the cause of additional defects, including frost attack, but interconnected pores may allow the transfer of moisture through the material. The degree of permeability of the material will influence its general characteristics.

Where cycles of wetting and drying occur the material may lose any water-soluble compounds present within it. This will expose a greater surface area to further effects, increasing the risk of deficiency. The moisture cycles will also cause movement of the material, which will adversely affect durability. Water absorbed into the material will generally add to its over-all specific heat capacity. This can mean that the temperatures of these damp materials will remain lower for longer. This may have marginal effects on its conductivity, but over large areas the risk of frost attack will increase.

Colour changes of materials are possible. This may be due to the leaching out of salts or the soluble salts of materials higher up the weathered face washing over them. The colour range can be extensive (Fidler 1988), and this will also be a measure of the differences in durability of similar materials. There are many types of natural stone, many having differing durabilities (Ashurst and Dimes 1977). The igneous types are generally less absorbent than the sedimentary types.

Deterioration of materials and water penetration

The deterioration of materials with age is not a process which appears to be capable of being stopped. The ravages of the weather are variable, and can make severe and unpredictable demands on any material. The general mechanism of ageing is essentially one of weathering. The wide range of diurnal temperatures, solar radiation, water, frost, and chemical attack can all occur to exposed materials. These will cause movements, wash away soluble matter, and increase the risk of frost damage. The exposed face is most at risk, and where this fails the process continues with the next underlying layer. A roughening of the surface in this way will increase the risk of further attack, as the exposed surface areas increase. Where the materials have durable faces these may fail because of two mechanisms. Firstly, the material at the face may behave physically in a different way to the underlying material, causing differential movement failure. Secondly, where the outer layer has failed, the underlying layer may not possess sufficient durability for its new exposure, causing premature failure. A similar problem can occur at the junction of different materials.

The small cracks and crevices formed by weathering will provide a harbourage for dust and debris. These may be places where moisture can remain trapped, fuelling any possible chemical attack or dampness penetration.

Water penetration and the deterioration of materials

The general process of frost attack was described earlier. For frost attack to succeed, some degree of porosity must allow water to enter the material's surface. High degrees of porosity may not necessarily be deduced as offering little resistance to frost attack. Likewise, low porosity may not be a measure of high resistance to frost attack. The traditional method of relying on the existing performance of porous materials may not be relevant for the new generation of insulated walls (*Architects' Journal* 1989). In the case of bricks, where small amounts of water are trapped in isolated pores and this water freezes, there may be little space which the water can expand into and this may cause the brick to be disrupted, blowing off part of the face. The incidence of freeze/thaw cycles of insulated walls will be more than uninsulated walls (*Architects' Journal* 1989).

The precipitation of soluble salts on the face of materials, or efflorescence, is generally a cosmetic defect. The space previously occupied by the salts in the material may now expose additional compounds to further dissolution. The pore network may be extended, creating additional scope for further attack.

Where water penetrates organic material this can create suitable conditions for the growth of fungi and moulds. These can break down the structure of the material causing discoloration and substantial loss of strength. The moisture content to support fungal attack is commonly assumed to be 20 per cent, although once established some rots can attack material with lower moisture contents. In the case of timber, high moisture contents are also associated with reductions in strength and dimensional instabilities. Where the timber is subjected to drying and wetting cycles the dimensional changes will stress the already weakened timber and may hasten further deterioration.

The laminated, or layered, forms of construction which involve facings, structure, infilling, and internal lining may be subject to a staged breakdown. The outer layers are most vulnerable to external influences, whilst inner regions may suffer from wind-blown water entry and/or

Fig. 13.7 A pattern of failure of the layered construction of an external wall.

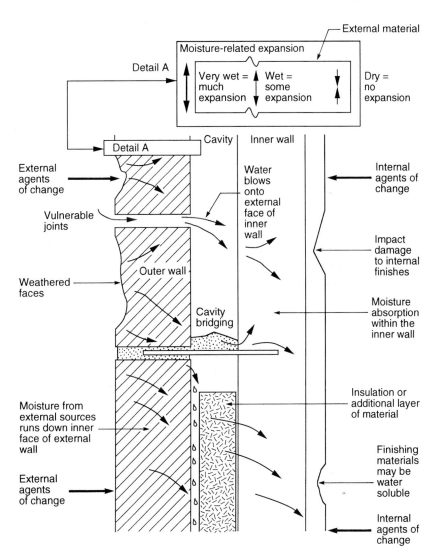

condensation. The inner linings and finishings are subject to the internal environment, also impact damage. A typical pattern of failure is shown in Fig. 13.7.

A similar process on a micro-scale occurs with the breakdown of thin film coatings. Outer layers can be breached by weathering or impact damage. This will expose the underlying layers to environmental attack, for which they may not have been designed. The adoption of thin film coatings brings with it design methodologies based on reduced margins. This fine-tuning of the envelope may make economic sense without consideration of the potential for minor defects to become major defects. The example of defective paint films to metals may mean that differential oxidation corrosion can continue under previously sound paint film.

Leakage characteristics of non-homogeneous composites

Where water can penetrate non-homogeneous composites the absorption rates are likely to be as different as the pore size distribution of the material layers. Large pores can fill quickly, whereas the smaller pores take time to fill, although due to capillary action water may move vertically and horizontally through them. The moisture content of the material will vary with depth of penetration of the moisture. This can lead to differential moisture expansion within the material. The interface between different materials can become a focus for these movement differences, as shown in Fig. 13.7.

Where different moisture contents exist, there is also the risk of differences in surface temperatures and induced temperature gradients through the envelope. This can occur across the face of homogeneous materials, inducing thermal stress, or across areas of different materials. This will place demands upon the accommodation of any joints between them.

The length of time taken for an exposed material to reach a standard moisture content will depend on many factors. These include rainfall intensity, orientation and the characteristics of the material. The drying-out phase is related to the latter factors, and the drying influence of the wind and sun. In general, the longer it takes a material to absorb water, the longer it will take to dry out. When the wetting and drying cycle of the material is considered in relation to the wetting and drying cycle of the ambient conditions around the material, it is possible for some materials to become progressively wetter at certain times, whilst at others it will be progressively dried out. This wetting process can be defect-critical, especially where materials such as timber and bricks are capable of absorbing considerable quantities of water.

Consequences of dampness in non-homogeneous materials

The consequences of water entry and dampness in non-homogeneous materials are accentuated by the influence of the interfaces and cavities between different materials. The directions of water migration through homogeneous materials will be influenced by pore size and gravity at the material scale, and by joints and water flow paths across its surface at the elemental scale. The placing of layers of construction to resist water entry brings with it the need to form effective joints between the layers. In the case of walls the interfaces will be vertical, allowing water to move through them under gravity. In certain situations water can pass through the outer skin of a cavity wall and run down its inside face. Where cavity trays are poorly positioned or constructed inadequately, water may pass through them to the inner skin. A similar effect can occur with other layered wall constructions as in Fig. 13.7.

The increasing use of filled cavities, in order to obtain the required thermal transmittance of the wall, presents additional scope for water

ingress. Insulation materials are commonly porous, and capable of holding significant quantities of water. As they become waterlogged they lose their insulating properties. Water can enter them from the outside by a variety of mechanisms, including driving through the outer skin. Even where an air cavity is retained, poorly positioned ties or insulation material can allow water entry.

The need for increased levels of insulation to meet energy conservation criteria will mean that where cavities are filled, the outer leaf will be subjected to more freeze/thaw cycles than an uninsulated wall (*Architects' Journal* 1989). Horizontal joints, common in roofing systems, may be physically below the available head of water, and also of sufficient size to allow capillary action to drive water through them. Timber supporting structures may then become damp, losing strength and allowing ponding to occur. This will increase the load and the material's inability to sustain it, causing severe overloading and even failure. This process may worsen progressively, giving a steadily increasing load to the structure.

The provision of internal drainage baffles in open jointed precast concrete panels is an acceptance that some water will gain limited entry. They will fail where the baffles are incorrectly lapped to horizontal flashings (Addleson 1982) which are insufficiently wide to accommodate the movements of the panels, or their fixing is inadequate and they simply drop out. Where the joints have been sealed as a remedial measure, water can be trapped behind them (Edwards 1986). This can flood cavities and break through into the inner wall or even the building. The external drainage of walls can fail at joints between materials and components, and be compounded by a lack of weathering. The junctions between impervious and pervious materials are areas of water concentration, and are commonly ignored as vulnerable areas (Marsh 1977). Defects such as this can lead to a long-term high moisture content of surrounding materials. Where they are of a non-homogeneous nature the additional problems of differential movement, surface temperatures and vulnerability to frost attack are present.

Permeability of minor cracks

Minor cracks in materials can allow water entry. These cracks may be so small as to be unseen, making detection difficult. The flow of water under capillary action may allow water migration to occur in a variety of directions. This may allow moisture to move up to and through ingress points. The influence of the surface tension effect between liquid and solid is critical and may change as the liquids dissolve salts from the solid. The problem of water migration through pores also manifests itself as rising dampness. It would be simplistic to relate the height that the dampness has risen to the pore size of the material. This ignores the above effects, and any effects due to covering materials, such as plaster, etc. The visual inspection

and surface probing of the wall alone can give misleading results (Seeley 1987).

Very small pores may resist water but allow the passage of water vapour. The use of breather membranes which allow water vapour to pass through them is an example of this. The ingress of water vapour can lead to increased vapour pressures in cavities, resulting in condensation when the dew-point temperature of the cavity air is reached.

Cracks and direct leakage

Where failure cracks occur in the building these can be direct paths for water entry. This can be considered as a secondary effect of the initial failure. Its consequences may be far reaching since the migration of water will affect materials previously unaffected by the crack. Cracking may pass through materials and components, or follow the joints between them. This can create complex paths which make the location of cause very different to the position of effect. Joint and material failure can occur, giving rise to a double failure. Sealants can embrittle and edges of concrete panels can spall due to corrosion of buried reinforcement.

The location of cracks within the building will also affect the amount of water passing through them. Those near to the top of the building may get less water delivered at generally higher wind pressure, whilst those at the lower parts of the building may get more total water delivered at pressures related to water run-off rather than wind pressures. Projecting eaves and cornices can throw water clear providing they are adequately weathered and throated. Smooth facades do not reduce water runoff, demanding increased performance from materials and joints.

Where cracking occurs below ground, the water pressures can be considerable. It can flood wall cavities and basements, and bring with it dissolved chemicals from the ground. Small cracks can leak in this way, even those difficult to detect visually. In the case of concrete, leakage may be symptomatic of a general failure of a material whose porosity is unacceptable.

Water channelling within the structure

When water has breached the primary defence, the remaining construction is generally more vulnerable to dampness. The breaches may occur in a variety of positions, for a range of reasons, although joints are common sites of failure. Where movement joints have been correctly constructed to penetrate the entire structure, any failure of the joint will provide a direct entry path for water. This can pass through a layered range of construction, soaking it as it passes. The interfaces between layers, as in Fig. 13.7, may allow water movement downwards when any gaps are large, and upwards when gaps are small. This will occur at all angles normal to the direction of water flow, and produce differing flow rates dependent on the

Fig. 13.8 Water channelling within the structure.

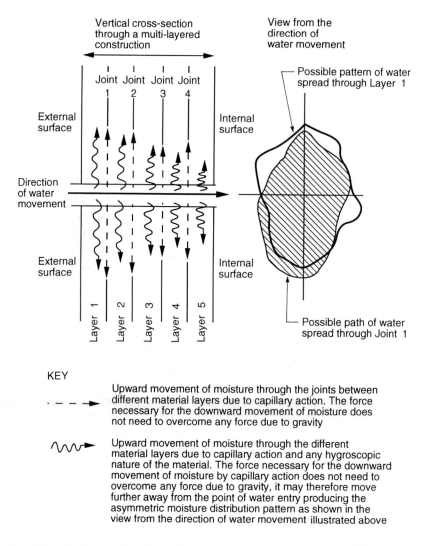

Vertical cross-section through a multi-layered construction

View from the direction of water movement

Joint 1   Joint 2   Joint 3   Joint 4

External surface

Internal surface

Direction of water movement

External surface

Internal surface

Layer 1   Layer 2   Layer 3   Layer 4   Layer 5

Possible pattern of water spread through Layer 1

Possible path of water spread through Joint 1

KEY

Upward movement of moisture through the joints between different material layers due to capillary action. The force necessary for the downward movement of moisture does not need to overcome any force due to gravity

Upward movement of moisture through the different material layers due to capillary action and any hygroscopic nature of the material. The force necessary for the downward movement of moisture by capillary action does not need to overcome any force due to gravity, it may therefore move further away from the point of water entry producing the asymmetric moisture distribution pattern as shown in the view from the direction of water movement illustrated above

physical constraints of the gap, and the materials. This is shown in Fig. 13.8.

This mechanism will force water into the building when the vapour and water pressure outside are greater than inside. When these pressure differences are reversed, some water vapour may escape, although the majority of water in the structure may remain since the pressure differences are likely to be in vapour pressure level, not water pressure level.

The degree of deficiency will be related to the resilience of the materials and components suffering from water absorption. Gypsum plaster can dissolve, pre-formed timber products can expansively disintegrate, and glass fibre insulation can sag. Masonry products are generally more resilient, although when damp they are more likely to suffer from frost

Fig. 13.9 The facade consists of different materials with variable runoff rates. Rainwater channelling is complex.

attack and pattern staining. Certain defects may occur deep within the structure, making access and inspection difficult. Any visual signs of distress on internal surfaces may not be directly adjacent to the failure of the primary defence. This may mean that areas of wall or floor may need to be removed in order to examine the extent of the defect. An endoscope may assist in this process, although where cavities have been breached and water movement through solid material layers has occurred, removal and exposure may still be required.

Additional features of the external skin

The frictional resistance of highly textured surfaces will reduce the water runoff rates during the initial period of a storm. Where light showers occur they may dampen the surfaces, become trapped in cracks and crevices, then evaporate. This will leave behind dust and any precipitated material. The appearance and ageing of the surface may change in this way. Storms sufficiently strong to cause water to flow over and down surfaces will wash

Fig. 13.10 Water channelling paths across the structure.

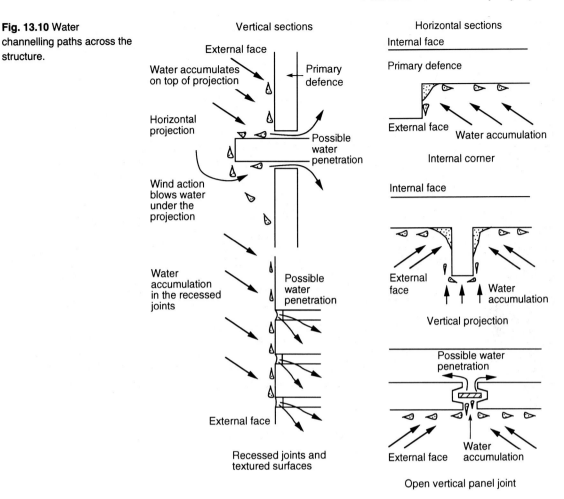

Vertical sections

External face

Water accumulates on top of projection

Primary defence

Horizontal projection

Possible water penetration

Wind action blows water under the projection

Water accumulation in the recessed joints

Possible water penetration

External face

Recessed joints and textured surfaces

Horizontal sections

Internal face

Primary defence

External face   Water accumulation

Internal corner

Internal face

External face   Water accumulation

Vertical projection

Possible water penetration

External face   Water accumulation

Open vertical panel joint

out these pockets, and create the channelling of water into corners and recesses. The position of recesses and internal corners is critical in this respect. Where a recess occurs in a generally flat surface, water will accumulate in it. This is seen in the movement joints in precast concrete panels, shown in Fig. 13.10. The relative impermeability of the surface will influence the amount of water moving across the facade. High winds can blow water upwards across the external surfaces. This can be stopped by projecting sills and transoms causing water accumulation at their undersides. This can lead to water ingress. Sills are commonly taken through the primary defence layer in cavity walls, and in the case of cladding systems may pass through the total system. These are obviously penetrations of the primary defence and may allow direct water entry.

The areas around openings can fail due to the poor integrity of the primary defence. Typical failures are shown in Fig. 13.11. The joints around

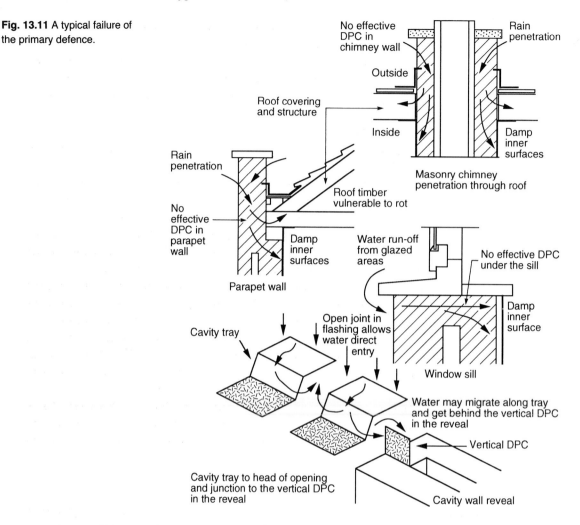

Fig. 13.11 A typical failure of the primary defence.

the openings should accommodate any differential movement whilst retaining a water barrier. Flashings can be poorly positioned or omitted (Addleson 1982, Davies 1985), allowing water direct entry. This will increase the risk of mould growth and dust adhesion to the damp inner surfaces. The position of these symptoms is usually adjacent to the opening, although failures in cavity trays at head level may result in water being discharged into the jamb of the opening.

Where walls are penetrated by horizontal elements, water may enter through them before the general areas of the primary defence are breached. See Fig. 13.11. These elements may act as effective water conduits because no barriers to water ingress exist. Alternatively, their jointing may be inadequate. Water may enter around metal pipes which pass through roofs and walls because the differential movement of the pipes and fabric have not been considered. Metal collars attached to the pipes can corrode, and butt joints between collar and pipe can fail.

The parapet wall is subjected to a most severe weathering process, and this can have implications for its physical integrity. Where the weathering process has attacked the exposed faces they are more likely to absorb water due to the texture and increased surface area. The junction between parapet and external wall may offer no barrier to moisture entry, allowing dampness to appear on internal surfaces and roof spaces. A typical failure is shown in Fig. 13.11.

Moving parts within openings can admit water. Their role as part of the primary defence is commonly carried out by thin sections of impermeable material. The tolerances of failure are therefore reduced, with increased emphasis being placed on meeting surfaces and joints. These can be disrupted by distortion of the components, which can take weatherproof joints outside their working range. Corrosion and thermal and moisture movements are all causes of distortion which can be increased by impact damage and heavy operation. Overpainting of timber windows can cause meeting surfaces to close up increasing the risk of water entry by capillary action. Where the fixings to large areas of lightweight claddings and curtain walling are inadequate, the whole area can flex in gusting winds. This will place mechanical demands upon the integrity of the cladding, and also upon its weatherproofing. Joints may open up or be stretched beyond their operating range, allowing water direct entry. Failed sealants may be as difficult to detect as the failed joints since the defect may only manifest itself during high winds. Glass can crack under these conditions, or in severe cases whole panels can drop out.

The influence of design The hygrothermal functional requirements of the external envelope can be simply described as an ability to discharge impacting rainwater and to maintain the stability of the envelope under extremes of temperature. These processes must be carried out whilst a physical and chemical compatibility is maintained between all the components and materials. The selection of materials for inclusion in the external envelope demands rigorous consideration of their weather-resistant properties commensurate with their expected life in the building. Many defects are associated with materials whose life expectancy may be definable, but whose practical application has assumed an infinite life expectancy. There is a need to examine the possibility of frost attack on all exposed materials. A survey of 510 defects found that 58 per cent were due to faulty design, and of those 295, 49 or 17 per cent were due to a faulty choice of material or component (Freeman 1975). The relationship between defect occurrence and poor maintenance is well charted, and this must also have implications for the effective life of any material or component.

Defects caused by inadequate design are usually assumed to be due to not following current design criteria. The selection of masonry units for the external skin of the building may be based on low permeability whilst carrying a risk of frost attack. The appearance aspects of design are vital to the construction of a successful building. Parapet walls without projecting or weathered copings, whilst giving clean lines are prone to attack. Typical design advice may be to consider the overall problem, rather than to offer detailed technical advice (Brick Development Association 1979) concerning cause and effect and the implications of choices.

Design and accommodation

The decision to select impermeable materials, components and joints may mean that the implications of defects associated with their failure are ignored. A total reliance on the impermeability of a material demands excellent performance at all times from the outer envelope, something which experience suggests will not happen. Where a more accommodating external envelope is adopted there may be a failure to carry through the possible water paths and movement patterns through the structure.

The design detailing and selection of materials/components for the external skin should balance the need for weather exclusion and adequate appearance. The conceptual nature of design will mean that technology is constantly running to keep up or else it will cause a stagnation of design, as is the case where the same standard details from the same standard texts are constantly re-used. When the water-exclusion side of the compromise fails, the building users get wet interiors, whereas poor appearance is no guarantee to defect-free buildings.

The influence of construction

For the construction process to produce quality construction adequate supervision is essential. Its importance is emphasised by the common practice of rewards and production assessment criteria being biased towards factors other than quality (Hinks and Cook 1988b). Where the tiered nature of supervision is incomplete, the resulting gaps may allow defective construction to be either accepted by default or simply hidden. An example is the lack of provision of DPCs at openings, which can be difficult to determine when the opening is sealed, although blatant cases of poor construction exist (Davies 1985). Material and constructional changes can occur without the approval of the designer. These changes may be driven by financial factors rather than a full appreciation of all of the technical implications. These implications must include the effects of any changes on the surrounding structure and materials.

Traditional construction

Traditional construction relies on the use of materials and methods which are simple and familiar to operatives and supervisors. The risk of changes with this type of construction may be considered minimal, since the general properties of the materials may be sufficiently similar to allow for extensive changes to be made without altering the performance of the structure. Changes from clay bricks to concrete or calcium silicate types may risk them being treated in the same fashion. Where a component material is changed this may place new demands on any movement joints between structure and component, which traditional jointing methods may be insufficient to cope with.

A failure to meet the agreed tolerance standards may leave water-entry paths in the primary defence. The general dimensional tolerance level of construction is wide, but even this can fail to be achieved. Indeed, buildings may be erected in the wrong place, and the possibility of errors of tens of millimetres is high.

A typical inherent defect in traditional construction is the water from the construction of the building. This accounted for 23 per cent of defects caused by faulty execution in a survey carried out between 1970 and 1974 (Freeman 1975). This may not strictly be the penetration of water through the primary defence, since it may have been introduced after its construction although the effects can be similar. Where DPCs and damp-proof membranes (DPMs) have been omitted or poorly constructed, there will always exist the risk of water entry. This was seen to occur in 12 per cent of the defects associated with the 1974 survey. In addition, 10 per cent of the sample possessed actual leaking joints.

Industrialised and system construction

Industrialised systems of construction were commonly based on traditional materials, although applied in a different way, placing new emphasis on accuracy and the interconnection of units. Many operatives were unaware of the implications of carrying out processes based on the now unacceptable workmanship and tolerance standards common for traditional construction. The poorly sealed joints or open joints where baffles have been omitted/ detached are a common defect of this form of construction. The interconnection of flashings around panels may have been lapped the wrong way, directing water into the building (Edwards 1986). There are similarities between this lack of appreciation of operation of the construction detail, and that associated with the placing of twisted wire ties upside-down in traditional construction. The supervision appears to be applied in a mechanistic way, ensuring that parts are arranged in the approved way, rather than giving an understanding of the consequences of poor construction to those carrying it out.

Trade specialisation

The traditional use of trades for certain parts of the work may assist in the excellence of specialisation, but makes the problem of quality coordination difficult. The meeting place of trades and activities under incentive schemes which are geared to production, may help in the covering up of earlier defects. The ongoing trade or operation appears to have no incentive to stop production in order to correct underlying faults.

Low maintenance regimes

The call for the effective maintenance of buildings has been made over many years, and included in many texts (Seeley 1987), but the unacceptable face of inadequate maintenance is still present. Where a low standard of maintenance exists, the integrity of the primary defence will gradually reduce. The point at which this allows water direct entry will depend on the hygrothermal resistance of the external envelope. A failure to maintain protective coatings on components may not lead to immediate water ingress, but begin the destruction of the component, which may lead to water ingress at some future date (Hinks and Cook 1989d). A failure to reseal mastic jointing in precast concrete panels may allow a slow build-up of water at the back of the panel. This may find a route into the building.

An initial and superficial appraisal of the decision not to maintain the primary defence may seem to have no consequence. Unfortunately, the results of this decision may not manifest themselves for months or even years.

Impact damage

The degree of roughness of use will affect the performance and durability of finishings and the structure. Sedentary users will make less impact damage on the building than those associated with heavy industrial processes. Doorways and windows are areas of high impact damage risk. Any impact damage can affect the fitting of the component into the opening, or destroy its weatherproofing integrity, allowing water direct entry. Broken panes of glass due to casual impact damage are common. This minor damage may lead to additional impact damage by building users who perceive the continuing existence of damage as a lack of interest by the facility/maintenance manager. The existence of damage to buildings may also act as a trigger to vandalism, whereas buildings which appear cared for are likely to remain intact (Underwood 1980).

The structure of buildings can also be affected by impact damage by the users. See Fig. 13.12. The forces required may be greater than those needed to damage finishings and components, but will depend on the nature of the structure. Brick walls can be impacted by moving loads and vehicles where no barriers exist. Any resulting water entry may be minimal compared with the structural implications, but this must be considered in relation to the use of the building.

Fig. 13.12 Progressive damage following vehicle impact on brittle lightweight cladding.

Occupancy patterns The way that people use their buildings appears to be continually changing, keeping pace with user and commercially driven expectations and technological advances. The domestic building has become refined into types of accommodation to suit particular markets, each having different occupancy patterns. These patterns will influence the hygrothermal performance of the building.

The time lags* and decrement factors[†] of the structure will move in ways which are affected by the occupancy. The changes will also affect the risk of condensation, and influence the ingress of water through the primary defence. Individual users may make a variety of demands upon the building

* The time taken for a change in Sol-Air temperature to be transmitted through the building fabric is termed the time lag. The Sol-Air temperature is obtained by adding together the external air temperature and the effect of solar radiation.

† The moderation of external heat flow rates by the structure of a building is termed a decrement factor.

in order to satisfy their particular needs. This may include considerable heating at all material times, allowing water and water vapour to be drawn into the building. Excessive ventilation may keep inner surfaces permanently cool. Commercial and industrial interiors will be occupied in a manner dependent on business efficiency, rather than the efficient use of the building. A continual occupation will make more demands of the structure, and increase the risk of impact damage. The provision of maintenance may be deferred where production is threatened. These factors may influence water penetration of the primary defence.

Thermal variations by the user

The need for energy conservation in the operation and use of buildings has meant that heating is commonly only provided during periods of occupation. The range of controls now available means that this can apply to individual rooms in buildings, and they can be retro-fitted to existing buildings. Where the heating provision inside a building is intermittent there will be times of major differences between inside and outside temperatures. This hygrothermal difference may act to draw water through the primary defence. Water stored in porous materials may act as a reservoir for this process, which may keep internal surfaces damp. A proportion of the heating of interiors may be due to the industrial, commercial or domestic processes going on within it. These may also act in a cyclic way, producing periods when large differences in temperature exist. High levels of insulation in walls, used as part of the drive for energy conservation, can lead to increasing cracking and movement above DPC level in masonry walling (*Architects' Journal* 1986c). This is attributed to the wide temperature difference that occurs across insulated elements, and may allow water through cracks in the primary defence. Solar gain will affect indoor temperatures, driving them up when structures are poorly insulated. Solar gain may be greatest through the roof, moving temperatures upward there before those in more massive walls.

A lack of appreciation of the need for adequate ventilation and heating of buildings by the users can lead them to make unacceptable demands of the building. A lack of understanding on the part of the specialist facility managers of the hygrothermal performance of their buildings may compound this problem.

High working moisture contents

The moisture content of the internal air in buildings is commonly greater than that found outside. This may emanate from the building users (*Architects' Journal* 1986b), or the processes taking place within the building. Where the air has a high specific humidity and a high temperature, the dew-point temperature of the air will also be high. The risk of moisture condensing on internal surfaces, or within the external envelope, is also

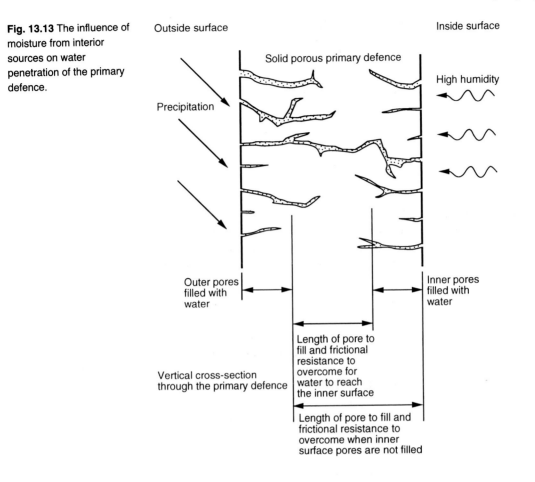

Fig. 13.13 The influence of moisture from interior sources on water penetration of the primary defence.

high. Where the external envelope is porous as with solid brickwork, this may become saturated from outside due to the weather, and inside from the moisture in the air. The filling of the inner surface pores with water may mean that water can more readily flow through from the outside, as shown in Fig. 13.13. This may simplify the ingress of water.

The positioning of insulation without vapour barriers will increase the risk of condensation water accumulating in the insulation. Insulants are commonly of open texture with large amounts of air space; they require large quantities of water to fill the pores sufficiently to allow water to pass through them. Where this filling is assisted by external and internal water, the time taken for moisture to pass through the material will reduce.

Sealing ventilating openings

The role of ventilation in the hygrothermal performance of the building is well documented (Burberry 1979). This was an important feature in pre- and post-war domestic buildings and although the levels of comfort may not compare with current values, the incidence of water ingress due to the influence of inner moisture was low. The open fire flue provided permanent ventilation, and insulation values were low. The high-humidity areas of kitchens, wash-houses and bathrooms were either primitive or well ventilated. The continuous occupation of dwellings was common and structures remained at even temperatures. This further reduced the incidence of moisture accumulation internally. The relatively modern need for high insulation values should have sharpened the need to consider ventilation. This did not happen in the now classic failures of insulated flat roofs and sealed precast concrete multi-storey blocks of flats. Where the need for ventilation was recognised and ventilators were installed, they were commonly covered over by the occupants. This appears to suggest that the short-term convenience of shutting out draughts is of more interest to the occupant than a long-term dampness problem.

Degree of intolerance to water ingress

Where the interior is very tolerant of water ingress the structure provides little more than a rude shelter from the weather. Many heavy industrial operations are contained within very basic buildings. As the use of the building becomes more sophisticated the threshold of tolerance to water ingress reduces. The degree of unacceptability of water entry can be assessed using several scales. The appearance of water dripping through the roof of a five-star hotel may be totally unacceptable for the guests, although its potential for ongoing structural deficiency may be small. Alternatively, the appearance of water dripping through a sealed cavity may be acceptable whilst its long-term effect may be considerable.

The relative 'healthiness' of the interior may be another scale used for assessment of water entry. Where water enters it may bring with it various organisms and chemicals which could affect the quality of the internal environment. If water may enter, it may be possible for similar organisms to escape from the building, which may have implications for specialised building uses.

The interior finishings of the building may be intolerant of water entry. They tend to be composed of less durable material, since their performance specification will be different to that of external materials. The water may dissolve them, as in the case of gypsum plaster, or cause the bond between them to break, e.g., wallpaper, paint films. Although they may suffer from a gradual reduction in appearance, they can be readily repaired. Indeed so readily that the original fault causing the defect may literally be papered over.

Insulation bridging

Poorly fitted pitched roof insulation may seal the ventilation paths around the feet of rafters. It may also act as a moisture bridge where no secondary

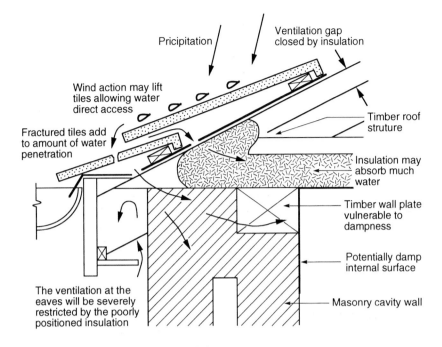

Fig. 13.14 Ventilation in pitched roofs: possible consequence of retrofitting insulation.

Labels in figure:
- Pricipitation
- Ventilation gap closed by insulation
- Wind action may lift tiles allowing water direct access
- Fractured tiles add to amount of water penetration
- Timber roof struture
- Insulation may absorb much water
- Timber wall plate vulnerable to dampness
- Potentially damp internal surface
- The ventilation at the eaves will be severely restricted by the poorly positioned insulation
- Masonry cavity wall

defence is provided behind tiles or slates, allowing water to enter the inner skin, as in Fig. 13.14.

Water- and air-permeable envelopes

With permeable envelopes the water is able to gain simple access to the pores during rain, but can easily migrate outwards when the vapour pressure outside is reduced. The materials are more likely to allow water to penetrate the full depth of the material under short storm periods, although the degree of admission will depend on the orientation and degree of exposure. The use of solid brick walls is no defence against water penetration in certain locations (BRE 1971b). Where cracks and open jointing exists this may allow water entry in a less controlled manner. It may be relatively easy for the water to enter in bulk, but it may only migrate out slowly as water vapour.

Material deterioration

The mechanism of material deterioration with age will involve the effects of water, snow, frost and chemical attack. Wind action will blow water and dry debris across the face of buildings and into cracks where their accumulation may cause problems. The breakdown of the exposed face of the material will increase its surface area, allowing more absorption to occur. Water trapped in the pores will increase the likelihood of frost attack (Taylor 1983b), which will continue the breakdown of the exposed face. The inner surfaces of certain materials may be more vulnerable to weathering, e.g., face-treated masonry and timber claddings, This will accelerate the deterioration of the material as the depth of attack increases.

Frost attack of porous materials can cause their complete destruction, although this would normally require a considerable amount of time to achieve. Water freezing will expand and where this is restrained the forces generated can be considerable. Porous materials can also be frost resistant since they have sufficient space to absorb the increase in volume, although the performance of individual materials is difficult to predict on this basis alone.

Capillary action will draw moisture through a porous material by an amount related to the size of the pores. This water movement will be assisted by gravity in downward migratory paths, but must overcome gravity when moving upwards. Water will move through porous materials in directions related to the internal structure of the material. This will produce results as shown in Fig. 13.8. Barriers between materials can be crossed since even small gaps will allow water to pass through them.

Extended water-discharge times

The available pore space within certain external materials can mean that substantial quantities of water may be held when the material is saturated. The saturation period will depend upon the degree of exposure and the rainfall. The time required for the water to escape from the pores will generally be longer than the time required to fill them up. This process will slow down the discharge times of water or water vapour from the permeable envelope. Since capillary action can continue long after the initial soaking of patches of external envelope, this may be pulling water into the structure whilst evaporation is pulling water back to the external air. Under intermittent rainfall patterns the ingress of water under capillary action can therefore remain virtually continuous.

Impermeable envelopes

Impermeable surfaces tend to be smooth, with few surface irregularities to trap water flow and add to the absorption rate. The reduced frictional resistance of water flowing over these surfaces allows wind action more readily to move water in many directions, not only downwards. This will influence the rate of water runoff and the paths of water flow over the external surface. These effects will be complicated by the introduction of different materials in the same facade: glass areas will have high runoff rates whilst their frames and sills will arrest and channel the flow patterns. These variations will influence the weathering pattern of the primary defence and its appearance. Gutterings and drainage paths can quickly fill with water, even under modest rainfall.

The impermeability of certain materials may include water vapour

Fig. 13.15 Possible hygrothermal consequences where impermeable envelopes are used as the primary defence.

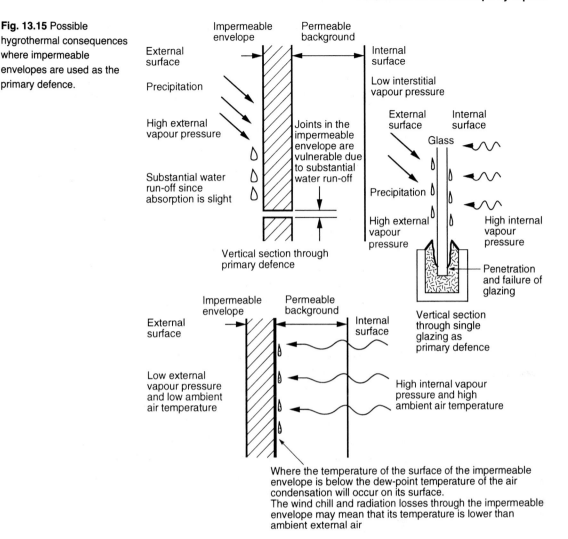

impermeability. This may cause it to act as a vapour barrier placed in potentially the coldest part of the external envelope, whilst recommendations are that it should be placed on the warm side of any structure (*Architects' Journal* 1986c). This may give rise to large vapour pressure differences between inside and outside, as shown in Fig. 13.15. The performance of joints and laps in this context will depend on their vapour resistance. Mastic seals tend to behave as the material, whereas laps and open joints will allow a more rapid equalisation of water vapour to occur. The rate of vapour pressure equalisation will vary depending on the incidence of joints and the overall vapour-pressure difference.

The provision of impermeable materials as outer coverings to the external envelope may seem to imply that inner regions should remain dry.

Fig. 13.16 A comparison of the positioning of impermeable panels on their vulnerability to breaches in the primary defence.

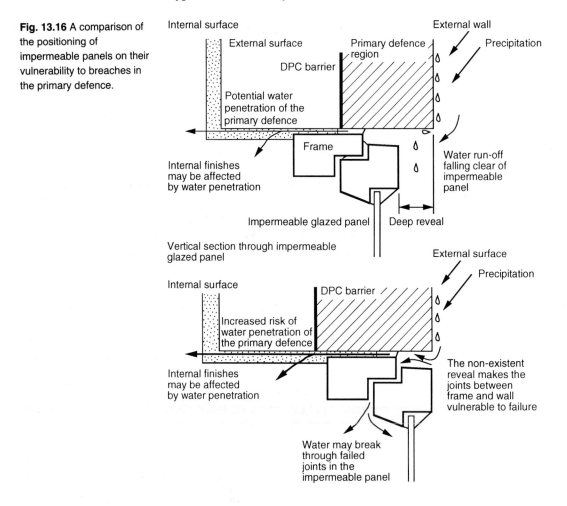

Unfortunately, the impermeable materials may be vulnerable to other defects, including fracture under minor loads. The amount of water passing through any cracks may be increased since the total amount of water passing over the surface of the material is a major factor in determining how much will pass through a crack or joint in its surface (Marsh 1977). Where impermeable coverings and panels are positioned on the outer surface of the primary defence they are in a more exposed position than those which are recessed. This may place new demands on their joints and seals which they may be unable to meet. A comparison is shown in Fig. 13.16. Where water is not thrown clear of the primary defence by sills and other projections it will add to the mass flow rate across the facade.

Water penetration of the secondary defence

Definition The secondary defence mechanism can be considered as that resistance offered by the inner part of the external envelope. This would normally lie behind any external material resistance. This may be a collection of layers as in the case of cavity walls, or innermost parts of a homogeneous primary defence, as in the case of solid walls. A failure of the secondary defence will normally follow a failure of the primary defence. This can lead to water penetration of the total defence and admittance to the interior of the building. Certain inner materials and construction methods are likely to be severely affected by the ingress of water since their performance requirements may not emphasise a high hygrothermal resistance or performance. Defects in the primary defence can often expose additional defects in the secondary defence. This can produce a domino effect of progressive failure.

Bridging and the cavity The development of the cavity wall was seen as an improvement over solid walls by offering an air void through which water could not pass. The primary defence of the half-brick outer wall was recognised as being of doubtful resistance in highly exposed positions (BRE 1971b), although the total wall was seen as providing a satisfactory solution for a wide range of applications. This will fail to be achieved where the cavity is not completely clear, as any solid bridging of the cavity allows water to pass from outside to inside (BRE 1982d). Constructional requirements will demand that the cavity is bridged around openings and sealed at high and low level. It is in these regions that breaches of the secondary defence commonly occur. Examples are shown in Fig. 13.17. DPCs may be inadequately lapped or positioned around openings. The construction of window reveals may include plaster and newspaper (Davies 1985). The cavity can act as a reservoir for water driven through the outer skin. This may accumulate in the cavity above flashings and bridges in the cavity and permeate through the remaining secondary defence. This process occurs more readily where no cavity trays exist. Water can penetrate the lapped joints of cavity trays since many rely on overlaps with no mechanical or chemical sealing. A similar problem of cavity flooding can occur with precast concrete panel construction where the erratic application of mastic sealant over open drained joints, without provision of weepholes, has allowed water to accumulate in the joint area.

Cavity ties The wall skins on either side of the open cavity must be effectively tied together in order for the total wall to behave as a composite element. The ties are zones of cavity bridging which link the primary and secondary defence lines. The flat and twisted metal ties each have some level surfaces where debris and dust can accumulate. Mortar droppings may also build up

Fig. 13.17 Potential breaches of the secondary defence which allow water penetration.

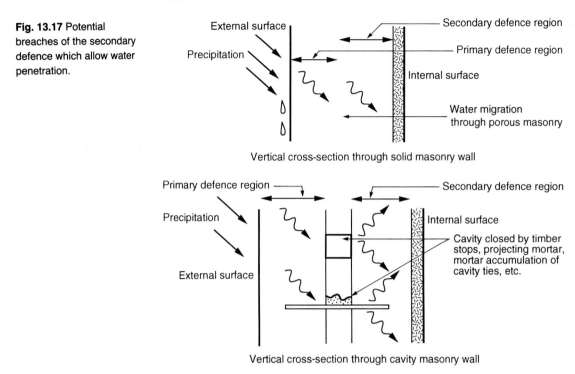

Vertical cross-section through solid masonry wall

Vertical cross-section through cavity masonry wall

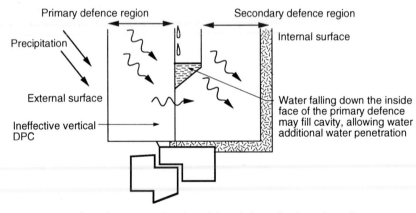

Vertical cross-section through head of opening in cavity wall

on the tie and together these accumulations may provide a porous bridge allowing water to migrate into the inner wall. This may lead to isolated patches of dampness, which can vary in accordance with the rainfall pattern. Where ties are positioned out of level the accumulations may build up to either side of the cavity, as shown in Fig. 13.17. The transfer of loading between the skins of the cavity wall may also be affected. Cavity walls which consist of concrete inner walls and masonry outer walls place demands on the provision of ties. In general, the risk of poor positioning increases, and with it the possibility of ties being missed out altogether. The checking/inspection required to determine the correct position is rarely carried out during the progression of the work, and becomes very difficult after it is completed. The penetration of dampness through concrete walls may take a considerable time, but a more immediate problem may be the corrosion of the tie. Timber-framed inner walls may allow for direct fixing of ties, commonly nailed to studs, which can accumulate mortar droppings. In this instance the dampness penetration may be more rapid and influence the decay of the timber.

The corrosion of metal ties will be influenced by the acidity of their surroundings and the presence of moisture and oxygen. The oxides produced are larger than the original metal and will tend to force open the horizontal masonry joint in which they are bedded (de Vekey 1979). This expansive force can cause spalling of concrete. Where cavity walls are closed around openings, the tie may directly penetrate any DPC between the leaves. Under severe conditions significant amounts of water may enter.

Insulation (application without understanding)

Insulation can be part of a composite or laminated solid wall, or applied in a variety of ways for roofs. Because of its open texture it may accumulate a considerable amount of water. This may come from a failure of the primary defence, or the condensation of water vapour from inside or outside. Its water-storage capacity may allow it to diffuse water over a wide area, a process aided by the capillary pores in the material. This is a similar effect to that shown in Fig. 13.8. Where the insulation abuts another solid material, a capillary gap may be formed. At times of high solar gain the primary defence may dry out quickly. The secondary defence will receive little solar gain and can remain at low temperatures, which will extend the drying-out period. Areas in permanent shade are more likely to become saturated and permit water penetration.

The commonly accepted classifications of 'Inverted', 'Warm' and 'Cold' forms of flat roof construction can also be applied to walls, as shown in Figs. 13.18(a), (b) and (c). This classification can be used to predict the incidence of defects from entrapped moisture from either condensation or poor vapour barriers. This may create conditions suitable for the existence of thermal pumping, with its subsequent effects on materials and finishes. This process enables water vapour to be drawn outwards through a poor vapour barrier at times of a low external temperature, only to be forced back inside during times of high solar gain and temperature. Where an impermeable primary defence exists, air trapped between it and the vapour barrier will generate a pressure dependent on its temperature. Water may drip through a ceiling in a manner similar to that from a leaking roof (Allen 1990). Industrial sheeted roofs can suffer from a similar defect (*Architects' Journal* 1985).

Fig. 13.18 A classification of insulation provision to walls and roofs. **(a)** Inverted wall construction. **(b)** Intermediate wall construction. **(c)** Cold wall construction.

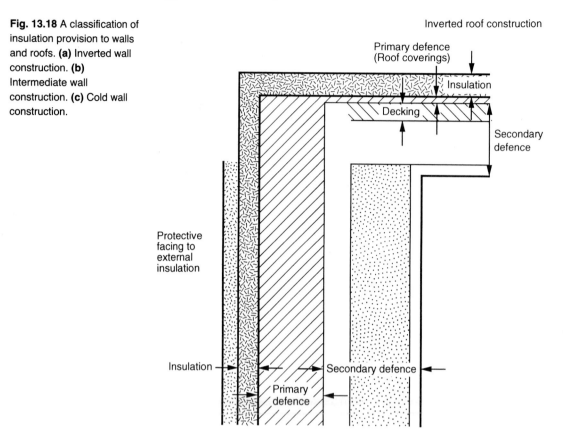

Mortar droppings The mortar used for masonry walling can provide a ready route for water through primary and secondary defences. The construction of vertical mortar joints, particularly deep joints of one brick thickness or more, becomes generally more difficult. The workmanship aspect becomes critical since it is rapidly covered over.

Cavities can be effectively closed by mortar exuding from brickwork joints. Where mortar from the inside of the outer skin projects into the cavity, it can allow water to drip from it onto similar mortar projections from the inner skin. The requirement for clear cavities is rarely achieved, and cannot be readily inspected for compliance. The common method of inspecting and clearing cavities at low levels may have some effect, although patches of mortar-filled cavity may exist anywhere within the wall. Patches of dampness can be associated with these solid wall areas.

Wider cavities are not considered structurally viable; they would also demand an increase in the workmanship aspect of lining up coursed units across large cavities. Where these factors set parameters for a maximum cavity width, the introduction of cavity insulation sets new parameters on minimum cavity widths.

Fig. 13.18 (b)

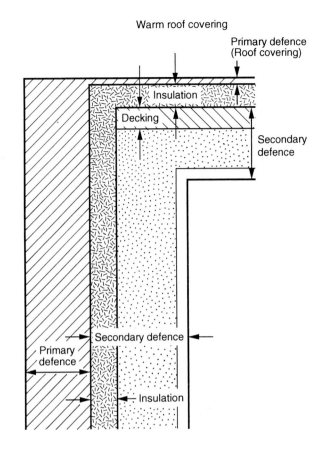

Warm roof covering

Primary defence
(Roof covering)

Insulation

Decking

Secondary
defence

Secondary defence

Primary
defence

Insulation

Solid walls Solid walls of brick perform as both primary and secondary defence lines. The only barrier to water penetration is the permeability of the materials, which may not be matched to the degree of exposure of the wall. These walls may possess rudimentary damp-proofing and parts of the structure can act as cold bridges. Driving rain will soak exposed walls and enter through weak points, commonly the joints. Accumulated water around openings is traditionally thrown clear by weatherings and drips. Erosion and impact damage may have weakened this line of defence by disrupting smooth surfaces and trapping water.

The link between ground floor and solid wall was reliant on either the continuation of the impermeable materials to stop rising damp or physical separation with air gaps. Butt joints between the DPCs and DPMs are common, although where the suspended timber floor was unable to ventilate away water vapour, defects could still emerge.

The meeting points of concrete-framed buildings with infill walls of cavity construction can be considered as isolated solid walls. The butt joints can be

Fig. 13.18 (c)

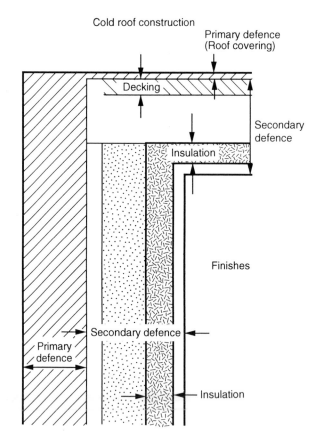

paths for direct water ingress, as shown in Fig. 13.19. The identification of the cause for the wetting of the inner surface should take into account the local weather, and whether parts of the concrete frame are acting as a cold bridge.

Precast concrete panels can be considered as the modern equivalent of a solid wall. Surface textures may hold moisture, and the porosity of the concrete may allow this to migrate as vapour and moisture through the panel. The durability of any reinforcement will be affected by this process, which may also cool the inner surface of the panel. This may encourage the growth of moulds on the inner surface.

Flooding of buildings   Buildings close to rivers or the sea, and those in low-lying areas, are vulnerable to flooding. This may be seasonal, associated with the neap tides and storms, or a random flash flood. This can occur after periods of drought have hardened the surrounding vegetable soil, causing a temporary

Fig. 13.19 Possible water penetration paths through solid walls.

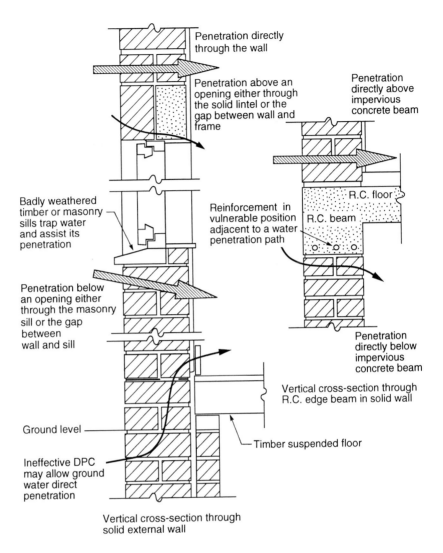

Penetration directly through the wall

Penetration above an opening either through the solid lintel or the gap between wall and frame

Penetration directly above impervious concrete beam

Badly weathered timber or masonry sills trap water and assist its penetration

Reinforcement in vulnerable position adjacent to a water penetration path

R.C. floor

R.C. beam

Penetration below an opening either through the masonry sill or the gap between wall and sill

Penetration directly below impervious concrete beam

Vertical cross-section through R.C. edge beam in solid wall

Ground level

Timber suspended floor

Ineffective DPC may allow ground water direct penetration

Vertical cross-section through solid external wall

impermeability. The rate of rainfall will exceed the rate of water absorption by the ground, and standing water appears. This may cover the low-level DPCs, flood suspended floor cavities, and lower the general air temperature around buildings. Where subsoil drainage paths become blocked, the water may back up connecting pipework, breaking joints, flooding gullies and spilling over walls. Inadequate sizing of pipes can cause similar effects at times of high rainfall.

Information regarding the risk of flooding from rivers and water courses can be obtained from river and local authorities (BRE 1989b), although any predictions of flooding are only likely to be as accurate as the last event. Where large bodies of water enter buildings they may bring with them suspended solids and chemicals aggressive to the building. Salt water can be particularly damaging (BRE 1973). Many internal finishings are water soluble and will be weakened by immersion in moving water.

The primary and secondary defences of the building can be penetrated by a wave of water breaking through doors and windows. The impact damage can be considerable, as seen during the failures of sea walls. Silt and mud can accumulate at lower levels and mark decoration.

The drying-out period of the structure will depend on its degree of saturation. The hydrostatic pressure associated with deep flooding can soak relatively impermeable materials. Deep cavities, basements and structural partitions may not be capable of free drainage when the flood subsides. This may complicate and extend the drying-out period. Material liable to fungal attack may be particularly vulnerable to extended drying-out times. Where timbers are adjacent to or embedded into structures which subsequently become saturated, the risk of fungal attack increases. Cabled and piped services demand special consideration (Seeley 1987).

Requirements of damp-proofing materials

Criticality
Preventing vertical moisture movement from the ground to the building interior is essential for the longevity of the structure, and to provide a healthy internal environment. Where buildings become damp this can cause a range of defects. Short-lived materials are generally vulnerable, and water-soluble sulphate solutions can cause disruptive expansion of cementitious material. Plaster may be water soluble. A general lowering of the ambient temperature and an increase in the relative humidity when coupled with moist, low-temperature surfaces provides ideal conditions for the growth of surface moulds, and increases the risk of insect attack.

The complexity of design, specification and installation, combined with the criticality of DPCs and the consequences of any inadequacies, mean that the subject requires close attention. As with many other building-related defects, this is often the result of the misapplication of existing knowledge rather than lack of knowledge itself.

Development of damp-proofing
Early traditional homes were generally built with thick solid walls without any damp-proofing. The first damp-proof layers to be frequently used were hard waterproof bricks bonded together in courses. The term 'damp proof course' developed from their use, and they are sometimes identifiable as bricks that are blue or differently coloured to those of the main wall. Alternatives included the use of slate, metal or bitumen damp-proof layers. All of these are capable of being waterproof, but have generally been superseded by modern plastics which are cheaper and more flexible as well as being strong. The term damp-proof course (DPC) has been retained.

The suitability of materials

The materials selected for use as DPCs and DPMs must obviously be highly resistant to moisture and water vapour, be inert to surrounding materials and have a durability comparable to that of the building structure. These features are difficult to assess. Aspects such as long-term performance under loading, ease of use, and appearance must also be considered. The materials should not tear or break in normal use or during installation. The choice of materials and their jointing will depend on whether protection is against upward, downward, or horizontal moisture movement. The exposure factor of the building will influence the design and the performance in use of the DPC.

To date, no standard gives detailed performance requirements for DPC materials, although BS 743 (British Standard 1970a) specifies the quality of most available types, some no longer in common use. This has been criticised as being out of date, out of touch and too generalised to be sufficiently useful (Duell and Lawson 1983). Materials are grouped in categories according to their flexibility:

Group A materials (e.g., bitumen, polythene and polypropylene, the most commonly used materials for DPCs) are flexible and are suitable for bridging cavities, stepped details, and dressing to complex shapes. Properly lapped or welted, these may be considered virtually jointless. Certain polymer-based DPCs are difficult to seal, although easy to lap. Also included in this category are metals such as copper, and lead, which requires protection against the corrosive effect of cement mortar. This is usually achieved by a protective coating of bitumen.

Group B materials are semi-rigid (e.g., mastic asphalt), and will resist water at high pressure. These are particularly suitable for tanking basements and retaining walls.

Group C materials are rigid (e.g., brick, slate) and capable of bearing high loads, but are the most susceptible to damage caused by building movement, and are not completely water resistant.

Care in the selection of the quality of a rigid barrier is essential, as is the matching of a suitable bedding mortar (normally 1:3). It is worth remembering that mortar is not a DPC and in these rigid applications can become the weak link in the barrier. A range of DPC materials is shown in Table 13.1 (Hinks and Cook 1988d).

The influence of detailing

Damp-proof detailing will be influenced by the construction method, as well as being practical to incorporate within the structure. For example, impervious panel-wall constructions can direct large amounts of water into localised areas, needing specialist DPC provision.

Table 13.1 Properties of common types of DPC materials

Type	B.S.	Weight/size	Joints	Comments
Lead	1178	19.5 kg/m^2 1.8 mm thick Code 4	100 mm	Expensive. Liable to corrosion when in contact with cement mortar Lapped joints to resist upward moisture movement Welted joints to resist downward moisture movement
Copper	2870	2.28 kg/m^2 0.25 mm thick	100 mm	Expensive May stain external surfaces Lapped and welted joint as lead
Bitumen	6398	A wide range exists 3.8–4.9 kg/m^2	100 mm	Brittle at low temps Extrudes under high load May be reinforced with metal or fibres
Polythene	6515	0.5 kg/m^2 0.46 mm thick	Lapped joint to be width of DPC	Black, low density No extrusion under normal loadings
Mastic Asphalt	1097 & 6577	12 mm thick	Homogeneous	Extrudes under loads >0.65 N/mm^2 Grit improves bonding
Slates	5642 & 3978	4 mm thick 230 mm long	Laid bonded in 2 courses in 1:3 cement mortar.	Liable to fracture External feature
Brick	3921	Standard	As slates	Engineering quality bricks with average water absorption

The relationship between damp-proofing and the structure

The distinctions between forms of damp penetration

Dampness can penetrate the building in many ways. Groundwater can enter as rising dampness; rain and other precipitation can cause penetrating dampness. Any deficiencies in the primary defence can allow water to penetrate the structure. All of these forms of penetration can produce damp interior surfaces, although the failure of any secondary defence is normally required for this to occur. The need to identify the mechanisms responsible for any particular failure is an essential part of any defects analysis.

Although the forms of dampness penetration can be conveniently packaged, any defect may be due to a combination of them.

Relationship between the DPC and the floor

The relationship between the DPC in the wall and DPC/DPM associated with the floor structure is essential for damp resistance. This will vary with the type of wall construction and the type of floor. The damp-proof membrane (DPM) is commonly of a different material to that used for the DPC. Although the detailing differs between solid concrete and suspended timber floors, the principles are constant. For the solid concrete floor, thick grade polythene (0.46 mm thick normally) is usually sufficient for a damp-proof membrane, since it does not carry a great load. It can be easily damaged during insertion, though, and this can be a cause of patchy dampness in the floor.

In order to resist water and water vapour pressure generous overlaps have to be made between the sheets of polythene. Where they meet the wall DPC, adequate jointing is required. This is absolutely critical for the exclusion of water and the avoidance of dampness inside the building.

Fig. 13.20 An example of poor lapping to joints in DPM under concrete floor.

Fig. 13.21 Possible defects associated with poor jointing of vertical and horizontal damp-proofing materials in a cavity wall at ground level.

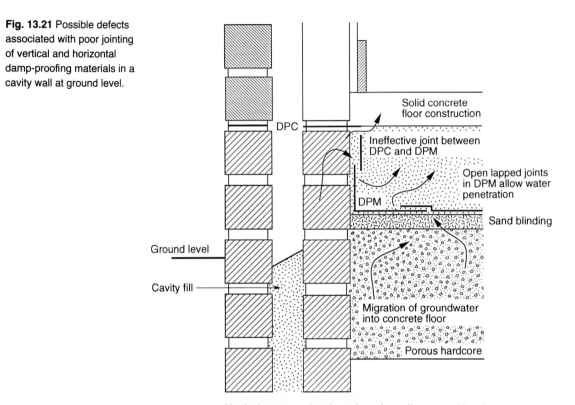

Vertical cross-section through cavity wall at ground level

Where this joint is defective, dampness is likely to appear at floor level adjacent to the wall. See Fig. 13.20.

Where the DPM in the floor is at the same level as the DPC in the wall, there is no need for a vertical DPC. Figure 13.21 shows the provision of a vertical DPC and possible defects associated with it.

With cavity walls this is no problem: the floor level is slightly higher than the inner-leaf DPC, but with the cavity acting as a vertical barrier, the level of the inner leaf DPC can be below the outer-leaf DPC. In solid walls this necessitates a floor level higher than ground level.

The influence of floor construction: timber

The important features of damp-proofing in the traditional suspended timber floor relate to the protection against damp ingress through the external wall to the underside of the timber floor joists. Groundwater could seep into the void beneath the floor, producing ideal conditions for fungal attack. Where the lower level of the floor void is lower than the surrounding ground, flooding is more likely. Newer buildings should not suffer from this defect since it would contravene the building regulations.

Traditionally, these floors rested on dwarf walls, although now with

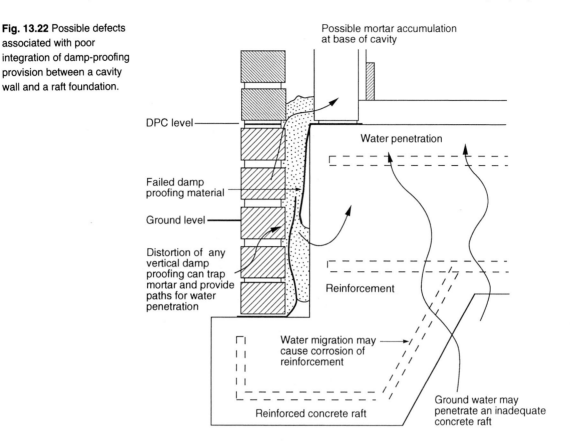

Fig. 13.22 Possible defects associated with poor integration of damp-proofing provision between a cavity wall and a raft foundation.

Possible mortar accumulation at base of cavity

DPC level

Water penetration

Failed damp proofing material

Ground level

Distortion of any vertical damp proofing can trap mortar and provide paths for water penetration

Reinforcement

Water migration may cause corrosion of reinforcement

Reinforced concrete raft

Ground water may penetrate an inadequate concrete raft

generally small-sized rooms, and the availability of joist hangers, these are less common. The levels of DPCs under the floor and in the wall may be different, although the ends of the joists should be above the wall DPC. Where a clearance between the end of the joist and the wall is less than 25 mm the joist may accumulate moisture and become vulnerable to fungal attack. It will be necessary to treat timber with preservative.

The influence of floor construction: concrete

Concrete rafts and reinforced slabs are advantageous in poor soil conditions and areas of subsidence. They enable the building load to be distributed evenly over a large area. Since the floor and foundation become a homogeneous structural unit, careful detailing of the DPM and DPC are needed. Problems arise at the edge wall detail because of the difficulty of constructing an integral DPC. Typical defects are shown in Fig. 13.22.

Due to the increased risk of movement the stepped DPCs at the base of cavities are vulnerable to cracking. Since raft foundations are particularly suitable for poor ground conditions, and these are typical locations where soluble sulphates are likely to occur, shallow rafts may be totally enclosed with a DPM in order to protect them.

Position of membrane and vulnerability to sulphate attack

Concrete ground-bearing slabs became very popular in the post-war period, replacing timber and its associated ventilation problems, and allowed a faster floor construction without the use of scarce materials. The position of the DPM is variable: the options are either above, below or within the concrete slab. The DPM below the slab can be damaged by the hardcore. The joints may be inadequately sealed or lapped. This can allow sulphate-bearing groundwater to attack the concrete. Indeed, the cocooning of reinforced concrete slabs with a damp-proof membrane may have been originally devised/applied for this reason.

Below-ground damp penetration

Original forms of construction

Early forms of basement construction were without any damp-proofing, and the structures were traditionally loadbearing brickwork, founded either directly on the ground or on (usually 1:4) concrete foundations (Adams 1912). Many of the Victorian basements were built in urban terraces at the time of the Industrial Revolution, and the constructional quality was variable. There is potential for the saw-tooth settlement of terraces built with semi-basements. The internal loadbearing structure of a basement, as it undergoes long-term deformation from the variable side loadings and upthrusts of a fluctuating water table, is not something to be tampered with unless imperative.

Methods of production

A common approach to the production of dry basements was to create 'dry areas', by building an outer retaining wall and leaving a small cavity, or larger open space, between this and the inner wall bearing the load of the building. Narrow dry areas (56 mm wide) were common, since this reduced the amount of basement excavation. This method is problematic, however, since the cavity can easily become filled with extraneous material including decaying vermin (Mitchell 1903). The smaller cavities were tied with tarred and sanded or galvanised iron ties, or canted glazed bonding bricks, and would be drained below the wall DPC. The glazed bonding bricks were considered superior to metal ties for structural purposes, and were of low permeability. Unfortunately, they are also likely to be less tolerant of differential movement.

Alternatively, the dry area could be opened out to 450 mm or more, for ease of clearing. These open spaces were used to provide natural light to basement storeys as well as provide natural ventilation. Access stairs to the basement were commonly sited in these areas.

Alternative damp-proofing methods

Damp-proofing at basement level was occasionally achieved using glazed stoneware slabs available in a number of lengths (about 50 mm thick) and perforated to act as both a DPC and ventilating brick. Other DPC materials used included sheet lead, slates set in cement mortar, and asphalt. With the provision of the open area it was possible to ventilate suspended timber floors in basements, although any failure in the ventilating system could lead to a rapid decay of the timber where basements are close to or below the water table. Dry areas should be drained, although backup and flooding problems with soakaways and sewer invert levels can occur.

Developments in tanking

By 1903, Mitchell was recommending that small dry areas be rejected, and an *in-situ* asphalt, or sheet asphalt tanking system be adopted. The first method involved spacing a 112 mm brick wall 20 mm away from the loadbearing wall, whose joints are raked out, and then progressively filling the cavity with liquid asphalt as work progressed (three courses at a time). This can produce horizontal joints between layers which may fail under severe water pressure.

Previous practice included rendering the external face of the outside (retaining) wall with Roman cement. This procedure was also used as a remedial treatment. The subsoils around the perimeter of cellars may have been drained with 50 mm agricultural pipes, covered with coarse gravel or faggots of brushwood (Adams 1912). These can become progressively blocked with fine silt, leading to a steady increase in water pressure to the basement sides.

The range of defects in basements

Basements can suffer from a similar range of defects to those experienced by the general structure. Settlement and movement problems associated with the ground are easily transmitted to the basement. Any cracking of the structure may be a ready path for groundwater ingress. The deterioration of the waterproofing properties of the basement can also produce the same effects, although visual inspection may be able to identify movement cracks. Under severe hydrostatic pressure water can move through lightly porous materials. Inner surfaces may appear sound, and yet permit general leakage. This can affect interior finishes and lower inner surface temperatures to points at which moulds and fungi can germinate.

Certain basement defects can be very difficult to remedy, involving external excavation and even underpinning and rebuilding basement walls and floors.

Leakage and discontinuity of tanking

The detailing of the tanking/DPC junctions needs careful attention if a continuous barrier is to be provided to the basement. This is similar but more critical than the external wall/ground-floor DPC/DPM junction. The hydrostatic pressure resistance of the basement depends on the thickness of the backing wall or floor to any tanking. This is assisted by the adhesion between the asphalt and the wall, and here adequate mechanical key and a sound background are important. Poor material quality and workmanship will obviously produce defective basements.

Characteristic basement failure

A particular characteristic of the failure of basement tanking is the appearance of symptoms remote from the source of failure. It is often difficult to locate the point of water entry accurately, and pattern drilling is sometimes needed. In general the defects in basements should be identified and solved before the cosmetic exercise affecting internal finishes is attempted. The option of constructing an inner dry basement within a failed basement must address and solve the inherent problems of continuity of water-proofing.

Historical perspective on restorative solutions

It is important to be able to identify where previous restoration or repair have been carried out, since this will probably have influenced the nature of failure and may restrict the options for its solution. Restorative damp-proofing around the turn of the century for flooded cellars involved draining the cellar by creating a sump hole. The outside walls were accessed and the brickwork allowed to dry out, followed by rendering. Brick floors were commonly replaced with 150 mm minimum of site concrete, and finished with a floated face of cement mortar. Among the tanking techniques developed, a method known as Callender's Bitumen Damp Course was also used for remedial work. Sheet asphalt was applied internally to the main wall, tied in using pockets, and clad with a half brick skin.

Ground-level penetration

Failure of the ground-level DPC

The ground-level DPC may fail where the wall is subjected to settlement. Where movement of the wall has produced cracking, this may break open the damp-proof layer allowing water a direct path into the wall and lower floors. See Fig. 13.23. The crack may accumulate dust and debris, which may add to any capillary action. The provision of impermeable materials as DPCs has also traditionally been associated with rigid brittle materials. Courses of bricks and slates bedded in rigid cement mortar can be effective

Fig. 13.23 The finish of the ground-level DPC has allowed rising damp to penetrate the wall and timber floor.

barriers to rising damp, although they will not resist water percolating downwards (Harding and Smith 1986). Unfortunately, they are liable to brittle failure which may leave large or capillary sized cracks. A reliance on a depth of several courses is no guarantee of continued effectiveness.

The change to a more flexible DPC material based on bitumens and plastics would appear to offer a better performance when accommodating minor movement. Unfortunately, these materials can be extruded out of joints. This is due to the applied loads, temperature cycles and the creep

Fig. 13.24(a), (b), (c)
Typical failure patterns of
ground-level DPCs.

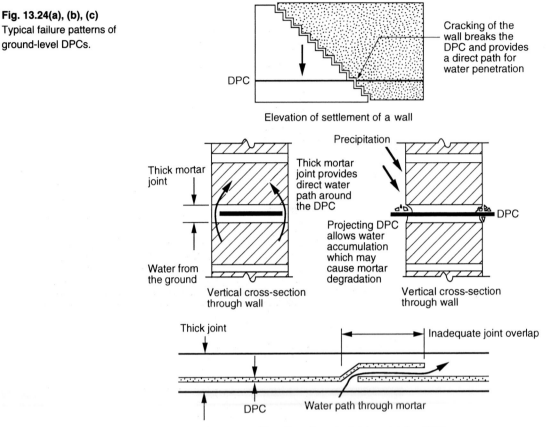

Elevation of settlement of a wall

Vertical cross-section
through wall

Vertical cross-section
through wall

Elevation of mortar joint containing DPC

properties of the damp-proofing materials. The force of exudation may be
sufficient to weaken mortar beddings, and the ledges formed can encourage
water into the wall. Typical failure patterns of ground-level DPCs are shown
in Figs. 13.24(a), (b) and (c).

The coefficient of friction of certain types of horizontal DPCs is sufficiently
low for them to act as minor slip planes. Where a horizontal plane of DPC
and linked DPM exists this effect may influence the whole building. The
amounts of movement are commonly small, but this will be influenced by
the loading from the building. The general effect will be related to the type
of construction.

**The role of inherent
faults**

The DPC will only perform correctly if it forms a complete layer throughout
the length and thickness of the wall. This was recognised as important for
the rigid DPCs where brickwork of the same or greater thickness than the
wall was used. Joints are vulnerable regions and where thin flexible DPCs
are used the joints are commonly lapped. These may not be sufficient and
allow water through, as shown in Fig. 13.24.

Fig. 13.24 (b)

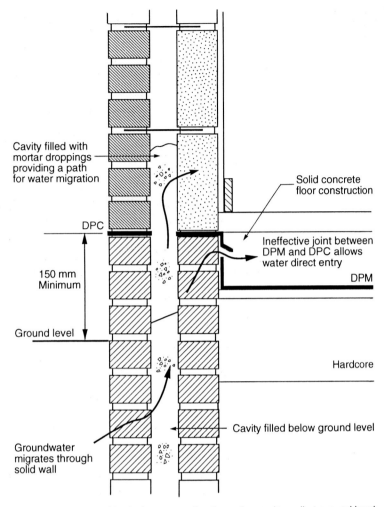

Cavity filled with mortar droppings providing a path for water migration

Solid concrete floor construction

DPC

Ineffective joint between DPM and DPC allows water direct entry

DPM

150 mm Minimum

Ground level

Hardcore

Cavity filled below ground level

Groundwater migrates through solid wall

Vertical cross-section through a cavity wall at ground level

Reductions in the length of lap, butt joints or gaps between lengths of DPC can all allow water to move upwards through the wall. These defects are commonly hidden within the mortar joint embedding the DPC, although the appearance of damp patches may help to identify the source of the defect. DPCs can be omitted altogether (Davies 1985). The adhesive taping of the joints to form an adequate seal can be difficult under site conditions. The influence of site workmanship on the production of inadequate joints will be related to the degree of supervision, since the joint is commonly hidden by ongoing construction.

In solid ground-floor construction the link between the DPC and the DPM can occur with minor differences in level. They are commonly different materials, with different characteristics other than their ability to stop water

Fig. 13.24 (c)

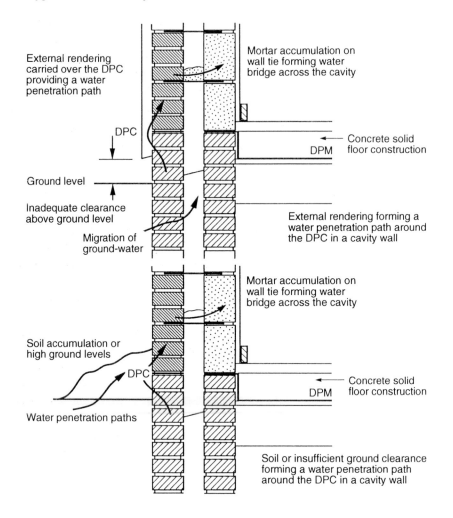

or water vapour. Any joints should be of the same vapour resistance as the materials.

Ground clearance of the DPC

The need for placing the ground-level DPC above ground level is a basic construction requirement, and has featured in most regulations and standards. The production of standard details showing the minimum clearance of 150 mm is recognised as giving reasonable performance (Hinks and Cook 1988d). Unfortunately this is not achieved in many instances, due to deficiencies in the construction process or subsequently.

Rendering can bridge the DPC, allowing water to migrate through it to parts of the wall, as shown in Fig. 13.24(c). The rendering may continue to deflect falling rainwater away from the wall, whilst permitting the passage of rising damp through its interior. In this condition it is more liable to frost attack. Soil may cover the horizontal DPC, either as part of gardening activities by the users or as made-up levels to entrances and ramps. Refer to Fig. 13.4. The soil will readily transport moisture to the outer surface of the wall and also assist in keeping the water runoff from the wall in contact with it. Vegetation may encourage the growth of external moulds.

The provision of weepholes at low level can assist in the damp-proofing of cavity walls, due to their influence on vapour-pressure reduction (Hinks and Cook 1988d). They provide a pathway for water splashed up from outside the wall to enter the cavity. Weepholes may become buried.

External influences on the effectiveness of the DPC

Water splash may only be significant at times of exceptional storms, or where weepholes are close to ground level. The critical distance of 150 mm between DPC and the ground should keep the amount of water splash to an acceptable level, although grass will be more absorbent to rain impact than paving slabs. Where entrances do not have a step above the DPC there is a risk of DPC bridging.

The provision of effective DPCs to door thresholds is recognised as requiring complex sealing (Duell and Lawson 1983), especially where the underside of the threshold is lower than the wall DPC. Ramps and stairways into buildings are also areas where the DPC can be bridged. The wall immediately below the DPC is in a vulnerable position, since it is likely to be damp for long periods. The risk of frost attack is high. The jointing above the DPC may also be vulnerable since water migrating downwards through the wall will be arrested in this area.

Where abrupt changes of wall direction and changes of DPC level occur, the damp-proofing joints become more complex to construct. Poor joints are rapidly covered up, and water ingress may take a considerable time to appear on inner surfaces.

Position of the DPC

In the case of external cavity walls the wall cavity must extend at least 150 mm below the lowest DPC. This allows free drainage of the cavity and should ensure that the cavity is clear down to a minimum ground-level clearance. DPCs in the external and internal leafs of cavity walls may be at different levels, with the cavity providing a vertical damp barrier. Some indication of this may be produced by the relative levels of the floor and external ground. The DPC can be located externally by evidence of a thick mortar joint between brick courses, possibly 25 mm. Where the mortar is incomplete, the actual DPC material may protrude. This can be due to the exudation of flexible material as discussed earlier, or be evidence of slippage of the wall.

Cavity bridging

If the DPC is carried across the cavity, it should slope towards the outer skin, and here care is required to prevent cracking. The constructional detailing at corners and junctions must be carefully considered to ensure continuity of the barrier (official advice concerning this appears to be lacking, although certain texts are helpful (Foster 1973)). There is considerable dispute over whether weepholes should be provided and whether they are of benefit (Duell and Lawson 1983), although in practice a recent survey showed that they are often omitted (BRE 1982d). It is of course essential that cavities (and weepholes) are clear of mortar droppings.

Rising damp   On walls this can reach a height of approximately 700 mm and abruptly stop. This dimension will vary depending on the amount of water present, the walling material and its porosity. The height could also be defined as the point at which the rate of rise of water in the wall is matched by its rate of evaporation from the wall. In this way similar materials may give differing heights of dampness dependent on the occupation pattern of the building.

A simple visual inspection of damp patches on walls at low level may show them to be due to other causes, e.g., condensation, or the existence of hygroscopic salts on the wall finish. In any analysis of surface dampness the hygroscopic moisture content (HMC) must be compared with the actual moisture content (MC) of the wall. Where the HMC is more than the MC, moisture is likely to come from the air; where the MC greater than the HMC it is unlikely to be coming from the air (BRE 1981f).

Cavity trays   Cavity trays should be provided with weepholes during construction. Drilling out afterwards is to be avoided since this invariably damages the DP tray and causes a build-up of debris in the cavity (BRE 1982e). It is important that mortar droppings are cleared out of the cavity tray area (Duell and Lawson 1983). The longer the cavity tray and the greater the height of wall discharging water to the tray, the more critical the weephole spacing becomes.

Damming-up of water at the cavity tray increases the risk of water penetration through the internal leaf, particularly with long lintels. There is evidence to suggest that water may drain outwards through the mortar joint above the DPC, but this is not always problematic (Ransom 1981). However, any fault in the damp-proofing will allow the internal leaf to become soaked, producing damp patches on the internal surfaces. Where open perpends are used as weepholes these may be adequately effective in slight or moderate exposures. Severe exposures may permit a net flow of water into the wall.

Insulated cavities   Carefully insulated cavities should not add to the threat of damp transmission and cold-bridging. However, blocking or sealing weepholes will render them partially or wholly ineffective. Where insulation prevents water from free draining from the cavity, this is unlikely to improve the thermal resistance of the wall or the durability of any ferrous cavity ties. Where the insulated cavity continues below the level of the DPC, the small gaps and open structure of the insulation may act as a bridge for groundwater to move up through the wall, by-passing any DPCs in the wall. Where insulation is retro-fitted to cavities, this may fill all voids, starting at the lowest level.

The depth of opening reveals

The major defect associated with reveals is their failure to keep out the weather. This internal corner junction in the external envelope must possess similar characteristics to the external wall whilst maintaining an effective seal between dissimilar materials. This is particularly important in the vicinity of the ground where the effects of groundwater and accumulated runoff from the facade meet. The penetration of vertical DPCs by ties and the reduced clearance of thresholds are potential routes for water penetration.

A common defect associated with depth of opening reveals is that of moisture gaining access to the back of the sills, thresholds and frames. This may be by direct ingress through the joint between the frame and the wall, or in the case of glazed units, the joint between the glass and the frame. The more exposed the opening, the greater the risk of rain penetration. Windows placed close to or at the external face may suffer in this respect. Smooth impermeable surfaces can concentrate water into the joint zone, placing considerable stress upon it. Where water is discharged from glazed or impermeable areas directly onto the outside surface of the wall, the accumulation of water may increase the risk of it penetrating the building, as shown in Fig. 13.19. Breakdown of mastic sealant may also be a major or contributory factor. Hard paving with a back fall to the building may prolong the time that the wall is damp.

Movement at the reveal

Movement due to moisture and temperature changes can stress the reveal junction, and may produce a wide range of differential movements. The provision of a contiguous DPC is essential, although its provision in certain constructional forms can actually make the junction inherently defective, as in the case of DPCs which return to the front of the wall, or are lapped to direct water into the building.

Tolerance and adaptation

The fit between components needs to be close to resist the weather, and yet large enough to absorb any differential movements. An overreliance on mastic sealant may simplify the construction, but will bring with it the need for detailed maintenance, in positions and situations to which access is not straightforward. The risk of vandalism to areas of sealant at low level must be considered.

The traditional construction methods did not rely on mastics as part of the hermetic sealing of buildings. The open joints allowed differential water vapour pressure between inside and outside to be equalised. Substantial depths of reveals offered further protection from the weather. In severely exposed positions, movable external shutters covered and protected openings. The user demand for the provision of daylight, access and an outlook at all material times has placed new demands on traditional technology.

Ineffective DPCs

An ineffective DPC may be compounded by the relative position of the door or window frame in the wall. Deep external reveals may place the window/wall joint behind the DPC weatherline, whereas shallow reveals may place the window in front of it. This can lead to direct entry paths. DPCs may be ineffectively positioned between structure and covering, as in the case of reinforced concrete frames and brick infill panels. Butt joints are common in this form of construction and the surfaces of the DPCs can become capillary pathways for water entry.

Damp penetration at roof level

The difficulties of inspection

Many problems arise with the durability and water-tightness of roofs. The roof is subjected to considerable wind, driving rain and a wide temperature range. It is often a difficult area of the building to inspect casually and maintenance may therefore be minimal. The degree of deficiency can be significantly greater than other easily inspected areas, and consequently defects may be more serious. The majority of faults with the roof are related to the integrity of the waterproof skin. There is some commonality in the faults arising with the flat and pitched forms of construction, independent of the materials used.

Many failures of roofs and guttering are evident only on detailed inspection. Many pre-1918 buildings can have complicated roof plans which increase the difficulty of inspection and maintenance. The inspection of pitched roofs by using binoculars from ground level can easily miss critical flat areas and those masked by chimneys and other obstructions. Although the inspection process for flat roofs tends to be easier than for pitched-roof finishes, major inspection problems do exist, and tend to be due to the lack of safe and effective access provision. This is essential if visual and physical inspection is to be carried out. Adjacent property may be used as static viewing platforms, which, although they may not offer total coverage, may be very useful for preliminary inspections. These difficulties tend to lead to poor maintenance, with inspections rarely carried out at the recommended intervals of two years (BRE 1982f). This leads to a growth of the defect, which results in these roofs tending to require either little maintenance or extensive repair work to major failures (BRE 1981e).

Any analysis of the complex mix of defects associated with flat roofs, in order to assess a particular problem, becomes a process involving detailed identification. The faults may interact with each other, or may mask other faults, further complicating the process of identification. Some of the more common faults with roof coverings are summarised in Tables 13.2, 13.3, 13.5 and 13.6.

The range of forms   The three standard types of roof construction are classified as:

(1) The **cold** roof – where the insulation layer is placed beneath the structural deck, and usually encloses an air space. This type of roof requires the correct positioning of vapour barriers and ventilation in order to be effective. Many older roofs were not provided with these features and this has given rise to a range of defects.

(2) The **warm** roof – here the insulation layer is still placed below the waterproof membrane; however, it is placed above any (warm) cavity. This produces a range of problems for unjointed multilayered systems.

(3) The **inverted** roof (or **protected membrane** roof) – here the insulation layer is placed above the waterproof membrane to act as a temperature buffer. It is normal to build in a weathering surface above the insulation. This protects the covering and reduces the occurrence of defects associated with temperature fluctuations. Since this is normally constructed on a concrete roof structure, its domestic applications are usually limited.

Examples of these standard roof types are shown in Fig. 13.25.

Large-scale unjointed   Large-scale unjointed systems as the primary defence mechanism may have
multi-layered roof   the advantage of providing both a water and water vapour barrier. Since
covering systems   these layers are easily recognised as vulnerable to water ingress the material choices are critical. The vapour pressure generated from inside the building can be greater than that existing outside at certain times. Where this pressure cannot escape, the layer may be severely distorted into blisters, since the times of high vapour pressure internally may coincide with high external temperatures, increasing the flexibility of the covering. The temperature-induced stresses are not localised at the joints as are those

Table 13.2 Defects associated with large-scale unjointed multi-layered roof coverings

Defect	Cause	Remedy
Crazing (surface cracks)	Dressing sand not rubbed in Poor solar treatment Water ponding	Because it is superficial, no action may be required. If appearance is important then take up and relay
Cracking (significant)	Differential movement around changes in direction and high stress regions Movement between asphalt and covering	Where water ingress has occurred, inspect the substrate before repairing. Adopt modified details as BSCP 6229
Blisters	Entrapped moisture from structure or within the building	Identify source of moisture. Cut out and repair. Lack of solar treatment will accentuate the problem
Ridging or cockling	Differential thermal movement between asphalt and sub-structure	If cracking is likely, then partial replacement is needed Where severe, removal and relaying required
Ponding	Symptomatic of: Differential movement Settlement Poor workmanship and/or poor maintenance	The provision of new falls may be constructed as part of an overall repair to the roof
General damage	Poor maintenance, workmanship and/or material Vandalism Roof traffic	Patching or repair. Consider replacement with more durable material. Provide protective measures
Chemical damage	Spillage from stored containers Plant and equipment	(as for general damage) Consider resiting the storage facility
Loss of chippings	Severe falls Wind erosion of poorly bedded chippings Inadequate depth of chippings	Consider alternative protective measures. Select larger sized chippings
Discoloured solar treatments	Surface film breaking down Oxide layer forming Repainting cycles too widely spaced	Repaint affected areas with more resistant material

Table 13.2 continued

Defect	Cause	Remedy
Water ingress	Cracking of surface layers. Inadequate or defective skirtings	Repair using approved details Avoid discontinuity of decking
	Failure around pipes penetrating the asphalt. Physical or chemical damage	Provide adequate collars
High temperature embrittlement	Overheating asphalt during the laying process. Poor or inadequate supervision	Remove all of the affected material Consider the liabilities of the parties concerned
Low temperature embrittlement	Low ambient and structural temperature. Unheated building. Cold roof construction	Where damage has occurred remove and re-lay Protect and asphalt with inverted roof construction

relating to layered systems. These may subsequently crack, allowing water direct entry.

A clear need for internal vapour-pressure reduction exists, and the provision of this through venting methods which penetrate the covering may itself allow water ingress through poor sealing and weathering.

The criticality of material choice is especially important when the introduction of new materials is considered. Adequate simulation of the performance in use of the material is an obvious prerequisite to selection, but appears difficult to achieve. Short-term testing may lead to short-term success and long-term failure. It is always unwise to make simplistic assumptions concerning defects. They must be set into a context which recognises that there is the potential for a variety of causes to interact to produce any failure. Because the systems tend to be unjointed, any movement characteristics can occur in three dimensions. Usually the third dimension is small in comparison with the length and width, therefore a failure to incorporate adequate area movement joints and the facility for the layer to move over the supporting structure can lead to cracking and water ingress. The movement joints are themselves critical areas, subject to embrittlement, detachment and impact damage which can allow water entry. The problem is accentuated where the thermal movement character-istics of the coverings are such that large distances are permitted between expansion joints.

Where changes of direction occur, e.g., at upstands, this will increase the height of the layer and concentrate movement stresses. Both effects can

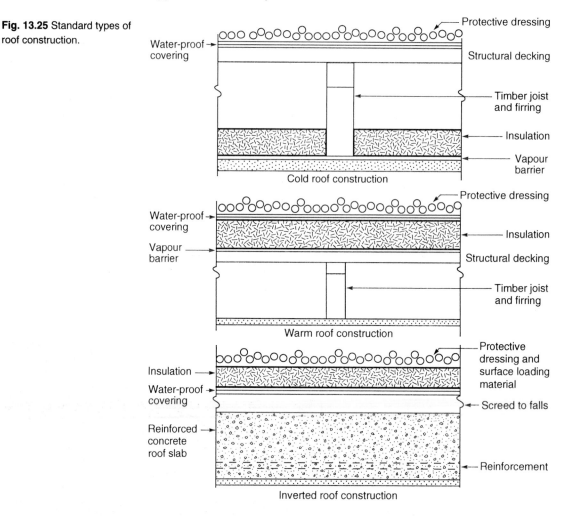

Fig. 13.25 Standard types of roof construction.

Cold roof construction

Protective dressing
Water-proof covering
Structural decking
Timber joist and firring
Insulation
Vapour barrier

Warm roof construction

Protective dressing
Water-proof covering
Insulation
Vapour barrier
Structural decking
Timber joist and firring

Inverted roof construction

Protective dressing and surface loading material
Insulation
Water-proof covering
Screed to falls
Reinforced concrete roof slab
Reinforcement

concentrate cracking, or lead to the detachment of the upstand from the supporting structure. Inadequately chased flashings may allow the passage of water, and lead to rotting of kerbs. The differential movement between the jointless layer and the surrounding structure can be compounded when the relative weights of construction are different. Deckings and supporting systems may be lightweight and prone to greater amounts of deflection than the surrounding loadbearing structure.

In layers which are sloping or near horizontal, the solar-induced temperatures can be considerable, causing ultraviolet degradation and embrittlement of certain materials. These temperatures may not reach those experienced during the laying of *in-situ* asphaltic materials, but any softening will make the material vulnerable to impact damage. The adoption of isolated solar reflective surfaces/finishes may lead to water ingress through cracking caused by the severe temperature differences between protected and unprotected layers or areas. Inadequate drainage provision of the layer, either through inadequate falls or blocked drainage paths, will lead to a range of defects similar to those associated with large-scale layered systems. The growth of moulds around ponding can drastically reduce the reflectance of any solar reflecting treatment, and may lead to cracking.

Fractures in unjointed surfaces

The suggested life span of a properly constructed asphalt roof is assumed to be 50 to 60 years, which makes it one of the most durable of all construction materials. Fractures in the asphalt surface can be failures that run through the material and allow water penetration. This may be the result of continual movement. Oxidation of the surface by ultraviolet radiation may increase deterioration. Cracks less than 3 mm deep are commonly assumed to be minor defects and can by treated by using a solar reflective surface. The real problem lies in assessing the depth of the crack. Cracks with a

Fig. 13.26 Asphalt distortion to side of low parapet wall gutter, producing a tear and hence a direct water ingress route.

definite line may be the result of structural movement of the roof, and the absence of expansion joints in the original structure.

In cases of persistent covering failure there may be a number of remedial layers of covering. These may create additional problems due to the different layers having different linear coefficients of expansion and contraction.

Where leakage defects occur they tend to be associated with discontinuity in the decking caused by movement, or inadequate attention to changes in the direction of the covering. In general, it is rare to involve the use of water testing for performance appraisal in these areas, although this can be a useful method to determine the cause of leakage defects.

Asphaltic flat roofs Asphalt becomes brittle at extremes of temperature, due to evaporation of the volatile component of the material at high temperatures, or the normal hardening of the covering at low temperatures. See Fig. 13.26. Asphalt is rather like paint and other applied finishes, in that it is tested and specified both in a liquid and a solid phase. When considering the constructional defects of the material, it is normally the solid phase that is referred to. This is because the inspection procedures are primarily concerned with the material as it exists, normally after a period of months or years. The need to assess the influence of poor workmanship must include any defects occurring early in the life of the finished material. In the case of mastic asphalt this must include the effect of the high temperatures experienced during the laying process. These are likely to be the maximum temperatures in the life of the material, and although the asphalt can cope with these temperatures quite readily, the effect on the underlying deckings must be considered. See Fig. 13.27.

Movement stresses and area effects It is well known that movement stresses may be induced by high temperatures, the degree of stress being related in an almost direct way to the temperature. This will mean that deckings with generally high coefficients of linear expansion may initially expand, then upon cooling contract. In doing this there exists the potential of imposing compressive stresses on the covering. Where a break layer of felt or other material has not been provided to enable the mastic asphalt to move independently of the decking, potentially harmful stresses can build up. This will not only occur during the laying phase of the material, but also during its life, when solar gain will induce additional temperature movements.

Temperature-related stresses act three-dimensionally, although the distances involved in the vertical direction mean that in many cases the effects in this direction are minimal. For most applications this is acceptable, but for the particularly vulnerable regions, such as upstands and skirtings, it

Fig. 13.27 Stress cracking failure of asphalt at jointing of metal trim.

should be considered. The overall roof area will influence the degree of movement, and if adequate control measures are not adopted, potential defects may occur. These may produce surface cracking. The incidence of cracking appears to be concentrated around junctions with surrounding structures, at projections and around changes in the decking structure.

Solar coatings   Although solar coatings are a relatively recent development, and some existing flat roofs and their failures pre-date this, the experience so far justifies their use. A variety of surface dressings are available to improve the solar reflectance of dark matt surface coverings. This can help to reduce maximum surface temperatures and therefore the thermal movement of the surface. These include basic stone chippings, solar-reflective paints, deliberate water ponding and the rubbing of sand into a freshly laid surface. The sand rubbing will break down the bitumen-rich surface of the asphalt, and provide dimensional stability as well as reflectivity. Although the sand rubbing can be effective for 'cold' roofs, it may be insufficiently reflective for other roof types and may contribute to their deficiency. Where there is an inconsistency of application a new series of differential movement problems may occur. The silvery appearance of some solar coatings dulls quite quickly, and this reduces their reflective protection. Where aluminium flake paints are used they tend to form a metallic skin with different linear coefficients of expansion and contraction to that of the base covering. This can cause splitting at the edges of the coating, as well as at changes of direction. Successful overcoating should also take into account the possibility of chemical incompatibility between some products.

Ponding Water ponding on the asphalt surface can induce surface crazing. The ponding may be caused by a variety of reasons, some of them related to the decking and roof structure. A typical recommended deflection limit for decking is 1/325th of the span. This requirement, whilst reducing the degree of water ponding, places upon the designer a need to establish the nature and duration of all loading to within a relatively fine tolerance. The critical loadings are likely to be encountered under unusual circumstances, be very variable, and therefore be difficult to predict accurately. The balance between uneconomic oversizing and flexible structures is commonly agreed without consideration of potential defects.

The moulds and other growths that thrive around the fringes of water ponding areas can drastically reduce the surface reflectance of solar coatings. The traditional stone chippings can help to trap surface water above the covering. Although this should increase the incidence of mould growth, the overall protective effect seems to remain high. This is in part due to the thermal mass of the chippings, as well as mould growth being concentrated within them. In order to retain the chippings during wind action, it is usual to lay them into a warm asphalt surface. This helps retention, but makes it difficult to remove them when remedial works are being carried out.

Water vapour and the hygrothermal link A compressible form of blister can be produced by moisture trapped underneath the unjointed covering layers. This may result from entrapped structural moisture or an incomplete or omitted vapour barrier. Recommendations for new work include the provision of proprietary ventilators in roofs of $10m^2$, since any moisture within the structure must be removed. The small-scale patching of the coverings is not always a satisfactory method for the repair of blisters, crazing or small cracks since it is difficult to ensure adequate bonding and continuity of the covering. This leads to secondary defects.

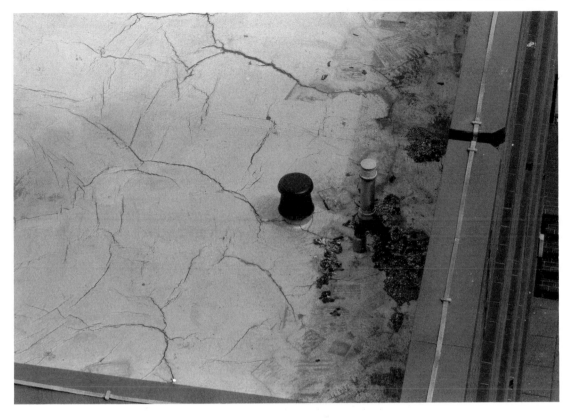

Fig. 13.28 Evidence of pattern staining associated with ponding and storage of bricks, also map cracking to retrofit reflective coating. Pipe penetrations present high-risk areas.

Pipes
Projections through the roof finish may provide ready paths for water entry. Pipes projecting through flat roofs can be provided with a sensible upstand which may still allow water in through gaps between pipe and upstand. Flashings cut into the vertical surface of the pipe above the termination of the skirting can fail due to an inadequate seal. The insertion of a metal collar to maintain a seal can fail due to its corrosion. See Fig. 13.28.

Decking and structure-related defects
Any unevenness or movement in the substructure can eventually manifest itself in the covering. Shear stresses can build up and cracking results. Although this takes generally longer for unjointed systems than for felt multilayered and jointed systems, the effects are similar. The provision of effective falls to the roof is an essential for water drainage. Flat roofs can rapidly become very flat roofs. This will allow water to accumulate, the rate of runoff to reduce, and waterways and drainage paths quickly to become blocked.

The historical problems associated with roof overloading have shown that there is a progressive link between the degree of deflection and the resultant load on the structure. The accumulation of debris will add to the problem of drainage provision since it can block waterways and increase ponding. Deficiencies in falls may be symptomatic of a failed decking or supporting structure.

Where the asphalt has not been allowed to move independently of the deck, ripples and ridging can result. With the movement away from the heavier forms of construction to those involving metal and timber decking, deflection values may be variable. Where metal decking systems are used in high-humidity environments, excessive deflection values may be associated with the corrosion of the decking. Chipboard decking can fail due to expansive disruption when allowed to absorb quantities of water. Total collapse is possible.

Surrounding structure-related defects

Where a lightweight decked asphalt roof is rigidly attached to a surrounding heavyweight structure, there is a risk of failure at the junction. The use of free-standing timber kerbs and metal lathing to achieve disassociation can fail due to water ingress causing rot or corrosion of the kerb. A lack of continuity of support by the kerb can cause sagging, leading to an increased risk of water penetration.

Parapet walls are regions that require adequate damp-proof detailing. These are frequently sources of water ingress, since they can provide a path for water which by-passes the impervious covering. See Fig. 13.29. The associated guttering provision may not provide an efficient route for the discharge of water away from the roof. During times of heavy rainfall the design parameters of the guttering may be exceeded, and since the effective depth of the gutter is a major factor in its carrying capacity, any reduction due to debris could be critical. Where the roof structure is of the lightweight type which requires isolation from the wall, the guttering system may act as a positive fixing. Any movements will be concentrated in the gutter region, where any failure would be critical.

Where asphalt has been tapered into a gutter, this can become an area for potential defects, since stresses are more concentrated and impact damage from access and inspection are more likely.

Large-scale multilayered and jointed systems

Leakage of the large-scale layered and jointed systems is a common fault in roof coverings. A range of typical defects is shown in Table 13.3. Defects are commonly associated with discontinuity in the covering, arising as a result of movement or inadequate attention to lapping and joint formation. Joints that allow water to flow into them are vulnerable to leakage. Where the covering is near horizontal the surfaces are also subject to a wide

Fig. 13.29 Rigid cement mortar has been used as a flashing between the roof and an abutting wall. Any discharge from this detail enters the guttering. There is no downpipe, however.

temperature range due mainly to solar radiation. This process is exacerbated by the incorrect positioning of insulation. Where this is placed directly under the layered systems their surface temperatures can reach high levels.

Temperature-related stresses can act two-dimensionally, producing surface crazing. This may progress into cracks through the materials allowing water direct access to the decking. Crazing and tearing can be made worse if a separating membrane between covering layers and structure is omitted. This can also occur in unjointed and multilayered systems.

Degradation and embrittlement of covering materials is accelerated by

Table 13.3 Defects associated with large-scale multi-layered and jointed roof coverings

Defect	Cause	Remedy
Cracking	Differential movement. Joints in substrate. Lack of consistent support. Severe changes of direction	If extensive re-lay, cut, and bond in a patch. Provide new sub-base. Adopt modified details as BSCP 6229
Small blisters	Expansion of volatile fractions in bitumen can occur with self-finished or mineral felt	If serious apply cut-back bitumen and grit
Large blisters	Expansion of trapped air and/or moisture. (May occur between felt layers or the decking)	Cut out and patch. If persistent, consider ventilation or heavy dressings
Ridging and cockles	Moisture movement of finish, sub-base or structure	Normal treatment is to re-lay. Site dressings may conceal potential defects
Lap or joint failure	Poor bonding. Severe tensile stress caused by blisters or ridging	Cut out and re-bond a patch. Simple re-bonding may be possible
Ponding	Inadequate falls or blocked drainage ways and areas	Remove and re-lay to falls. Additional layers of built-up covering may be sufficient
General deterioration	Incorrect choice of covering. Lack of surface dressing. Poor/inadequate falls. Lack of regular inspection	Re-lay if deterioration is severe. Extensive site dressing can be used to restore exposed surface
Loss of chippings	Severe falls. Wind erosion of poorly bedded chippings. Inadequate depth of chippings	Consider alternatve protective measures. Select larger-sized chippings
Dents and rips	Physical damage	Patch, paying attention to bonding. Area may be subject to traffic
Sticky or semi-liquid surface	Chemical damage	Remove source of contamination. Remove affected covering, patch or re-lay
Water ingress	Joint failure. Inadequate or defective skirtings. Incorrectly weathered joints. Failure where pipes, etc., penetrate the felt Physical damage	Mechanical fixing around pipes, rain-water outlets and skirtings, may be required
Splitting parallel to metal eaves trim	Poor fixing of eaves trim	Remove trim and replace with welted felt drip as BPCP 6229

ultraviolet light, whilst the process is compounded by oxidation. Inadequate bonding between layers can produce blistering of the layered surface. This may be due to poor temperature control during laying, by trapped moisture from inside the building, or dampness driven out from the drying structure. These faults may not always produce immediate leakage, although the additional strain on the coverings may cause progressive joint failure.

Penetrations through the layered systems can be problematic, since they are invariably associated with skirtings and upstands, which are liable to vertical movements and capillary effects. These deficiencies may allow water direct entry through poor sealing between the covering and the penetration. Ponding can allow water to flood into joints above the covering level, and impose new loads on the structure.

Ridging of the layers can occur when the substructure moves, stretching the covering layers, and then contracts, pulling them into ridge shapes. This process can tear the layers where fatigue cracks occur at the top of the ridge. The junction between layered systems and the surrounding construction are sites of potential water ingress. This may be due to differential movement stressing corners, or pulling out upstands. Parapets may provide direct entry paths, as shown in Fig. 13.11, and the associated guttering may flood into roof spaces at times of heavy rain, due to blockages, poor falls or inadequate size. Many problems occur with the placing of damp-proof courses in parapets (and their omission). Often DPCs are omitted from the underside of coping, and throatings to the underside of copings may be omitted. Parapet gutters may be defective. The light rendering to brick parapets, common with 19th-century buildings, is liable to take up water through surface cracking, allowing water penetration into the building.

Failure can occur at the edge of flat roofs due to differential movement between the general roof area and its edge. Where the edge is laid on metal or timber inserts its movement characteristics will be different to those of the general roof area and cracking at right-angles to the edge of the roof can occur. Carrying the roof slab through to the rear face of the outer leaf, or continuation to the outside surface of the wall (often disguised with mosaic or brick slips), can also cause water travel. Impact damage can penetrate the covering layers and may be due to pedestrian traffic, vandalism or falling debris.

Movement stresses and materials

Movement stresses may be induced by the generally high coefficients of linear expansion of many of the bitumen-based materials. Glass fibre and polyester types generally have a lower expansion coefficient, and so have fewer thermal movement problems. These temperature-related stresses act two-dimensionally, producing surface crazing and cracking. Even with moderate roof areas, this can be critical if suitable control measures are not adopted. Defects tend to occur around junctions with surrounding structures and at projections, where stresses are concentrated.

Crazing

Although the internal environment makes a contribution to the temperature drop through the structure, a major contributing factor is solar radiation. Degradation due to ultraviolet radiation can produce hardening of the coverings, which means that the material no longer has the necessary tensile strength to resist even minor stress; consequently any movement of the deck or the structure can cause tensile failure of the surface. This process generally proceeds concurrently with the oxidation of the material, making it difficult to determine the exact cause of the defect. In general, the covering without solar protection is likely to suffer to a far greater extent from ultraviolet degradation than from the effects of oxidation.

Hard and compressible blisters

Inadequate bonding between the three felt layers will produce hard blistering of the roof surface. Poor bonding can be caused by poor temperature control during laying, also by trapped moisture arising from using wet felt, by high vapour pressure inside the building, or damp driven out from the drying structure. These faults do not always produce immediate leakage. An example of the blistering of an asphaltic roof covering is shown in Fig. 13.30.

Fig. 13.30 Blistering to mastic asphalt

The empty, compressible form of blister can be caused by moisture trapped between layers, also by poor workmanship, such as inadequate pressure during application. Materials distorted during storage may produce defective finishes also. Any moisture within the structure will exacerbate the production of blisters by the renewal of internal water vapour pressure.

Ridging

Ridges in the covering material are almost certainly the result of unevenness or movement in the substructure and they will probably follow the pattern of joints in the substrate. Roof deck panels expand and contract during temperature cycling, and these dimensional variations in the deck can cause the covering to stretch, and then ridge up on cooling. Repetition of this increases the size of the ridge and leads to fatigue cracks at the ridge. Similar expansion ridges occur around isolated fixed points such as pipes and parapets. The problem is most severe with warm roof details, especially if the separating membrane is omitted. The insulation layer below the waterproof covering holds the temperature of the covering high, and make it more fluid. Where the roof is trafficked, failure can occur rapidly.

Large-scale single-layer jointed systems

The single-layered systems do not form continuous coverings extending over large areas of the external envelope. Fixings through a one-layer system could easily form a direct path for water entry. Therefore the materials may be of a high relative density, or in the case of lightweight materials, extensive fixings or interlocking methods to resist negative wind loading are used. A reliance on weight of covering is shown by the application of continuously supported metal roofing. General characteristics of a selection of the metals commonly used as continuously supported roof coverings are shown in Table 13.4.

Table 13.4 General characteristics of metals used for roof coverings

Name	BS No.	Relative density	Thermal movement (*10 per °C)	Tensile strength (N/mm²)	Melting point (deg.C)
Lead (milled)	1178	11.3	29.5	15–18*	327
Lead (cast)	None	11.3	29.5	12–15	327
Copper (annealed)	2870	8.94	71	216–355	1083
Zinc (rolled)	6561	7.14	23*	139*	419
			40**	216**	

* Parallel to the direction of rolling
** Perpendicular to the direction of rolling

These materials can allow water penetration where the supporting structure is liable to deflection under the imposed loadings. This may strain joints or tear the material. Ponding on flat surfaces may add to the imposed loads, exacerbating this effect. Where surfaces are to be trafficked, the risk of impact damage increases, and the loadings become difficult to quantify, since they may change with time. Minor indentations may subsequently harbour dust and debris, becoming sites for gradual chemical attack. The use of duckboards will remove the traffic, but may cause concentrated loads from the duckboard fixings. The treatment processes for timber may be incompatible with some single-layer systems.

Expansion and longevity

Metal fatigue can occur, particularly at points of curvature due to the daily expansion and contraction of the covering. Timber decking which has not been laid in the direction of the roof fall can impede water runoff due to the possible shrinkage movement of the timber.

A major factor involved in the longevity of metal roof coverings is the behaviour of the metal surface when exposed to the atmosphere. The concept of oxidation was discussed earlier under metals generally. The minor variations in atmospheric oxygen, when considered over the life of roof coverings, mean that differences on the top surface are minimal. The underside is more likely to experience oxidation variations. In general, the metals used for construction, and roofs in particular, are of types which produce a relatively dense oxide layer. Common defects associated with metal roof coverings are shown in Table 13.5.

The higher the temperature, the higher will be the oxidation rate. This would, with impervious oxides, offer increased depth of protection.

The process of alloying can be used to enhance the oxidation of the parent metal, or to improve its mechanical characteristics. Most roofing metals contain some alloys for workability as well as durability reasons. This text can only serve as an introduction to the subject of the behaviour of metals, and for more information standard sources should be consulted (Taylor 1983a, Everett 1975). The oxidation layer can break down owing to impact damage and environmental pollution. The continual removal and formation of oxides will reduce the thickness of the covering and help to concentrate failure stresses.

Creep

Metals can suffer from creep. This time-dependent strain may become a significant factor when considering water ingress since the time scale can be considerable, and inspections may be difficult. Lead is particularly vulnerable, whereas zinc is more resistant. Creep may open up joints and pull the covering away from the surrounding structure. In severe cases the material may tear, although for this to happen steeply sloping surfaces are

Table 13.5 Defects associated with large-scale single-layered jointed roof coverings

Symptoms	Cause	Remedies
LEAD		
Splits and cracks	Insufficient thickness	Patch if isolated
		Temporary solution
		Replace with thicker sheet
	Underlay softening, causing lead to stick, restricting thermal movement	Re-lay using correct underlay
	Incorrect joint spacing	Welding of isolated faults
		Replace decking and lead where severe
Dents and cuts	Physical damage	Consider removing cause
		Thicker sheet code
		Patch with soldered joints
Edges lifting	Wind action	Increase fixings
		Increase lead code
Surface marking loss of metal	Chemical attack (electrolytic corrosion), from external sources	Re-lay avoiding direct contact with dissimilar metals
Sugaring to underside of metal	Corrosion from internal sources*	Re-lay after removing the risk of moisture
White streaks (corrosion)	Concentrated water flow	Re-lay and divert and/or divert water flow
Movement of lead	Creep of the lead	Provide extra fixings, taking care to reform around rolls
		If severe, consider alternative material
COPPER		
Patina or verdigris	Oxidation, the amount of atmospheric pollution affects rate of patination	Consider how to improve appearance
		No remedial treatment required
Corrosion of fixings, external	Adjacent to cathodic metal, e.g. aluminium	Remove from the cathode
		Provide copper fixings
Corrosion from internal sources	Condensation, which may cause failure of organic substrate, e.g. timber	Re-lay using adequate vapour barriers and durable sub-strata
	Chemical incompatibility with elements in bitumen felt	Remove underlay and provide new inert felt to BS 747, Type 41 A(II)
Cracking around seams and rolls	Over-working causing localised hardening	Remove and replace if severe
		Redressing if faults are minor
Water ingress	Single cross welts provided on pitch less than 45 deg.	Reform to provide double cross welts
		Provide adequate sealing
	Corrosion from adjacent cathodic metal	Remove cathodic metal
		Ensure the electrolytic link is broken
	Movement of sheet edges due to wind action	Remake failed joints
		Consider providing thicker sheets or smaller bay size
Dents and cuts	Physical damage	Consider removing cause
		Provide thicker sheets
Edges lifting	Wind action	Increase fixings
		Provide thicker sheets
ZINC		
Surface corrosion and pitting	Atmospheric pollution	May not be suitable material for heavily polluted environment
		Consider replacement with alternative material

Table 13.5 continued

Symptoms	Cause	Remedies
Internal surface corrosion	Water draining from adjacent cathodic metal roof or fittings	Remove cathodic material
		Re-lay with compatible material
	Condensation	Re-lay, providing vapour barrier
	Corrosion from acidic timber decking	Re-lay using well seasoned neutral timber decking
Water ingress	Leaking welted lap	Reseal joint
	Insufficient clipping	Provide additional clips
	Splits and cracks	Replace defective sheets
	Capillary action under cappings	Remove rubbish adjacent to cappings
	Nailing through cappings and sheets	Replace sheets
Dents and cuts	Physical damage	Consider removing cause
		Consider providing thicker sheets
Edges lifting	Wind action	Increase fixings
		Provide thicker sheets

*The risk exists of water being drawn in through the joints and rolls from outside. Particularly where a warm deck roof construction has been adopted

normally required. Water running over impervious surfaces will produce concentrated flows through any open joints or cracks.

Joints and jointing   The primary joints in the covering are either parallel to the direction of water flow, known as laps, or perpendicular to the direction of water flow, welts. Failure of any of these joints will allow water entry behind the primary defence. Laps can fail where cracks have been induced by the work-hardening process of forming the joint. Sharp corners, as with rectangular zinc rolls, are areas of stress concentration and vulnerable to cracking. Laps can be made into seams where the material possesses sufficient rigidity. Copper roofing commonly uses this method of jointing. These can be damaged by impact loading either being opened up or bent over to lie in the path of water runoff. In this position they can allow water direct entry since the open part of the seam can become a capillary pathway. Welts are likely areas of leakage where falls of the covering are shallow. Water can move through the joint by capillary action, and may pond on the upstream side of the joint as a result of its height. A similar problem can occur with standing seams, as shown in Fig. 13.31.

A failure to provide adequate flashings and soakers at the abutment between covering and surrounding structure can leave direct waterways through the covering. The differential movement problems of fixed upstands may cause these to pull out of the surrounding structure.

Fig. 13.31 A standing seam in an aluminium-covered roof has been flattened by impact damage. Dust accumulations have provided a site for the growth of moss which itself becomes a chronic source of dampness. A risk of further traffic damage from the nails exists.

Corrosion Corrosion can penetrate metal coverings. This may be due to the anodic reaction of the covering material, when linked through an electrolytic cell to a metal lower down on the electrochemical series. Copper is cathodic to a number of other common metals, whereas zinc is strongly anodic. Oxidation which results in porous films or substantial volume changes, will accelerate any perforation of the covering. Moisture and differential oxygen levels at the back of the lead coverings has caused an increase in the number of cases of condensation corrosion since 1974 (Murdoch 1987). It is possible for a difference in pressure to exist between the outside and the back of the covering. This could allow water to be drawn through joints and start corrosion cells at the back of the covering. The supporting and backing structures to single-layer jointed systems may be chemically incompatible with it. This can occur between certain bitumens and copper.

The accumulation of organic matter in gutters and channels may generate acids which can attack the coverings. The joints are particularly vulnerable because of stress concentration caused during cold working. Drips formed at the transverse junctions of coverings can be a source of water entry. Where the height of the drip is insufficient, capillary action may draw water upwards through it. Since there are many joints and junctions at drips there are also many routes for water entry. The underlapping covering layer should be rebated to a depth equal to the thickness of the material, otherwise ponding can occur behind the double thickness of material.

Small-scale lapped systems

The provision of a covering system based on small units will of necessity involve a large number of joints or overlaps. The units keep out the water because they overlap other surrounding units, in both directions. These overlaps are critical areas for water ingress and most covering systems are based on an assumption that water generally flows downhill. This may not be true during high winds, as can occur in areas of severe exposure or extreme height. The provision of adequate lap to the coverings is normally based on the degree of slope of the surface and its exposure.

Angle of creep

Water will enter the small butt joint between covering units. This water will fan out on the underlying surface of coverings, describing an angle of creep (King and Everett 1971). The steeper the slope, the narrower the angle, allowing smaller unit sizes to be used. Where shallow slopes occur water may run into the butt joints of the underlying courses of units and breach the primary defence. This effect is shown in Fig. 13.32.

Pargetting is the pointing of the underside of the tiles to prevent water penetration. This may be due to an excessive angle of creep, poorly nesting units, or excessive wind induced movement. Generally, any examples of this will be in poor condition nowadays.

Covering or cladding

The point at which coverings become claddings is difficult to define on certain building shapes. Small-scale lapped systems can be used as vertical claddings, where a range of decorative effects is possible. The restrictions on laps and sizes may be considered less onerous, since angles of creep are reduced, although water may still gain entry. Wind action and the slight movement of the units contributes to this. The supporting system for small-scale units is commonly battens. Where these are not provided with counter-battens to lift them clear of any additional backing structure, water may accumulate on the upper surface of the batten, leading to rot or damp penetration of the structure. Vertical flashings around openings in vertical coverings must be correctly lapped and linked to the general existing damp-proof barriers of the opening.

Fig. 13.32 Relationship between angle of creep and lap to slope and size of covering units.

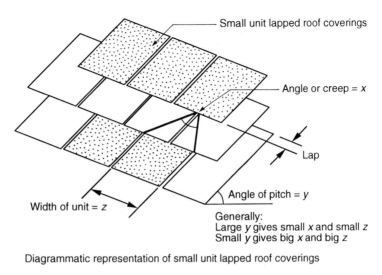

Small unit lapped roof coverings

Angle or creep = x

Lap

Angle of pitch = y

Width of unit = z

Generally:
Large y gives small x and small z
Small y gives big x and big z

Diagrammatic representation of small unit lapped roof coverings

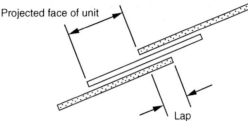

Projected face of unit

Lap

Cross-section through small unit coverings

The units are normally constructed from virtually impervious materials. The permeability that does exist is likely to increase with time as the material ages. Slates are generally very thin and this enables them to follow closely the shape of the roof; unfortunately they can be relatively light and once disturbed by wind action possess good aerodynamic properties.

Location and leakage

Where the location of the unit in the mass of all the other units is incorrect, water may gain direct entry. Location failure can occur when fixings, such as nibs, nails or pegs, fail. Nibs are commonly an integral part of the unit and require adequate hooking over supporting systems. Nibs can snap off when embrittlement and slight movement concentrate stresses at the foot of the nib. This will allow the unit to slide out of position, exposing the underlying structure or any secondary defence. Where no secondary defence exists, the nibs to roofing tiles can be inspected from inside the roof space.

The traditional fixing to tiling battens using nails or pegs of oak or iron can become defective as a result of recurrent condensation on the underside of the roof covering. This can be associated with reduced ventilation between the tile and the underlay, which in turn can lead to increased susceptibility to frost attack. Newer types of breather felts are thought to offer better control of condensation, they could therefore be used for remedial work. The problem increases as the pitch reduces. The use of adequate lapping and support of the underlay felt, particularly at the eaves, is essential if water ingress is to be avoided.

Positive fixing　　Nailing of the units will involve the risk of corrosion. This risk can be assessed in relation to location and consequences (BRE 1985d). Nail failure may contain an amount of mechanical wear due to the movement of the unit. Where one unit moves out of position, this may have an effect on the remainder of the units, causing distortion and reduction of laps. The partial removal of edge support for isolated units may not admit significant quantities of water provided secondary defences are satisfactory. Defects with the fixings may occur as a result of the continual condensation on the underside of the roof covering. Nailing positions are always covered by (two) other tiles and so replacement of tiles requires support from the adjacent (usually lower) tiles.

Domino failure　　Large-scale removal of units, as in the case of severe wind damage, may also breach the secondary defence. This is likely to occur in areas of high negative wind pressure causing the units to be pulled from their supports and thrown off the roof. A domino effect of rows of units being removed can occur. Verges and ridges are particularly vulnerable. See Fig. 13.33.

This mode of failure can occur to other coverings, even those of high relative density and significant positive fixings.

Critical edges　　The junctions between the unit coverings and the surrounding structure are vulnerable to water entry. A range of different construction details are used. Where the coverings meet a vertical obstruction a butt joint will pass through the covering. Where this joint has not been protected with flashings or soakers, water can enter. Flashings from brittle materials, e.g. mortar, may crack owing to differential and structural movement. A failure to adopt effective treatment to all projections through the covering will lead to similar failures. Gaps around chimneys and pipes may produce damp inner walls and trigger corrosion.

Unit coverings made from impervious materials can discharge large quantities of water to the surrounding structure. Where the height of upstands is inadequate, or joints in flashings have failed, water may enter.

Fig. 13.33 Loss of tiles from an exposed open verge/eaves junction caused by wind action. The progressive or domino effect is obvious.

Drainage ways and gutters can accumulate debris washed from the coverings. These can be cementitious in the case of new concrete units. Gutters can become blocked, discharging water either down the face of the building, or in the case of parapet gutters into the roof space.

Other problems occur as a result of the sagging of the roof over the party wall. The brick wall expands upwards, and the timber rafters and wall plates shrink. Similar effects may be produced by spreading roof trusses.

Organic alternatives Organic material, e.g. timber and thatch, can be used for small unit coverings. These are capable of absorbing significant quantities of water whilst remaining relatively impermeable. Permeability will generally increase with age, placing an increasing demand on the secondary defence. The rough surface associated with the degradation of cedar shingles may mean that considerable water absorption will occur. This will have implications for the integrity of the total primary defence, and perhaps where the covering is part of a total timber building, the stability of the structure.

Since they are both plant forms, timber and thatch are water-vapour permeable, whilst resisting the passage of large water molecules. The considerable intercellular space within timber can also act as a reservoir, in much the same way as it would whilst growing. This facility is provided at a cost of reduced thermal resistance and an increase in weight. This latter factor can be critical, particularly where the roof structure has been designed on a traditional/rule-of-thumb method, where joint stresses can rapidly be exceeded. Failures of this type are commonly associated with a combination of effects, since water absorption may be due to heavy rain, perhaps associated with high wind load. Thatch has considerable thermal insulating properties. This is generally due to its intercellular space being empty. A physical failure of the material will normally eliminate this effect, compounding the original defect.

Both timber and thatch coverings carry an inherent design 'defect': they both perform poorly in a fire, and special procedures are necessary before they are adopted into new construction. Birds' nests and rodent harbourage are a problem with thatch, which when sodden will support vegetation. Fungi and moulds may also find suitable conditions for growth in thatch.

Loading and impact failure

The risk of impact damage to sloping unit coverings is mainly from falling objects. These may be propelled by wind action, causing cracking or total impact failure. Falling pieces of masonry or other structural items may be due to weathering and age. Many of the materials used for small unit coverings are relatively brittle, and where this increases with age they become progressively more vulnerable to impact damage.

Changing clay tiles or slates for modern concrete tiles may overstrain existing roof timbers. Where they have not been replaced or upgraded, roof distortion may occur. The horizontal component of the extra weight may also cause problems with the walls.

Drainage

The problems of punctured valley or parapet gutters and blocked downpipes will provide a focus point for water entry to the roof and/or building. Where incorrect depths have been provided, a lack of regular cleaning will increase the risk of overflow. Timber support systems can fail, since they are as vulnerable as the main roof timbers. Projecting features, and their associated joint with the roof covering, are also vulnerable areas. The weatherproofing provision at the rear of chimneys is a common failure zone. The area is difficult to inspect, and may have been designed as a flashing, rather than as a gutter. Additional problems may arise at the junction between a tiled roof and a wall. These joints require to be flexible, and where cement render is used failures will occur unless there is additional protection from metal soakers and/or regular renewal/maintenance.

Deficiencies in tiling   Continuing manufacture of clay tiles ensures availability, although due to the general price level, there is significant pressure to re-use old tiles. Where re-use has occurred any already badly weathered or inherently defective tiles are potential sources of water ingress. Where single lap tiles are used for roof pitches less than 40 degrees, water entry through the primary defence may occur. This is due to the increased angle of creep allowing water to enter directly through the vertical butt joints between units.

Tiles generally   The range of types of roofing tile is considerable, and this is being extended to include patterns suitable for restoration applications. Where tile mixtures are employed, a failure to assess their individual fixing and laying requirements may have produced latent defects. Because of the relatively thin cross-section of concrete tiles, the maximum aggregate should not exceed 4.75 mm. This can be readily checked by breaking open the tile. This is also a useful method of defect diagnosis, where faults, including those listed in Table 13.6, can be examined. A breakdown of the normally uniform internal structure is common with most defects.

Acidic atmospheric pollution will slowly etch cement away from the upper surface of the tile, and this can accumulate in gutters and downpipes.

Repairs/replacement of concrete tiles will modify the appearance of the roof. These generally lighter patches may be indicative of additional defects in the roof other than tile failure. Repairs of single tiles are rarely satisfactory and may themselves become sites for future defects. Where roofs have been repaired, the access ladders may have caused cracking or premature failure of the guttering or edge detail.

Deficiencies with slates   Although deposits are available, UK slate production only accounts for 1.2 per cent of the roof-covering market, with the majority of slate production concentrated in one quarry, Only 2 per cent of all quarried material is suitable for roofing purposes, and most of the splitting work is still done by hand. Slate is imported from the EC, and although generally of good quality it can be variable. Evidence of satisfactory performance in hot and dry climates may be misleading for UK use (*Architects' Journal* 1987a). The traditional specification that the slates, 'should ring true when struck', together with the need for the grain to run longitudinally, provides useful guidance for site inspection. Owing to the inaccessibility of the slating battens, replacement of damaged slates can be problematic. Lead tacks are used, although these can fail as a result of wind action and age. The slates tend to become brittle with age, and this increases the likelihood of breakage of adjacent slates during any replacement process.

Table 13.6 Defects associated with small-scale lapped roof coverings

Defect	Cause	Remedy
Roof units loose and/or slipped	Corrosion of fixings	Replace fixing with durable alternative
	Lack of fixings	Provide fixings, using Al, Cu, or stainless steel, as BS 5534
	Severe wind loading	Replace, check the design with regard to CP3 This can also cause wear around the nail hole, causing failure
	Deterioration of tile nibs	Replace individual tiles. Where nibs are not integral, they may be replaced
	Deterioration of tile battens	Remedy condition for fungal attack Replace defective battens
Sagging of roof	Distortion of roof timbers, due to long-term movement	Remedial work will only be necessary when structurally unstable Take long-term measurements Provide permanent propping
	Roof timbers deteriorating (fungal and/or insect attack)	Remedy cause of attack Proceed as for defective timber Minor cases may be treated with timber patching
	Corrosion of metal fixings	Re-make joints with non-ferrous fixings Patching of gussets may be required
	New roof covering heavier than original	Remove and replace with lighter covering
Roof spread causing cracking	Shrinkage of wall plate	Make good around wall plate
Roof spread causing cracking and outward movement of wall	Underdesign of roof timbers	Provide additional structural members in roof
	New roof covering heavier than original	If severe replace by lighter covering
	Deterioration of timber (fungal/insect attack)	Remedy timber attack, replacing timber where attack is severe Metal rods, plates and adjusting nuts can be used to pull walls back to the vertical

Table 13.6 continued

Defect	Cause	Remedy
Lamination and spalling of clay tiles	Frost action	Short-term replacement is the precursor to total roof replacement
	Low roof pitch	Run-off times are extended Consider replacement
Lamination of natural slates	Atmospheric pollution	Most indigenous slates are resistant Check country of origin Replace separate units although the whole roof may eventually fail
	Frost	Rare. Slates of poor quality Replace
Deterioration of cedar shingles	Fungal attack, due to being located in an area of high rainfall	Remove shingles and replace with preservative treated type

Implications for inner timber structures

The difficulty in replacing structural timbers damaged by water ingress or related defects arises principally because of the need to relieve loading during the operation. There may also be a general reluctance to remove or replace parts of visible elements such as roof trusses, which may have historic value in certain older buildings. The injection of low-viscosity epoxy resins, polyester resins or glass fibre into the timber voids, together with the use of metal dowels, can relieve the timber of the structural load whilst maintaining its appearance. In order for this technique to be effective, the temperature, humidity, cleanliness and timber moisture content must be controlled. The porosity of the timber must also be determined, and this can be variable. In addition, the localised stiffening of the timber may cause movement stresses, and the change in thermal conductivity and vapour permeability could lead to condensation inside the timber at the interface between treated and untreated parts, possibly initiating new decay.

Alternative techniques exploit the high permeability of decayed timber to inject preservative into it. The diffusion of borates is a typical technique. Treated rods are inserted into known damp areas, so that the borates are carried in solution through the timber. This targets the treatment, and where the rods remain in a stable condition, the treatment can have considerable longevity.

High risks and consequential deficiency

**Identification of high
risks**

High-risk areas occur where primary and secondary defences against water penetration coincide at the external surface of the building; also at points in the external skin and wherever the transfer of functional responsibility for water exclusion occurs, such as where roofs discharge the water across flashings or into gutters then from guttering to downpipes and finally into drains. Some examples are shown in Figs. 13.34, 13.35 and 13.36.

Fig. 13.34 Failure of rainwater discharge system. This can occur directly through ingress through roof coverings, by direct bypass of the gutterings, through loss from blocked guttering or at various stages of the downpipe and drainage system.

Fig. 13.35 Fractured or incomplete rainwater pipes allow unsightly and destructive damage to facades.

Blockages in chimneys The primary and secondary defences can be directly breached by blockages in masonry chimneys. These structures must by definition penetrate the outer envelope and the method of exclusion of water must not hamper the flow of the products of combustion within the flue. Although there is a case for providing chimneys with cavity walls, most are of solid wall construction, and this further complicates the methods required to stop water penetration.

Fig. 13.36 Indirect bypass
of guttering leads to chronic
or prolonged dampness in
fascias. This is exacerbated
by flush eaves detailing
which ensures continual
contact with damp rendering
or walls.

The amount and disposition of the solid particles of combustion products
will be related to the fuel used and the construction of the chimney. Many
of the chemicals produced by fuels are aggressive towards cementitious
materials. The sulphurous types are particularly aggressive. When in
solution, these chemicals can attack the mortar joints and linings of flues.
Figure 13.37 shows a typical sulphate attack of a brick chimney. This can
alter the shape of the chimney with a lean away from the prevailing wind.
Mortar joints open up and this will admit more moisture to accelerate the
process. This in turn will affect the integrity of any DPC provision to the
chimney, leading to further water ingress.

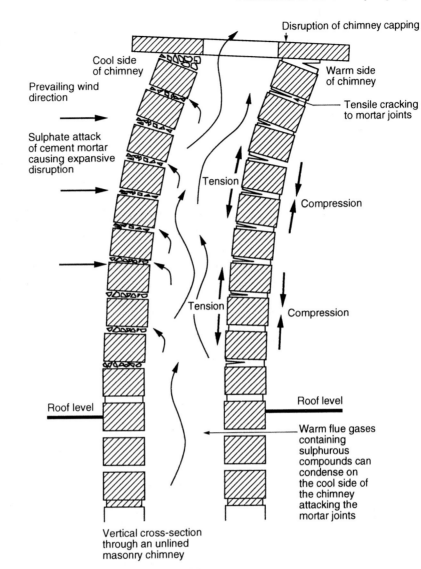

Fig. 13.37 Schematic representation of sulphate attack of brick chimney.

Disruption of chimney capping

Cool side of chimney

Warm side of chimney

Prevailing wind direction

Tensile cracking to mortar joints

Sulphate attack of cement mortar causing expansive disruption

Tension

Compression

Tension

Compression

Roof level

Roof level

Warm flue gases containing sulphurous compounds can condense on the cool side of the chimney attacking the mortar joints

Vertical cross-section through an unlined masonry chimney

Alterations to construction

Alterations to existing buildings may not be a qualitative match. The haphazard out-buildings and extensions, many dating from times of relaxed planning requirements, may cause breaches of the primary and secondary defences of the original building. This may be due to the different standards of construction of the original building and any extensions. The meeting point of the new and old construction may be a butt masonry joint or a nailed metal or wire tie fixed to the existing wall. The vertical joint may offer a water penetration route through the wall, particularly where it is of half-brick thickness. The toothing of new brickwork into existing is made more difficult because of the changes of brick size. A failure to find suitable sizes can mean a structurally and aesthetically poor joint.

The differences in ground loading of extension to existing building may cause differential settlement. This will strain any positive joint between them and may lead to cracking. Where any weatherproofing is provided to seal the joint at high level, this may need to accommodate subsequent movement.

The use of rendering to the extended wall area may breach the DPCs where these are at different levels. This covering over of potential defects complicates the identification of their causes. This is particularly true where a building has been extended and altered many times.

A range of water-ingress problems at high level involve defective roofs and flashings. These may be difficult to inspect. They may also be difficult to see, which may partially explain why they are defective. Simple breaks and omissions are readily identified, but valley and parapet gutters may only fail during periods of high rainfall. The drainage of new or extended roof areas into existing guttering may cause overloading. Blockages can occur in distorted or irregular guttering. The intersecting shapes of sloping roofs demands careful construction of flashings. Where this has not occurred, water may flow into the roof space. At ground level and below, the integrity of the DPC may not be achieved. Where different ground-floor levels occur the clearance above ground level of the DPC may be insufficient to stop water splash or direct ingress. Basement extensions demand waterproof construction, but this may be difficult to link to existing basements, which may rely on thick walls to keep out the water. The use of special seals to construct waterproof joints demands high workmanship.

Bridging of secondary defences

The inner skins of cavity walls may be behind the secondary defence of the cavity, although where the cavities are filled they may play an increasing role in preventing water ingress. The demands of this role will increase with the number of potential water paths. Poor joint filling of blockwork may allow water to accumulate, as well as offering little resistance to penetration. The practice of shell bedding demands that the mortar shells are well sealed. Concrete columns and beams which cross the cavities may do more than create cold bridges. Paths through the secondary defences may be created, as shown in Fig. 13.19. The production of cold bridges is commonly followed by the production of damp or wet and cold bridges. This condensation of water vapour from inside the building may migrate out through the primary defence during hot weather. This can meet any water migrating in from outside, speeding up the water ingress to the building.

The ability of water to flow under capillary action means that even small porous paths can lead to water ingress (Gratwick 1968). Mortar droppings and other waste material may provide suitable paths. Joints around DPCs may provide capillary paths, as may those of cavity trays. Water can build up in the tray and flow through lapped joints into the inner surfaces of the wall (Duell and Lawson 1983).

Flexible secondary defences may fail under roof coverings where wind action may cause their failure, e.g., the tearing of sarking felt. The open joints in coverings should not be laid to admit direct water entry, but isolated alterations and repairs may not provide this protection. The work of birds and rodents may also breach this secondary defence.

Maintenance of coverings and flashings

The maintenance of coverings and flashings is commonly carried out when a defect has progressed to such an extent that water has entered the building. This may not occur until flashings and upstands have been pulled apart by the differential movement of roofs or structure. Access to the upper reaches of buildings is essential to assess the performance of coverings, flashings and upstands. The use of binoculars may be useful but is no real substitute for close access. Physical inspection may accidentally cause some impact damage, and all necessary safety precautions must be carried out. The inspection of flat roofs is particularly important following the extensive range of defects associated with them (Eldridge 1976). The recent corrosion of metal roofs (Murdoch 1987) would seem to suggest that even the traditionally durable materials require regular inspection.

The degree of failure of the flashing may mean that only sufficient water enters to wet but not penetrate the primary or secondary defence. The wetting and drying cycle may persist for some time before water actually emerges inside the building and is seen. It may be that water will never enter the building to an extent that can be detected visually. This tolerance of marginal failure will depend on the type of construction. Lightweight forms may have a reduced tolerance, whereas multi-layers offer an increased tolerance.

Where the edges of flashing or upstands are exposed to high wind speeds, as is common at roof level, the action of the wind may open up failed joints. Continual flapping may cause failure of the flashing material.

Deterioration of materials

The quantification of the frost resistance of new materials can be difficult, since this varies with exposure and is difficult to simulate accurately. Where materials have a proven history of adequate performance, care must be taken in selection since minor alterations in position, orientation and their relationship to other materials can have serious consequences. During the laying of masonry units frost protection is essential for the maximum strength of the mortar and for its long-term durability. Faces attacked by frost can spall away, with mortar faces generally more vulnerable immediately after construction. Weaker, more flexible lime mortars are more liable to attack than the stronger cement types. Lime mortars set slowly and develop strength fairly slowly, increasing the likelihood of frost attack. This defect is generally evident in the early life of the building since underlying mortar layers are more likely to be frost resistant as they develop long-term strength.

Clay tiles on roofs are exposed to severe frost effects, including the night-time heat loss problem. There is some concern that clay tiles used 50–60 years ago appear now to be coming to the end of their natural life. Where pressed types are used these can delaminate, whereas the traditional hand-made pressings have lower strengths, but appear to have a greater hygrothermal tolerance.

14 Penetration of the envelope by gases

Gas penetration generally

Air and gases

A number of gases may penetrate the external envelope. The movement of air through the structure can be beneficial when it brings with it low relative humidity and high oxygen levels. Although this may produce draughts it will keep internal relative humidity levels under control and assist in drying out damp structures. The air can also be the carrying medium for fungal spores, dust and debris, which can have an adverse effect on the building envelope. Their accumulation can lead to a range of defects, including timber decay, low surface temperatures and excessive weathering of external surfaces.

Low temperatures and moulds

The movement of low-temperature external air into the envelope will reduce its temperature, and can lead to the incidence of internal or interstitial condensation. Where internal air has a high relative humidity, due to the occupancy, condensation is more likely. Stagnant air pockets in the external envelope can create suitable conditions for the growth of moulds, particularly where they have high moisture contents or are in ambient conditions of high relative humidity (BRE 1985e).

Explosive and toxic gases

Explosive and toxic gases can also penetrate the structure. The hazardous conditions generated inside the building may have little effect on its structure. This can alter in the case of explosive gases, e.g. methane, which can also be harmful to any occupants. The seepage of radon from the surrounding ground is another example of gas penetration. The Institution of Environmental Health Officers estimate that between 50,000 and 90,000 dwellings have radon concentration levels that require some action being taken in a few years (Thorn 1990).

Condensation mechanism

Incidence of occurrence

The incidence of condensation in houses in the UK has been estimated by the Building Research Establishment. The report suggested that 1.5 million houses are affected seriously by condensation and a further 2 million suffer slight condensation. Condensation is likely to occur in virtually any building where an amount of moisture is added to the air by internal operations/occupants, since 'cold' regions are always likely to be present. It is therefore the magnitude and distribution of the defect that is important.

Mechanism of condensation

The essentials of the mechanism of condensation revolve around the differential capacity of air for moisture according to its temperature. At higher temperatures the humidity capacity of the air is markedly higher than at low temperatures. Relative humidity is an expression (as a percentage) of the ratio between the vapour pressure of the air alone and the total vapour pressure when it is saturated. Air is saturated when it is holding its maximum amount of moisture at a fixed temperature. It is not possible for the air to absorb any more moisture unless its temperature is increased.

The interrelationship between the physical properties of the air can be calculated by the use of formulae (Smith, Phillips and Sweeney 1987) or shown diagrammatically on a psychrometric chart.

Likely locations

Warm, damp locations in buildings, frequently the kitchen or bathroom areas, can produce high moisture-loaded atmospheres. When air is sensibly cooled its relative humidity will increase until it is totally saturated. The temperature at which this occurs is termed the dew-point temperature, and is unique for a range of air/moisture mixtures. This can occur at relatively cold surfaces within the room of origin such as single glazed windows or tiling at wall or floor level. See Fig. 14.1.

Since a reduction in temperature is needed for condensation to occur, the provision of insulation within any structure will also provide additional potential sites for condensation. This may be within the structural envelope, or within a moisture vapour porous material. This is termed interstitial condensation.

The influence of air flows

Where there is a sufficient stack effect or ventilation within the building to carry the moist air away from the source, it may condense some distance from the origin of the moisture. This may be a remote, poorly heated corner of the building. This transmission of the moist air throughout the building is also assisted by the pressure developed by the increased amount of

Fig. 14.1 The evidence of paint damage may be related to the severe condensation on the glazed area.

moisture vapour of damp air. This means that it is likely to displace the colder air as it moves towards it. Consequently the moist air movement through the building enclosure is dependent on the available ventilation routes. Depending on the vapour porosity of the structure, the humid air may disperse into it.

Contributory factors

Primary production factors

The contributory factors affecting the occurrence of condensation are primarily those which dictate the production of moisture. These include appliances such as dishwashers or washing machines, showers, (unvented) portable heaters and animal or human occupants. The home sauna can be considered a high condensation risk area. Indoor drying of clothes is problematic and obviously tends to occur mostly at the cooler times of the year. This is when there are also particularly damp external conditions, which themselves encourage condensation. Industrial interiors may have open steam baths, seed germination rooms, open warm-water tanks and other moisture emission sources. The growth in the number of indoor swimming pools and leisure centres with extensive changing/showering facilities has created a range of potential condensation problems.

Occupancy and moisture production

In this context the amount of condensation will depend on the number of people using the building, the processes going on within it and the pattern of usage. There is an important link here with the heating regime and the physical characteristics and response of the structure to temperature change. This will be a particular problem where almost instantaneous high air temperatures can be achieved before the temperature of a heavyweight structure can catch up. Well-insulated lightweight structures which have no effective means of controlling/reducing the moisture content of the air can also be problematic. The prevailing weather conditions will also have an influence of the occurrence of condensation.

Condensation and changing lifestyles

The widespread incidence of condensation in domestic buildings is a by-product of the mid- to late 20th-century lifestyle. It can occur in a number of different building types from a number of eras. In the thermally heavyweight structures of the early 20th century there is a distinct possibility of the structure becoming out of phase with the external environment. For example, a sudden change from dry cold weather to warm humid conditions may find the construction wanting, and condensation patches will emerge (Seeley 1987). This gives rise to the potential for inter-stitial condensation. In the times when these buildings were constructed this was not particularly important since the convention was to ventilate thoroughly. This was partially because of the era and also because of the lack of tightness of the structures anyway. In a sense then condensation is the by-product of the tightening up of the building enclosure.

The contribution of construction materials

In addition to this there has been a move away from the permeable heavyweight materials that would have absorbed some of the dampness and buffered against surface effects. The modern materials are dense, highly finished materials, commonly designed to be impervious to moisture and wipe clean – particularly in the areas of the building most prone to surface condensation occurrence. Condensation is therefore more likely to occur and the quality of internal finish is less tolerant to slight deficiency. In particular, the move from the absorbent plasters and permeable paints has exacerbated surface condensation.

Patterns of use   An important feature is the use of the building. In households where the house is unoccupied for long periods of time and then subjected to extreme changes in environmental conditions, such as heating, cooking and washing, the structure is particularly prone to become out of phase with the internal environment. Condensation is likely to occur. This can be contrasted with the use of buildings in the early part of the century where dwellings were rarely unoccupied and any heating was usually more moderate and consistent. It was likely that the structure could approach thermal equilibrium and to some extent be in harmony with the ambient external environment.

Consumer durables   There are other major factors included in the equation that have led to an increase in the incidence of condensation. These include the increase in the number of consumer durables which produce moisture in particular, the move to carrying out high-temperature washing inside the house, and the use of tumbler driers and showers. In conjunction with the extra production of moisture in the air by mechanical means there is the not inconsiderable production of moisture from the normal activities of people, although it is perhaps too early to say whether occupants are more or less active than in previous generations.

Insulation, ventilation and condensation   In the mid 1970s the oil crisis precipitated the widespread reassessment of energy requirements and the use of insulation on a widespread basis became confirmed. At the same time the ventilation levels acceptable in homes were reduced and people began to live in tighter enclosures. Double glazing was installed to increase the thermal resistance of the building envelope and a side-effect of this was that the ventilation rates in homes were markedly and generally reduced. Indeed, the 'draught-proofing' of openings is seen as an important part of any energy-saving plan for a building. Unfortunately, the problems of condensation associated with this measure do not appear to be adequately stressed. Two common methods of overcoming the problems of condensation are to open a window or to turn up the heating system. Both methods will increase fuel usage to maintain a standard internal ambient temperature.

The removal of chimney breasts   The removal of chimney breasts is a move which has continued throughout the 20th century. It is common to remove chimneys from the design of domestic buildings, to block them up or remove them completely. This creates a possibility of the sealed chimney filling with stagnant air, which can lead to condensation. Also there is much reduced ventilation of the moist internal air since the open flue would have been a major influence on

Fig. 14.2 Tell-tale of previous chimney. Note also pressed tile roof of a different era to the main structure of the building.

its natural exhaust ventilation. See Fig. 14.2.

In addition to the reduction in the ventilation provided by chimneys, there has been a reduction in the ventilation provided by windows in the external envelope. The modern window is much more efficient at limiting air movement. Indeed, the newer standards and specifications appear to be placing increased emphasis on air and water exclusion. The regulatory criteria for the provision of ventilation openings may bear little relevance to the actual ventilation opening used by the occupier. The adoption of draughtproofing as a means of reducing the cost of heating and minimising the discomfort of draughts may have all but closed off air movement and eschewed the general air change conditions in the building.

The misapplication of thermal insulation

The satisfactory performance of a building depends on its overall thermal character rather than merely the levels of insulation and the disposition of any insulation material. Indeed, it is the misapplication of thermal insulation that has been responsible for many of the hygrothermal problems occurring in the last 25 years or so (Allen 1989). There are a number of related reasons for this, and most of the defects associated with its misapplication involve moisture or thermal movement problems. There may be secondary aspects of health and ventilation also. It is recognised that correctly applied insulation can have benefits for many buildings. The application of the science of heat flow and psychrometrics to the technology of insulation is essential for satisfactory building performance. This has been seen to fail in a number of ways, including the omission of vapour barriers, blocked or inadequate ventilation and the bridging of cavities (*Architects' Journal* 1986c).

In some instances the misapplication of the insulation may be involved directly in the fault; in other cases it acts as an agent. Frequently the relationship between the material and the existing structure is not given due regard. The insulation of walls to keep the structure warm can have several side-effects. There is firstly an increased propensity for condensation occurrence on the colder surfaces. The provision of insulation is likely to affect the pattern of distribution of condensation around the building. Highly insulated areas will act to concentrate heat flow through areas without insulation, the cold bridges. Before these areas warm up they may cool down the moist air below its dew point and cause condensation.

Positioning of thermal insulation and defects

The positioning of thermal insulation can make a significant effect on the response of the building to its internal and external climates. Examples of roof construction are shown in Fig. 13.25. The examples of the warm and cold roof details as used in the last 40 years can produce a range of different response patterns. These show an evolutionary trend of placing the insulation further and further way from the inner surfaces of the external envelope. The trend appears to have been driven by a succession of defects centring on condensation adjacent to the insulation. The positioning of insulation anywhere within an external envelope will give a similar thermal transmittance value, but may produce a range of potential defects. The provision of rooms in the roof space of dwellings demands special consideration with regard to the positioning of insulation (Pitts and Yeomans 1988). A failure to ventilate the insulated sloping roof/ceiling may lead to condensation problems. This is complicated by the relative narrowness of the construction at this point.

Cold construction

Where insulation is placed in the region of the secondary defence close to the internal finishes, this will be the region where the greatest temperature

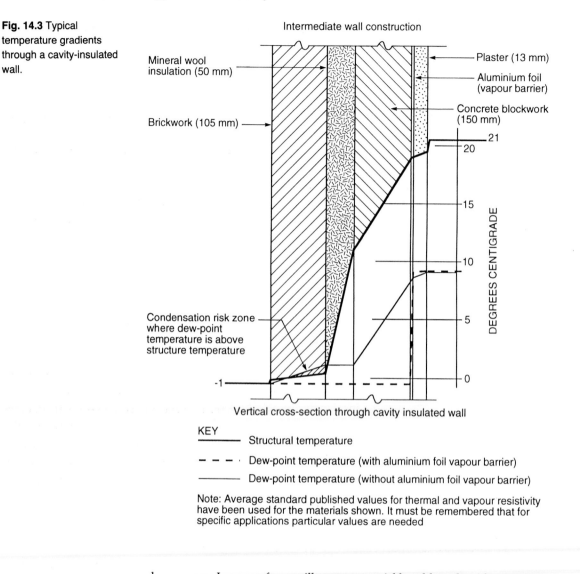

Fig. 14.3 Typical temperature gradients through a cavity-insulated wall.

Intermediate wall construction

Mineral wool insulation (50 mm)

Plaster (13 mm)

Aluminium foil (vapour barrier)

Concrete blockwork (150 mm)

Brickwork (105 mm)

Condensation risk zone where dew-point temperature is above structure temperature

DEGREES CENTIGRADE

Vertical cross-section through cavity insulated wall

KEY
————— Structural temperature

– – – · Dew-point temperature (with aluminium foil vapour barrier)

————— Dew-point temperature (without aluminium foil vapour barrier)

Note: Average standard published values for thermal and vapour resistivity have been used for the materials shown. It must be remembered that for specific applications particular values are needed

drop occurs. Inner surfaces will warm up quickly, although without vapour control the risk of interstitial condensation increases. Outer and primary defence regions are likely to experience temperature changes in tune with the external environment. A typical temperature gradient through a cavity-wall construction is shown in Fig. 14.3.

Warm construction   The placing of insulation close to/within the primary defence allows the secondary defence and finishings to move towards those temperatures common in the interior of the building. The outer regions of the primary defence outside the insulation are likely to experience marked temperature differences. Low external temperatures will keep outer surfaces colder for longer and may lead to increased frost attack. High external temperatures will similarly keep external surfaces warm.

Buffered or inverted construction

Insulation totally outside the primary defence can be vulnerable to impact damage. In this position it will buffer the external envelope from major external temperature swings. This effectively creates a warm external envelope. The massing of the structure will influence the time taken for this warming to be achieved since it will extend the time lag of the structure. Rapid and intermittent heating of the interior coupled with high relative humidity levels may produce inter-envelope condensation, although this form of construction appears to offer a reduced risk (*Architects' Journal* 1986b).

Surface condensation

Surface condensation

Surface condensation is relatively easy to identify, although it is important to distinguish any symptoms from those of rising or penetrating damp from an external source. The two may occur in combination, particularly in older properties which do not have a damp-proof course. On paintwork the appearance of condensation may produce a range of coloured stains, including pink or purple. The occurrence of surface condensation is associated with surfaces within the enclosure that are at a lower temperature than the dew-point temperature of the air.

Cold bridging and condensation occurrence

The occurrence of surface condensation is particularly likely to occur in areas of the building enclosure exposed to high heat transmission, termed cold bridges. These are common around openings in the structure where poor constructional detailing may lead to a closing of the reveal around the opening which has a higher potential heat transmission than the remaining structure. Examples of cold bridges are shown in Fig. 14.4.

Although predictable (Burberry 1988), the calculations are based on steady heat flow rather than the transient conditions critical for the occurrence of condensation. Another common cause of cold bridging in the structure is the lintel details recommended in textbooks of the 1950s and 1960s, which show concrete lintels bridging the cavity and creating a path for heat transmission. Some of these are still being included in current textbooks. Reinforced concrete framed structures may provide cold bridging where floor beams and columns are placed in the external envelope.

Fig. 14.4 Examples of cold bridges in the external envelope.

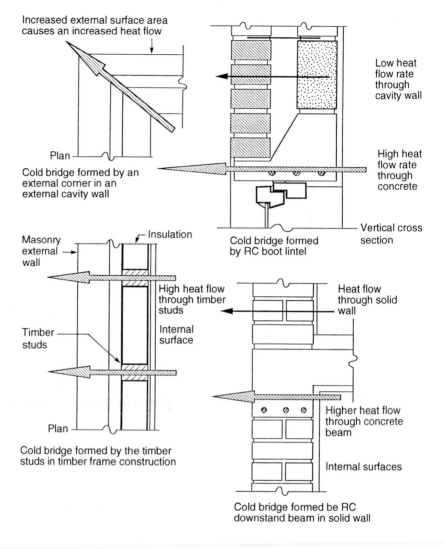

Increased external surface area causes an increased heat flow

Plan

Cold bridge formed by an external corner in an external cavity wall

Low heat flow rate through cavity wall

High heat flow rate through concrete

Vertical cross section

Cold bridge formed by RC boot lintel

Masonry external wall

Insulation

High heat flow through timber studs

Internal surface

Timber studs

Plan

Cold bridge formed by the timber studs in timber frame construction

Heat flow through solid wall

Higher heat flow through concrete beam

Internal surfaces

Cold bridge formed be RC downstand beam in solid wall

The significance of condensation

Condensation rivals rising damp and rain penetration as a significant dampness problem. A significant occurrence of condensation is reported in the system-built concrete structures of the 1950s and 1960s. Some of these constructional designs were imported from the continent and were designed for a distinctly different climate than ours. The Scandinavian example of highly thermally insulated buildings is effective, firstly because they have recognised the need to conserve scarce natural fuel resources and secondly because of the general uniformity of their seasonal weather. Where these systems were directly imported into the rapidly variable UK climate the thermally heavyweight construction easily became out of phase with the rapid ambient temperature fluctuations.

Interstitial condensation

Location of interstitial condensation

The insulation of previously permeable structures with impermeable cavity fills produces a significant change in the nature of the external envelope. It changes overnight from being a moisture transmission path to a moisture-resistant barrier. The important feature of this is that the barrier is not at the surface; rather it is within the structure. In such cases the moisture-laden air is free to permeate through the permeable inner leaf of the wall until it meets the insulation layer. Here there is a drop in temperature associated with an impervious barrier. Condensation will only occur where the temperature gradient in the wall is below the dew-point temperature of the air within it. This is shown in Fig. 14.3.

Occurrence of interstitial condensation

The occurrence of interstitial condensation is by no means restricted to walls. There is ample evidence of the occurrence of condensation in the cold-roof designed flat roof of the 1960s. A significant contributory cause of interstitial condensation is the absence of vapour barriers. This is a frequent constructional problem, and was rife in the early constructions of cold flat roofs and some forms of timber-framed buildings. Where the vapour barriers are punctured by fixings or impact damage their effectiveness is reduced. Although these may be small areas they may still permit an accumulation of moisture. Thermal pumping can also operate through these poor seals.

Porous materials and interstitial condensation

When interstitial condensation occurs within a porous insulation material, the condensed water will be absorbed. This will reduce the thermal resistance of the material as well as create the potential for dampness problems within the primary and secondary defences. These defects may take some time to fully manifest themselves, during which time they may trigger additional problems.

Implications for the structure

The implications of this for the structure are that the building envelope becomes damp and transmits heat well, and there is the moisture and temperature gradient within the structure necessary for bacteriological or fungal growth. There is a general risk of an unhealthy environment.

Surface and interstitial: the condensation commonality

The defects associated with surface condensation are concerned with moulds and the attack of surfaces and finishes. The interstitial sites of condensation can cause structurally-affecting dampness and marked reductions in thermal transmittance.

Both effects are physically common, with water condensing out of the air. The main difference is the position of the dew-point temperature of the air in the envelope. The dew-point temperature is likely to be reached at or near the insulation, except where very high relative humidities exist internally. This sweeping across the envelope of the position of the dew-point temperature will influence the range of potential defects. When close to the outside, weathering and frost problems are more likely. Inner positions may soak insulation and where inner surfaces are wet, hygiene problems can occur.

Identification

Distinguishing condensation and damp penetration

The distinction between condensation and dampness caused by externalities is important, since the remedial actions are markedly different. In the case of surface condensation the quick solution may be an absorbent paint with a fungicidal additive. The long-term solution may involve a change in occupancy pattern of the building. The point at which a lifestyle is incompatible with and creates defects within a building is not easily defined.

Indicative patterns

Examine the dampness for any evidence that it is a pattern staining that may indicate wall ties creating a damp bridge. In such cases the patches of dampness will be localised and evenly spaced, about 450 mm vertically by 900 mm horizontally in modern properties or a little wider spanning in the older structures. It may be possible to establish the surface effect by using a protimeter to analyse the dampness of the structure just below the surface, although readings can be complicated by the action of deliquescent salts at the surface. These will absorb and evaporate water depending on the ambient conditions.

Evidence of rising damp

Examination of a wall suspected of suffering from condensation-related defects should also include examination of evidence for rising dampness. See Fig. 13.23. This usually produces a seismographic pattern starting at floor level and spreading up the wall to a distance of up to 0.70 m. High-level dampness is unlikely to be rising damp, but may be penetrating dampness from a failure of the external envelope or concentrated water discharge. Internal corners can be sites of cold bridging and stagnant air pockets; this may result in surface mould growth and dampness. The possibility of damp air movement through a building to a suitable site with the correct dew-point temperature for condensation to occur should not be overlooked. Air movement can occur through floors, and internal and external walls.

Leakage of the structure

Check also in cases of high-level dampness for possible leakage of the roof at eaves level, or any blockage or cracking in the downpipes. Problems with the placing of cavity insulation batts in modern houses can channel the water across the cavity and concentrate it into the inner leaf. In such cases the dampness may be related to the external environment and appear after rainstorms or in persistently damp weather. The need to examine all possible causes of dampness in order to determine the true cause is the ideal model to be applied to all building defects.

Distinguishing features of condensation

It is unlikely that condensation will produce distinct or sharp-edged damp patches, compared with leakages (Seeley 1987). In new buildings during the drying-out period there is a possibility that similar symptoms to those of surface condensation may occur. Seeley notes the difficulty of convincing houseowners that this period can extend to three years (Seeley 1987). The occurrence of dampness in corners may be exacerbated where these are in externally exposed positions, since they will lose more heat. Impervious coatings may produce a film of water or even generate droplets. In such instances there is a definite health risk of mould/fungal attack indoors. Alternatively, it is possible to estimate the likelihood of occurrence of condensation using temperature and humidity measurements (BRE 1985b).

Routes for interstitial condensation

Water accumulating within the structure could have moved in horizontally from outside through failed primary defences. Failures at roof level can permit vertical movement of water into the external envelope. Both can produce symptoms very similar to those of interstitial condensation. Rooms where a high relative humidity is likely may also be surrounded by an envelope with a similarly high moisture content.

Radon

Radon in buildings

Radon is a feature of the background radiation levels. It is a dense gas released as uranium-238 (present in the ground and traditional building materials) decomposes. Radon is radioactive, and a risk to health (Bennett 1987) has been long established from mining incidents. Across most of the UK the background radiation level is relatively low and the risk of cancer is correspondingly reduced.

Entry into the building

The gas passes through any permeable layers in the structure from the ground and into the internal spaces, particularly from the floor. The dissipation of the gas within the internal spaces will of course depend on the ventilation rate, and where this is low there is the possibility of build-up and increased exposure of the occupants.

Radioactivity mechanism

Radon embedded in the materials such as concrete and bricks is a primary source of radioactivity. As it decays it produces penetrating gamma rays. This radioactivity will pass through building materials and the occupants are directly exposed. The decaying radon gas releases polonium, another carcinogen. The mechanism involves dust carrying the polonium being breathed in and settling in the upper lungs. Alpha rays released from the collecting polonium damage the skin and produce cancer.

Findings by scientists at Bristol University link radon to 2500 prostate cancer deaths each year, and 200 leukaemia deaths per year (Chevin 1990). The levels of radiation involved are many times greater than those allowed for radioactive discharge.

Construction factors

Factors that will increase the risk are floors to houses which have no barrier to the radon gas, exacerbated by any features of design or use that minimise the ventilation rate. Common examples include insulated walls, which have a reduced permeability, and draught-proofed openings or double glazing (Hollis and Gibson 1986). Dense concrete floors with a low vapour permeability or floors with effective vapour barriers should be of minimal agency to radon transmission, but this will be affected by any cracks and openings in the floor slab. Consequently, it can become the biggest single source. The external background levels in the air will also mean that some gas will be introduced into houses which themselves are not contributory sources. The materials used in the construction of traditional building products may contain uranium and decay to release radon and radioactivity directly into the house.

Obviously the nature of the ground under and surrounding the building will be important. High uranium concentrations occur in the Highlands and Cornwall, and the presence of significant concentrations of radon and its by-products is likely.

15 Deterioration in coatings

Paintwork generally

External paintwork can make a real contribution to the durability of buildings. Its importance is therefore much deeper than merely presenting a good appearance, and deficient paintwork is commonly indicative of deeper problems. Paintwork is heavily reliant on thorough and appropriate maintenance, however. This is coming under increased pressure due to the increased costs of labour and materials. In response to these pressures property owners and managers may extend the maintenance period past the traditional five-year cycle of stripping and repainting. The potential of the modern environment to damage coatings and building structure alike has imposed a greater criticality on paintwork. Durability is no longer just a function of the product, the preparation of the substrate and its application.

The immediate consequence of deferred maintenance is diminished protection, leading to progressively greater damage to the property. Furthermore, the cost of renovatory work may be increased, since paintwork in an advanced state of decay is more difficult to prepare and remedy.

The durability of any paint covering can be related to the quality of preparation of the substrate, the correct choice of paint and its application. The temperature and humidity conditions during application of the paint are important, as is cleanliness of tools. A key difference between paint products is the type of thinner used. The properties of some of the major types are shown in Table 15.1.

Paints vary in their tolerance of adverse conditions and application methods (May 1986). Common problems caused by inappropriate choice of paint arise with incompatibility between the new paint and the substrate or existing coating. Although the trade is less likely to make these mistakes, the DIY approach may inflict secondary defects.

The external timber details requiring most frequent repainting are window sills, and the lower parts of frames and sashes. End-grain timber, such as occurs on window sills is very often neglected when it comes to applying preservatives or painting. This part of the window is quite critical, since the endgrain timber acts like a straw and will draw in water very efficiently. The other sites on timber sections where the paint system tends to fail prematurely are usually at sharp edges. A range of typical defects

Table 15.1 Comparison of the properties of the common types of paint

Properties	Type	
	Water-based (emulsions) and Acrylic* (emulsions)	Solvent-based Oil paints and alkyd based[†]
Flammability	Non-Flammability	Some types flammable
Fumes	Low	Medium
Cleaning	Simple	Elaborate
Recoat	Within 24 hrs	Depends on the volatile evaporation range
Drying	Not rapid	Can be rapid
Build	Low	Medium to high
Durability	Moderate	Very durable
Corrosion	A risk	Little risk
Adherence	Poor to oily substances	Poor to friable and oily backgrounds
Colour-fastness	Weak colours	Ultraviolet degradation can affect certain types
Permeability[‡]	High	Low
Flexibility	Moderate	Moderate to high

* Acrylic emulsions are claimed to be flexible, durable and give a permeability which can allow moisture/vapour pressure equalisation and not permit soaking of timber

† Polyurethane-modified alkyds are relatively brittle and may not cope well with external conditions where significant thermal and moisture movement of the substrate occurs

‡ Microporous paints have been developed to produce an 'ideally' permeable external coating. These may be water or solvent borne. Overpainting will desroy this property. They generally weather by erosion making maintenance straightforward

associated with the painting of timber is shown in Table 15.2.

There are many causes and effects of paintwork deterioration, not just related to the product and its application, but including the conditions in use of the painted surface.

Paintwork: background to defects

Defects in paintwork can arise from a plethora of causes, which commonly occur in combination. The main agents of decay are exposure to sunlight and rain. Pollution in the atmosphere and a range of instabilities of the background can also occur. Additional water-related problems will arise during the life of the coating if condensation occurs behind the paint, or if surface defects allow water to penetrate the surface coating on a large scale. The range of defects occurring with paints are widespread, producing a variety of symptoms. See Fig. 15.1.

Table 15.2 Painting defects on woodwork

Painting defect	Possible causes		
	Preparation of wood	Application	Exposure
Blistering	Damp or unseasoned wood; liquid or vapour beneath coating Knots not properly treated	Paint poorly applied	Excessive heat
Peeling Poor adhesion Flaking	Damp or unseasoned wood Surfaces not properly cleaned, powdery or friable	Paint poorly applied, particularly the priming coat, or omission of it or use of unsuitable primer Too long between coats	
Irregular cracks	Damp or unseasoned wood Paste or size left on wood	Hard drying paint applied over soft coatings	Excessive heat
Chalking Powdering	Early coats in system may have failed to satisfy porosity of substrate	Could be incorrect or unsuitable formulation	Lengthy exposure, possibly severe
Insufficiently opaque ('grinning') Colour uneven	Surfact not cleaned	Paint poorly applied Undercoat too thin or uneven	
Resin coming through	Knots not properly treated Very resinous knots not cut out	Too much resin for knotting to seal	Resin softened by heat
Delayed drying uneven drying	Surface not properly cleaned Residues of paint removers. Painting on creosote without sealing	Finish applied before undercoat completely dry	Damp, cold or frost
Discoloration	Painting on cresote or bitumen without sealing Surface not properly cleaned Knots not properly treated Stains from various species of timber, either as solid, ply or veneer	Possible use externally of colour intended for internal use	Lengthy exposure to bright sunlight Chemical attack

Table 15.2 *continued*

Painting defect	Preparation of wood	Application	Exposure
		Possible causes	
	Pigment dyestuffs, generally reds or maroons, melt when burned off, are absorbed into the wood and discolour the paint system, unless sealed		
Poor gloss	Alkaline materials left on wood	Paint poorly applied Still, cold air in unheated rooms	Damp, fog or frost; lengthy exposure
Bloom–hazy or white appearance on gloss		Gas or paraffin heating during drying	Condensation
'Sinkage'– patchy low gloss		Unsuitable undercoat Finish applied before undercoat sufficiently dry	
'Sheeriness' –uneven gloss		Paint not properly stirred. Failure to maintain a wet edge Vigorous brushing of matt paint in dark colours	
'Wrinkling'– loss of gloss	Can also be caused by excessive retention of preservatives or their solvents, owing to insufficient drying out	Paint dries too quickly Skin-drying of paint film Mixing resin-based paints with oil and other paints Adding unsuitable solvents	Excessive heat

Flaking, peeling and cissing

Flaking occurs with the loss of adhesion between the paint layer and substrate, producing a breaking away of small particles. All or some of the paint layers may be affected. Flaking is commonly an external problem, which tends to arise with hard-drying paint types, or any paint applied to a friable surface. Timber backgrounds can cause early flaking of paint because

Fig. 15.1 Advanced damage
to paint coating in
high-exposure area.
Evidence of overpainting an
oil-based gloss with water-
based emulsion as a
makeshift repair.

Fig. 15.2 An example of severe loss of covering and peeling to an external metal door. Note the concentration of peeling on ledges.

of entrapped moisture, and this problem is particularly likely at the joints which will be the sites of maximum movement. Timber that is highly grained or well weathered is likely to produce an unstable background. As differential movement across the grain or within the weathered timber occurs, this will set up surface stresses which will destroy the integrity of the painted surface. Layers may flake off following the pattern of the grain, so illustrating the nature of the stresses. Problems may arise at the interface between the painted surface and the substrate, as shown in Fig. 15.2.

Paints bond less effectively to non-porous surfaces, and some oily hardwoods such as teak are troublesome. Flaking and other bond-related faults occur more easily with these surfaces. The problems with flaking and detachment from metallic surfaces may be related to residual millscale, a galvanic coating, or simply excessive thermal movement of the substrate.

Peeling is a large-scale form of flaking produced by differential movement. It can be caused by the paint layers on their own, as a reaction to other (background) paints, or between the paint coating and the substrate. This is distinct from cissing, where layers 'roll off' the background. Cissing is produced by the incomplete wetting of the background and therefore patchy adhesion (Melville and Gordon 1979).

Bubbles and blisters Bubbles in the painted surface can also occur with hardwoods. These are the result of entrapped moisture, air or other gases, or from resins and crystalline salts. This can be termed 'blistering' and affects the dried film. It is worse with the dark-coloured coatings which reach more extreme temperatures. The formation of small blisters or bubbles in the wet paint surface can produce pinholing of the surface. Energetic mixing of the paint or vigorous application froths the paint, and small bubbles are created in the wet surface. These dry more rapidly than the surrounding film and the bubbles become entrapped in the finished coating.

Overpainting Overpainting existing coatings rather than removing them can produce a range of problems. See Fig. 15.1. These include incompatibility, poor base surface quality or integrity, cleanliness or extent of preparation. Different types of paint may cause differential stresses, and the additional weight of the new coating may overstress an old, friable base. Particular problems arise between old and new coatings of oil- and water-based formulae.

Checking and crazing Poor-quality paints may develop 'checking' in the form of Vs or 'crows-feet' shaped cracking. This is very fine and itself does not immediately threaten the protective properties of the film, but left unattended can lead to 'crazing' or 'alligatoring'. Crazing defects arise most frequently with paints based on or containing bitumen. They occur when hard coatings are applied to soft backgrounds. As these dry they shrink and distort the background. The addition of excessive driers can produce the same effect. Patterned cracks form in the dried paint coating, but the defect does not always penetrate the film completely. The cracks are raised above the surface, producing an alligator-skin effect.

Erosion and chalking Mildly-eroding surfaces exhibit 'chalking'. At this stage, this may not significantly affect the protective properties of the film, but is characteristic of older or over-thinned paints. The binding agent is oxidised, accelerated by sunlight. Interior paints used in exposed locations are susceptible, as are highly pigmented spirit-borne paints (Johnson 1984). The unbound pigmented filler lies loose on the surface, spoiling the appearance. The effect is an easier paint to maintain and restore than those which crack or chip. Highly porous surfaces which were incorrectly primed produce similar effects as the background sucks moisture out of the drying paint film. Paints containing coarse pigment produce rough flat surfaces. If these are abraded they 'polish up' to form a patchy appearance defect.

Application defects   'Wrinkling' or 'rivelling' occurs as fine corrugations in the film shortly after paint application. It does not necessarily produce a fault in the covering quality, unless interfered with in the semi-dry state, but is unsightly and collects dirt. Uneven paint application will produce this. Edges of surfaces are liable to thick paint coatings as paint gathers, and these may run and rivel. Paint quality and skill in application are important instruments in preventing this. In extreme cases 'sagging' will occur. Paints which flow particularly easily can produce sagging, and coating thickness must be carefully controlled.

'Bittiness' occurs when dirt is incorporated into the surface from the atmosphere, or dirty tools are used, or if the paint is not properly strained. Gloss surfaces show the problem most; flat finish appearances are quite tolerant of dirt. Other surface appearance defects arising during application include 'brushmarking' (or 'ropiness'). Severe brushmarking produces weaknesses in the paint film, and a loss of durability.

Sprayed paints with poor flow characteristics (too viscous) tend to create surface pimpling. The problem is less frequent with brushing. This 'orange peeling' effect may not be problematic unless the surface finish is critical.

Certain masonry paints contain fine masonry aggregate. Although this is held in an emulsion capable of supporting the weight of aggregate, this only applies over a short time span. Continual application from a well-mixed batch which is allowed to stand will apply progressively less of the aggregate to the surface. The characteristics of the covering will change together with its dried appearance. Certain masonry paints have the ability to fill minor cracks in coverings.

Response to the environment   'Blooming' of a paint film is a chemical response to the environment. It has a whitish appearance most noticeable on dark-coloured films, and occurs in damp atmospheres with the loss of gloss. Its occurrence is exacerbated by pollution. Oil-based paints are generally affected, particularly those with a high proportion of resin. This is similar to sweating of the paint surface.

Paints are commonly assumed to last 5 years or so. This is highly dependent on the specific circumstances. Sloping or horizontal surfaces collect water and suffer greater erosion and pollution damage than vertical. Severe pollution or marine atmospheres all deteriorate painted surfaces more rapidly. Elevations facing south or west are affected by sunlight and wind most. Once deterioration to paintwork starts for any reason, it can progress very rapidly. Frequent inspection is advisable.

Rapid deterioration occurs where water can collect on surfaces (such as window sills), or becomes entrapped within joints or by bad design detailing. End-grain timber is especially vulnerable to water uptake: the rate can be several hundred times greater than that of lateral wood surfaces.

Colour variability

Paint colour variability is problematic, but does not necessarily indicate any deficiency in protection. Incompatibility with previous coatings can cause bleeding. For example, bituminous paints are the most problematic bases, and can produce brown 'staining' through overpainting. Untreated knots will produce similar effects. Where the previous coating is visible through the overpaint, 'grinning' is produced. Painting applied to ferrous metals may develop rust spots and stains once in place. Soluble salts in porous backgrounds will also affect paint films as they emerge from the substrate. Incomplete mixing, old or poor-quality paint can produce a variation in colour density or tone. 'Floating' may result, as the constituents separate out; fading may even occur in the tin.

Coated surface defects

Paints that do not properly dry out will remain 'tacky'. If the paint surface is rubbed down prematurely the paint may lift up and produce a 'balling' effect at the surface. This can be difficult to cut down for recoating. Such problems arise also from lifting of the paint, produced by incompatible layers. The commonly cited problem of 'saponification' is the soapy development of paint attacked by damp alkaline backgrounds. Walls containing cement, or timber stripped using an alkali paint remover, may produce a softening or even liquefying of alkyd paints and bleached patches in some emulsions. Where an alkali paint remover has been used, the residues can attack the paintwork and produce saponification. 'Kiss' marks on fletton bricks due to the stacking pattern when firing can be difficult to cover and may produce a rectangular patchy appearance.

The hygrothermal function

The principal function of external paintwork is to control the entry of water into timber (*Architects' Journal* 1986d). Wood alters its moisture content to approach equilibrium with the environment, and this produces disruptive changes in size. The changes are not the same in all directions: more expansion occurs across the grain than along it, since it is an anisotropic material.

If timbers are continually changing in moisture content this can give rise to deterioration caused by splitting and distortion. The grain may become obvious and unsightly, too. The extent of moisture-related movement timber undergoes will depend on the particular wood; however, this can be significant enough to cause problems with inelastic paints. Hence an important property of paintwork is extensibility without cracking, so the paint skin remains intact. Otherwise, the wood and the paint move differentially, causing flaking and an incomplete coating. Once the coating is deficient, moisture control is lost, and the situation worsens rapidly. Repainting may become essential in order to preserve the wood.

The way in which the paint allows water vapour to pass through it will determine how naturally the timber can move, and how quickly it can approach equilibrium with the environment. An impermeable coating allows complete control of the moisture content of timber, which could theoretically eliminate problems with rot if applied to completely enclosed dry timber. However, the completeness of the skin cannot be assured. There is a distinct likelihood that any moisture entering the timber will become trapped beneath the coating. Whilst rot and possibly frost damage occur in the wood (*Architects' Journal* 1984b) the pressure of entrapped water will cause the remaining paint to blister or peel off.

A paint that is too permeable will control the moisture content of the timber poorly (the problem with traditional stains). This encourages the excessive movement mentioned earlier. Correct permeability is an important function of any external paint. The permeability of a coating varies widely from the low-solids stains to the traditional finish of several undercoats and a gloss finish over a primed surface (BRE 1977b).

Masonry background faults

Faults with the product choice or background preparation are the common sources of failures of masonry paints. Application to unsound backgrounds also causes problems. The stability of any substrate is essential. Friable masonry cannot easily be over-painted successfully. The resulting poor adhesion will cause the paint layer to fail prematurely (Graystone 1984). Masonry paints may protect the substrate from the environmental pollution that would damage it. The cause of failure of these paints is usually moisture within the structure during painting or rising damp and condensation. All of these will affect the quality of adhesion. Separation of the substrate and paint can cause problems with water retention in the gap between the two, and this can accelerate any degradation.

The paints will require to be immune to the alkali environments when in contact with cement mortar and concrete. Their resistance to efflorescence may also be suspect, although this is usually a transitory phenomenon occurring as salts in the burnt clays are brought to the surface of bricks during the early wetting–drying phases. Only in prolonged and severe cases is it serious or more than an appearance problem. Less porous paint films will resist the passage of the salt solutions and may be damaged.

Toxins

It should be noted that Victorian paints relied heavily on lead, and although the Lead Paint Regulations of 1927 reduced the permitted content, paints used for at least the first half of this century still had a significant proportion of lead in them.

16 Deterioration in timber

Material-induced instabilities

Material-induced instabilities in timber arise predominantly from its vulnerability to fungal decay or insect attack. This usually produces a loss of structural strength and dimensional integrity. There are a number of possible mechanisms involved, which may operate in conjunction. Wood deterioration is a potentially significant problem, since timber and timber products amount to approximately 45 per cent of all the materials used in traditional buildings and 67 per cent of all materials used in timber-framed buildings.

In verifying the nature and extent of any damage, it will be necessary to establish the type of timber and the agent(s) of decay. This requires familiarity and a certain amount of intuition gained by experience. Specific texts dealing with these topics at great length and in excellent detail are available (Hickin 1981, Coggins 1980).

Fungal attack

Fungi

Timber fungi are primitive types of plant which grow on and in wood. The fungal spores are produced in such vast numbers that it is safe to assume that they will be present in any building at some stage (Gibson and Lothian 1982). All fungi follow a typical growth cycle, and the degradation caused to building timbers is merely an extension of their natural role. An extensive root system, or mycelium, is usually developed to remove nutrients and oxygen from the cellulose and hemi-cellulose of the timber. This causes a breakdown of its chemical structure. As the fungal root system elongates and forces through the cellular structure, it is physically destructive. The compressive and tensile strength of the timber is reduced, and physical distortion occurs. Obviously it is important to assess the stage in the life cycle correctly, since this indicates the relationship between potential and actual damage caused. However, the range of fungi producing similar basic decay symptoms makes it difficult to establish the exact cause unless the latter distinctive stages have been entered.

The risk of attack

The risk of fungal attack depends largely on the species of timber and the proportion of sapwood. Species such as European redwood are extensively used in the construction industry for first fixings and are particularly susceptible to fungal attack. Others such as cedar have a natural resistance which enhances their durability. Timbers which have become decayed through fungal attack are more susceptible to insect attack, particularly from the deathwatch beetle or weevil.

The likely incidence of sapwood attack increased in the latter half of this century as the use of faster-growing coniferous timber became widespread. This was both a response to the depletion of the forests and an effort to minimise cost whilst matching the expanding demand. To counter this increased vulnerability there is now widespread mandatory pretreatment of timber with preservatives poisonous to fungi and insects. This was less commonplace prior to the early 1960s (Gibson and Lothian 1982) and represents a threshold in terms of the types of timber and decay likely to be encountered. The timber used half a century ago was mostly better quality than now (Melville and Gordon 1979), principally because of the restrictions placed on sapwood and the nature of growth allowed by the smaller demand. Very dense heartwood rots very slowly generally, but where the timber was poorly treated, as was common in the post-war constructions, there is a great likelihood of wet rot. Indeed, the susceptibility to wet rot also appears to have increased post-war (Melville and Gordon 1979).

Natural to artificial durability

The shift in emphasis from natural durability to quality of preservative treatment and retreatment after working is critical to the performance of modern timbers. This is exacerbated critically by the relatively recent trends in stress grading and design which result in reduced cross-sections of timber with higher allowable proportions of sapwood (Gibson and Lothian 1982). There is less safety margin remaining for fungal or insect attack.

Conditions and surroundings

The occurrence of fungal attack depends on the surroundings, the type of fungus present, and the characteristics of the wood. If these conditions are all suitable, attack will commence and continue until a significant change occurs or the supply of timber is exhausted. When attack is suspected without ambiguous symptoms it must be established whether the environmental conditions are likely to support it. Any evidence of chronic dampness, particularly in conjunction with poor ventilation and a normal internal temperature range, indicates a potential site.

In addition to the environmental conditions, any biological evidence of rot and the overall physical soundness of the wood should be examined. Not surprisingly, many of the locations prone to rot are the most difficult to survey, so this may involve opening up and inspection of concealed timber.

Wood damaged by fungus may sound hollow and dull, and be an abnormal colour (Mika and Desch 1988). It will yield easily under the pressure of a sharp instrument such as a pen-knife, and tend to break across the grain. The chronic dampness required for continued growth will be readily identified by a moisture meter.

Integrity and residual strength

In addition to establishing the extent of timber involved, the state of integrity and residual strength must be assessed. The effects of decay in any structural timbers are obviously potentially serious and its appraisal should be treated accordingly. For instance, bonding timbers are a critical element in the construction of solid walls. Their failure through rotting will produce symptomatic buckling and instability in the wall, a more serious secondary effect.

The consequences of the rotting of timber in structural members are not restricted to the location of occurrence. The transferral of loads as key components fail will produce other secondary or even tertiary failures. It is most critical to assess the degree of structural redundancy, which in modern buildings is likely to be significantly less than their overdesigned forerunners.

Progressing, dormant or terminated attack

It will also be necessary to assess whether the fungal attack is current and ongoing. This may involve monitoring for activity. It is particularly important to distinguish between the different types of fungi, since their significance for the structure and the methods of purging vary. In addition to the plain occurrence of fungal attack, there may also be a chaining effect. Where fresh timber is in contact with infested wood the fungal attack will spread. Where detailed appraisal is required a mycologist should be consulted.

Contributory factors

External assessment of the building will give an indication of the general state of maintenance and any obvious sources of moisture entry. Note can also be made of any characteristics of the site that make it more susceptible, such as any slopes or signs of a high or fluctuating water table (Gibson and Lothian 1982).

At ground level, the likely sources of frequent or continual dampness are associated with faults in the design or workmanship of the damp-proof coursing. Bridging of the damp-proof course may also arise from the use of the building such as neglect of the adjacent ground, or following the building of extensions. The ground level may interfere with any underfloor ventilation intended from airbricks, assuming they were incorporated.

Exposed soil beneath a suspended timber ground floor may be sufficient

to create a suitable environment for deterioration of the floor timbers. This source of moisture and vegetable growth should now be unlikely in relatively new work where the local authority/by-laws/building regulations specify oversite concrete.

Other circumstances affecting floor joists arise from their contact where they bear directly onto walls, or are built into thin solid walls or otherwise damp masonry. A similar rot scenario can occur with timber lintels (pre-1900). A symptom of rot in the ends of built-in timber floor joists may be excessive vibration. In such instances it will be necessary to lift floorboards and inspect the perimeter. With thick walls the zone of water penetration under normal circumstances may not be deep enough to allow wetting of the timber. Cracks in rendered walls will encourage the entry and retention of moisture, especially in areas of high exposure. In solid wall construction this can be a significant source of damp penetration. These cracks do not have to be large or numerous to supply sufficient water for fungal growth, and may be concealed by external fittings.

Water discharges

Discharge of significant amounts of water from faulty rainwater handling services or flashings around openings can cause breaches of otherwise sound construction. Porous constructions may be continually or recurrently wet. Blocked or rusted gutters may allow the eaves timbers and fascia panels to become saturated. In such instances fungal attack is unlikely to be the only symptom; there may be discoloration of the external skin of the wall and evidence of frost or sulphate attack.

Gaps and exposure

Wet rot is likely to occur in external joinery exposed to prolonged damp conditions. Apart from direct contact with the ground, window and door construction is vulnerable. Rot will occur in the region of water uptake. Water can become trapped in the exposed joints and below perforated paintwork. Capillary gaps around the frames where the mastic filling has perished, or open joints where it has been omitted, hold water and are susceptible to driving rain. Particularly vulnerable areas include end-grain timber to sills and the lower portions of window and door frame components. Water traps may be indicated by dirt collection, and may be concentrated on the elevations exposed to driving rain. The absorbency of the exposed timber may produce conditions for rot despite the more extreme temperature range externally.

Problems from roof coverings

Similar problems arise from damage or neglect to roof coverings, whether of the unit type or continuous membranes. Design detailing can be the root cause or an exacerbating feature. These elements of the building envelope

are generally relatively poorly maintained because of the difficulty of casual inspection and access. Common rotting problems with the roof structure arise from parapet and valley gutters to M-roofs with defective linings or blocked outlets (Pryke 1980).

Internal leakage

Internally, similar problems may be caused from long-term leakages in plumbing. Care is required in analysing this defect, since the emergence of water very often bears little relation to the location of its source. Joists to upper floors are much more rarely attacked by dry rot than at the ground floor. Acute locations include damp, poorly ventilated cupboards, and where there is leakage of rainwater, such as in the roof space.

The role of ventilation

As a whole, the degree of ventilation in buildings has changed markedly with the demise of the chimney. This difference in natural ventilation between properties of the different eras seems to be a significant factor in itself. Frequently a reduction in ventilation is used in an attempt to minimise heating requirements. Aside from the direct health implications of this there is the increased risk of rot internally.

Timber in flat and pitched roofs is a common location of decay. In some high-quality houses built early this century the eaves construction was closed by design. This reduces ventilation within the roof space and encourages the stagnancy required for fungal growth. The eaves construction may also be susceptible to water ingress and wet rot problems from overflowing gutters or incomplete roof surfaces.

Vapour barriers and fungal attack

Many of the problems with timber rot internally arise from the omission of vapour barriers (Taylor 1983). This can be a significant contributory cause of fungal attack in the roof space, particularly if there is insulation blocking the eaves ventilation routes. Condensation produced by ordinary use of the building will collect in the roof space. This is exacerbated by insulated roof spaces without vapour barriers. As the warm moist air passes up through the insulation and its thermal gradient, it undergoes a thermal shock and the water vapour condenses. The roof spaces immediately above humid enclosures such as bathrooms and kitchens are obvious focal points.

Condensation can occur at ceiling level, and or rafter level depending on the positioning of the insulation layer(s). Where the condensation coincides with poor ventilation, virtually ideal conditions for surface fungal growth can be created, concealed by the insulation. In addition, surface moulds will occur in poorly ventilated buildings suffering condensation (BRE 1985b).

In flat roofs frequent problems have arisen with cold roof details used widely in the 1960s. Here the omission of a vapour barrier meant that the

occurrence of condensation and rot within the roof void was rife. To compound this, the older cold-roof detailing to flat roof construction was almost always unventilated. The loss in strength of the timber structure produces secondary drainage and waterproofing faults. A vicious circle may occur, with the roof eventually failing completely.

Identification of fungi

It is essential to be able to identify correctly the type of fungus present. The differentiating criteria for fungi are the colour and scale of the mycelium, the fruiting body and its spores. A second opinion or specialist advice may be required for detailed analysis or where unusual conditions exist.

There are a number of wood-rotting fungi that damage building timbers, and these are frequently divided into the rather ambiguous categories of wet and dry rots. Both dry and wet rots will attack hardwood and softwoods (Ragsdale and Raynham 1972), and are by far the most frequent fungi to attack building timber. Neither the wet or dry rots will grow in a dry environment, although the dry rots will not grow in saturated conditions.

True dry rot

Serpula lacrymans (formerly *Merulius lacrymans*) is the exclusive form of dry rot found in building timbers. It is a brown cubical rot that attacks the cellulose and lignin in wood, and progressively renders it into a dry, frangible powder. *Serpula lacrymans* is widely considered as the most serious of the wood rots, because of its structural significance and pervasiveness.

Dry rot fungus is more likely to attack floor timbers at ground level or below. Flat roofs are also susceptible. The rot requires a stable, damp atmosphere. Poor ventilation is a key factor, so vital areas to check are concealed timbers, such as the framing to partitions and panelling. Internal plasterwork accommodates the movement of the decaying timber poorly, and cracking may appear. There may be distortion of the panelling under its self-weight. The surfaces of panelling and skirting boards may become slightly uneven. By the time such damage becomes an obvious symptom the attack may be very severe.

The surface appearance of the rotting timber depends on the stage of attack as well as the environmental conditions. In still air and in damp conditions the rot develops a fluffy, cotton-wool appearance. In drier conditions the surface appearance is more greyish, similar to felt. There may be bright yellow patches where there is light. This is the mycelium, which may perish and disappear as the timber supply becomes exhausted, leaving the hyphae only. It is further characterised by the presence of hollow strands (hyphae) which search out and transport moisture. An example of dry rot hyphae is shown in Fig. 16.1. The hyphae vary in size from 1 mm to 10 mm in diameter, and will cross or penetrate non-cellulosic materials in their path.

Fig. 16.1 Characteristic dry rot hyphae strands on brickwork. (Reproduced by kind permission of Rentokil Ltd).

The fruiting bodies (sporophores) develop in the final stages of the life cycle and are identifiable by their white perimeter. These produce spores by the million for the redistribution of the fungus. The surrounding timber may be coated in a fine dusting of the red spores which are easily dislodged from the fruiting body. In still air the fruit bodies may exude water and this corresponds to the name *lacrymans* which is translated as 'crying'.

A frequently cited musty or mushroomy smell occurs at an advanced

stage in poorly ventilated enclosures. The timber becomes brownish-red and is of excessively low density. It is common for the surface of the wood to exhibit a deep cross cracking, cubical effect. The rotted wood shrinks as it loses its cell contents. There may be internal collapse of the timber element, producing an undulating surface. Advanced decay may remain hidden, however, and culminate in structural collapse. Indeed, there may be no symptoms obvious to the building user of the existence of dry rot.

Damp brickwork may also become infiltrated and provide a site for fungal development or dormancy. It is obviously essential that such locations are identified during inspection. The examination for dry rot requires extensive and precise inspection of the wood. Any survey should be extended a considerable distance beyond the limit of obvious attack, if necessary to include the whole building.

There is some dispute as to whether the *Serpula* spores can transcend dry periods in a dormant state. The late H.E. Desch (Desch and Desch 1970) suggests that a period of two or three months at a moisture content below 16 per cent is sufficient to kill off the spores. However, it is most unlikely that the whole of the infected timber within a building could drop to such a low moisture content for a sufficiently long period to produce this, since the infected areas around the mycelium are usually kept moist. Hence the spores appear remarkably resilient.

Wet rots The wet rots are less pervasive and generally less harmful than dry rot, but can become widespread in excessively damp conditions such as leaks or heavy condensation. They require a damper atmosphere than the dry rots for initial growth, which is likely to be localised. Opinions about the ideal moisture content for growth vary from the range of 24 to 28 per cent (Desch and Desch 1970) up to about 50 per cent (Taylor 1983). For fungi generally, however, moisture contents in excess of about 40 per cent by dry weight exclude oxygen from the fungus in the wood, and the process becomes starved. In addition, temperatures in excess of about 40°C are usually sufficient to destroy most fungi (Taylor 1983). The required moisture content for optimal growth varies slightly between different types of fungi. Generally a minimum moisture content of 20 per cent is critical for germination of the spores (Gibson and Lothian 1982, Taylor 1983). Continued supply is needed for the transport of the fungal enzymes into the cellulosic timber and the dissolution of the resulting glucose for return to the fungus. Moisture contents below 18–20 per cent are widely considered to be secure (Gibson and Lothian 1982), and properly seasoned timber examined under normal internal conditions in modern houses should have a moisture content around 12–15 per cent. Fluctuations in moisture content may drive fungi into dormancy.

The most frequently occurring wet rot is *Coniophora puteana* (formerly

Fig. 16.2 Wet rot of timber mullion, showing breakdown of timber.

Coniophora cerebella), also known as cellar fungus. It produces a cubical failure in the wood that is difficult to distinguish from the superficial symptoms of dry rot. *Serpula lacrymans* may be identified by the hyphae if no fruiting bodies remain.

The wet rot darkens the affected timber, which also becomes very brittle and easily powdered. It may crack parallel to the grain. It is frequent for the deep timber to decay more, since it is least sensitive to wetting and drying cycles. There may be a skin of sound timber. Fig. 16.2 illustrates wet rot in a timber mullion.

The initial signs of attack in painted timber may be a bulging of the coating caused by backpressure. Wet rot may be less structurally significant than dry rot, although longitudinal cracking may develop.

In the latter stages a visible fruit body develops which may be olive green or brown (Hutchinson, Barton and Ellis 1975). This will produce and release the spores for transportation to new sites. The sporophores are rarely found indoors, but outside they may occur on timber holding water. They are usually smaller than the dry rot. In concealed timbers with poor ventilation wet rot may emerge as dry rot. The increased tendency to draught-proof may also be contributory.

Less frequently occurring rots are *Poria vaillantii* and *Poria xantha*, both of which attack softwoods.* Some fungi are indicative of the onset of dangerously damp conditions rather than problematic themselves. These include sap stain and elf cups.

Poria vaillantii

Poria vaillantii (also *Poria contigua*, and now referred to as *Phellinus continguus*) requires very damp conditions and is only likely to be found in saturated ground locations. In its advanced stages it produces a similar visible effect to that of dry rot, and it is categorised as a dry rot rather than a wet rot. Hence it is sometimes referred to as white pore fungus. The attack is usually more localised than dry rot and this may be a useful distinction. Note also that the spores are lighter in colour than the *Serpula*, and in the dry state the fruiting bodies remain leathery (instead of becoming brittle as the *Serpula lacrymans*). It is obviously very important to distinguish between *Poria vaillantii* (sometimes also referred to as *Fibroporia vaillantii*) and *Serpula lacrymans*, because of the different degrees of remedial work involved. The *Poria xantha* strain favours warm humid atmospheres, such as have become much more common in the modern home. Spores in the air settle on the timber and lie dormant until the environmental conditions become correct. The humidity and temperature needs will be particularly well met in the thermally heavyweight buildings which encourage condensation.

Sap stain

Sap staining is produced as the sap content of the timber structure is drained. The timber generally remains structurally sound, but may develop coloured streaking caused by *Ceratostomella* fungi (Desch and Desch 1970). This is frequently a blue or blueish-grey, and is most obvious under light coloured paintwork. It penetrates the wood deeply and is caused by the hyphae within the cell structure of the wood.

Other staining effects may be produced by different fungi. In particular, one form of incipient decay produced by fungi also produces a streaked

* Desch discussed a range of less common fungi at some depth in Desch and Desch (1970).

staining to the timber, but is structurally significant. This has been noted with imported (unseasoned) timber where the prolonged damp conditions of the bundled wood produce the correct conditions for attack.

In addition to the mechanisms of sap stain and conventional fungal attack, there is the possibility of chemical or bacteriological involvement. In such cases the presence of moisture in the timber catalyses a chemical reaction, producing staining to any timbers in contact with ferrous metals.

Although sap stain can be distinguished from fungal attacks on the basis that its immediate structural relevance is limited, there can be symptomatic similarities. The attack usually occurs in sap-rich wood, the type most susceptible to fungal attack. Hence the discovery of true sap stain may be a precursor to and early warning of other fungal attack. Where the conditions are suitable for blue stain to occur there is a significant risk of wet or dry rot occurring. Heavily stained timber may already be infected by a separate fungus.

Insect attack

Insect attack generally

Insects cause much of the other damage to building timbers. The common term used to describe the damage is woodworm. The main external symptoms are the flight holes formed by the adult insects as they leave the wood. These have characteristic spacings and size which are indicative of the types of insect responsible. A summary of the main attack characteristics of a range of wood-boring insects is shown in Table 16.1.

The larvae produce worm holes as they burrow through the wood, converting the cell walls or contents to frass by extracting and absorbing the glucose. This causes the majority of the damage. There may be some of the frass or bore dust visible, but the effect on the wood is only evident on sectioning. Frass indicative of the species of beetle attack is left in the bore. The size and shape of the exit holes are key symptoms of the type of insect attack. Whether the attack is continuing is obviously important but a difficult detail to establish. The common beetles causing damage in this country are the house longhorn, the powder post, the (common) furniture, and the death-watch beetles. A full inspection of the building will be necessary to ensure the identification of all affected timbers.

House longhorn

The house longhorn beetle (*Hylotropes bajulus*) is one of several species described as longhorn beetles. Most thrive in tropical conditions only, but occasionally mating of house longhorn beetle adults occurs in temperate climates, followed by infestation. They attack the sapwood in softwoods only and are a particular problem in northern Europe (Desch and Desch

Table 16.1

Species	Timber Sap or Heart	Fungal decay	Shape of transverse section	Larval gallery Straight or Meandering	Stained or Clean	Frass Quantity	Frass Description	Flight hole Shape	Flight hole Size
Anobium punctatum	Mostly S but sometimes H	Not present or very slight	Circular	M	Clean	Moderate	Cigar-shaped pellets. Gritty feel	Circular	About 1.5 mm
Xestobium rufovillosum	S and H	Always present	Circular	M	Clean	Moderate	Bun-shaped pellets	Circular	About 2.5 mm
Ernobius mollis	Outer sapwood where bark adhering	Absent	Roughly circular, sometimes oval	M	Clean	Moderate	Bun-shaped pellets present of two colours according to whether bark or sapwood has been eaten	Circular	Up to 2.5 mm in bark
Ptilinus pectinicornis	Sapwood	Absent	Circular	M	Clean	Moderate	Fine particles only. Silky	Circular	About 1.5 mm
Euophyrum confine / *Pentarthrum huttoni*	S and H	Present; often wet rot	Indistinctly rounded	M	—	Small	Fine in texture round granules	Narrowly oval; ragged margin	About 1.1 mm

Species column detail:
- *Anobium punctatum*: Wide range of softwoods and hardwoods
- *Xestobium rufovillosum*: Hardwoods oak, chestnut; rarely softwoods if adjacent
- *Ernobius mollis*: Spruce, pine, larch, fir
- *Ptilinus pectinicornis*: Beech, elder, maple, oak, alder, plane, sycamore, elm, hornbeam, poplar
- *Euophyrum confine* / *Pentarthrum huttoni*: Softwoods and hardwoods

Table 16.1 (contd.)

	Species	Timber Sap or Heart	Fungal decay	Larval gallery Shape of transverse section	Larval gallery Straight or Meandering	Stained or Clean	Frass Quantity	Frass Description	Flight hole Shape	Flight hole Size
Lyctus sp.	Wide pored hardwoods eg: oak, ash	Starch rich sap	Absent	Circular	M	Clean	Copious, sometimes bursting undamaged outer skin	Fine particles silky feel	Circular	About 1.4 mm
Hylotrupes bajalus	Softwoods but recorded in some hardwoods such as oak, alder, poplar	Sap first but heart at late stage	Absent	Oval	Straight in early infestations later meandering	Clean 'ripple' teeth marks	Copious, often bursting veneer of unattacked surface wood	Fine particles (rejected) with well formed cylindrical faecal pellets	Oval	Up to 6 mm long axis, but flight hole made by males often very small
Phymatodes testaceus	Usually oak but chestnut and ash recorded	Outer sap	Absent	Flattened oval and broad channels	Mostly with grain	Clean	Large quantities sufficient to loosen bark	Roughly cylindrical with one or two constrictions; irregular ends	Oval	About 6 mm long axis
Clytus arietis	Usually beech and oak but fruitwoods etc., not uncommon	Under bark until half grown then sap	Absent	Oval	Meandering	Clean	Moderate	Fine particles and loose aggregates	Flattened oval	About 5 mm long axis
Nacerdes melanura	Usually softwood, occasionally oak. *Eucalyptus* recorded	S and H	Always present	Large irregular channels	Usually with grain	Clean	Moderate	Soft and wet with fibrous shavings	Ragged oval	About 5 mm long axis

Table 16.1 (contd.)

Species	Timber Sap or Heart	Fungal decay	Shape of transverse section	Larval gallery Straight or Meandering	Stained or Clean	Quantity	Frass Description	Flight hole Shape	Size
Bupestis aurulenta	S and H	Absent	Flat oval	Mostly with grain	Clean	Tightly packed in arc-like pattern	Irregular spherical	Oval	Up to 5 mm long axis
Sircidae wood wasps	S and H	Present but not usually detectable	Circular	Curved	Clean	Gallery tightly packed	Coarse	Circular	Up to 8 mm
Scolytidae PLATYPODIDAE *Ambrosia* beetles	S and H	Presence shown by staining	Circular	Very straight	Stained	Nil	—	Absent from sawn timber	—
Species Many hardwood species, more rarely in softwoods									

Note: the description of the wood damage caused by SIRICIDAE is equally applicable to the melandryid beetle *Serropalpus barbatus* which is common in Europe but only occasionally introduced into Britain.

1970). Attack on hardwoods is less likely. Damage may occur during storage of the timber or in place.

The house longhorn larvae feed below the surface of the wood, leaving a shell of apparently sound timber. Damage may occur without prior symptoms, although there may be a surface bulging with established attack. Generally, attack is less frequent than by the furniture beetle. Any infection must be cut back to sound unaffected timber. The attack may be structurally serious enough to require replacement of members, particularly in roof spaces. Other areas of recorded damage include floors and internal joinery.

There may be a long life cycle of up to ten years but more generally five to seven years. Consequently there is significant internal damage to the timber before the adult beetle emerges, sometimes this causes total disintegration (Lee 1982). If the surface is broken, frass resembling flints mixed with wood dust is likely to be released. The adult beetle is frequently 20 mm long, and leaves large flight holes (3–9 mm) which are oval and widely spaced, but not uniquely characteristic.

In this country the consistently high temperature conditions required by the house longhorn beetle are rarely met. Continuing infestation is unlikely, and appears to occur mostly but not exclusively in the structural timbers in roof spaces. Areas where evidence of the house longhorn have been found are restricted to the Home Counties – north Surrey has suffered frequent occurrences, and a trend has emerged in Essex (Desch and Desch 1970). There is some evidence of attack in Hampshire and Berkshire also, and that it may be spreading there. These areas appear to have relatively high summer temperatures, which may continue/increase as a result of any global warming.

The discovery of house longhorn beetle infestation has resulted in the mandatory pretreatment of roof timbers in new constructions in these areas since the early 1950s in local bylaws and the 1965 Building Regulations. No infestation has been found originating in timbers older than 50 years.

Powder post   Powder post beetles (*Lyctus*) rely on the starchy content of the wood. There are several species producing similar effects, and it requires a careful and close inspection to distinguish between them. The *Lyctus* beetle is a temperate species and usually attacks only the sapwood of certain hardwoods, in particular elm and oak, also ash and chestnut. The attack mechanism involves the starchy content of the sapwood. The very fine-grained timbers are resistant, as are those which have little starch content. Damage usually occurs rapidly at the sapwood edges of timber members, and the initial attack may be completed by the involvement of the common furniture beetle.

The life cycle is between one and two years, with emergence of the adults usually in the spring and summer months. The frass is particularly fine and smooth and the completeness of destruction below an intact skin is a useful distinguishing feature when discovered. The timber locations where attack is likely include oak flooring, and particularly any wood that is high in sapwood content. Older timbers appear to develop an increased resistance to attack.

Common furniture beetle/woodworm

The furniture beetle (*Anobium punctatum*) has a widespread infestation and is probably responsible for most of the damage to wood in the UK. Staveley and Glover (1983) suggest that it is probably present in about 80 per cent of pre-war houses, and a higher proportion of pre-1914 houses. They also suggest a significant infestation in post-1950 houses. Desch and Desch (1970), on the other hand, suggested that the risk is often over-emphasised.

The damage is produced by the larvae as they burrow through the sound wood, producing a cream-coloured, gritty, cigar shaped frass. The life cycle is anywhere between two to five years and the adult beetles emerge from the wood in the summer months through small circular flight holes (1.5 mm). The occurrence of infestation is high in cheap softwood carcassing timbers (roof and floor timbers, and particularly ground-floor joists) and as the name suggests, old furniture. The beetle also traditionally infests the hardwood timbers of period properties. Timber which has been affected by damp is more likely to be attacked by furniture beetle. Furthermore, although the woodworm will commonly infest sound timber this is unlikely in very dry conditions. The attack is predominantly in the sapwood and the proportion of this largely dictates the extent and potential structural severity of the damage. Attack may occur cyclically and produce a gradual loss of strength.

Death-watch beetle

This insect (*Xestobium rufovillosum*) is frequently found in old hardwood elements of the timber structure, such as oak, particularly if there has been some rot. Chestnut and elm may also be affected. Infestation is usually restricted to the older period properties, but damage to adjacent softwoods has also been recorded The adult beetle is between 6 and 8 mm in length and may live up to ten years, although cycles of 5–6 years appear to be more common. It leaves larger flight holes (2–3 mm) than those of the furniture beetle, and the oval discs of frass are intermixed with wood dust.

Infestation commences in the cracks and crevices of the timber. The damp and poorly ventilated conditions necessary for fungal attack are also ideal for infestation by the death-watch beetle, consequently this often follows fungal attack. The mature larvae are reasonably resilient to dry conditions in the order of 12 per cent moisture content. The damage generally starts in wood exposed to moisture ingress, particularly the sapwood. This is frequently associated with structurally significant timbers under leaking roof surfaces or behind eaves guttering. The attack may spread from the initially affected sapwood to involve all the timber, creating sizeable hollows within the timber section. As with any discovery of attack to structural timbers, an urgent analysis of the residual stability and strength should be undertaken. Consequently, the attack may be expected to be structurally significant. Common locations affected include the ends of built-in joists and tie beams, and wall plates in contact with damp masonry. This has a similarity with fungal attack.

Infestation is widespread in England and Wales, although the more severe climate of Scotland and northern England effectively limits its spread.

Wood-boring weevils
These insects (*Pentarthrum huttoni, Euophyrum confine*) attack wood and plywood that has suffered or is suffering from damp and established fungal decay. This commonly includes the timber in contact with damp masonry, such as wall plates or built-in timber joists. Below-ground-level timbers are generally susceptible, and the comments regarding the need for oversite concrete to protect against fungal attack are relevant to the weevil also.

The insects are characterised by their prominent snout and the very small flight holes (less than or equal to 1 mm) they leave in the timber. The frass is comparatively small. The damage the beetle produces is similar to that of the furniture beetle, and it totally destroys the integrity and structural strength of the wood.

Bark borer
This insect (*Ernobius mollis*) infests the sapwood of softwoods. It is also found in waney edge timber. Attack occurs frequently in roof timbers and floor joists. It is important for the defects analyst to distinguish between this and the furniture beetle, whose flight holes are similar in appearance, since the bark borer produces no structural damage.

Pinhole borer (Ambrosia beetle)
Other borers include the pinhole borer, or shothole borer, or ambrosia beetle (*Scolytidae platypodidae*). The infestation occurs in the forest and may affect both softwoods and hardwoods such as hemlock and oak. The seasoning process kills the insect and the special ambrosia fungus it creates for the larvae to feed on. Consequently, any evidence will not represent an active attack. Whether the timber is suitable for use will depend on the extent of attack, and this should have been assessed at the time of incorporation into the structure.

Other causes of attack

Damage to cellulosic materials may also occur from rodent attack, or higher plant damage such as caused by ivy and moss. The acidity of the moss and lichens also will damage metals, and the root penetration into porous materials may produce moisture damage below the surface. In contrast there is empirical evidence that the presence of ivy creates a boundary layer at the wall surface which buffers against extremes of temperatures.

Inspection

Inspection methods   If the eradication of insect attack or fungal infestation is to be effective the inspection and analysis requires to be very thorough indeed.

It is common for exit holes in floorboarding to be on the underside only and the removal of boards will be necessary in order to inspect them thoroughly.

Live larvae or adult beetles may be found in active infestation, and this may also be indicated by a scattering of bore dust around the area of flight holes. The months when inspection is most likely to reveal activity are April to June for the death-watch beetle, June to August for the common furniture and house longhorn beetles, and this whole period for the powder post beetle. It may be necessary to analyse samples of the timber using X-rays for evidence of live larvae.

17 Thermal behaviour of the building

Overview

Energy crises

The first escalation of energy prices since the oil crisis of 1973 has brought the issue of thermal insulation to the forefront of building design and refurbishment. The increased awareness and sensitivity over the direct and secondary implications of energy production and consumption have drawn into the public and political arena the logistics of depletion and conservation techniques. At the individual building level the sheer cost of the heating resource is a direct incentive to minimise energy usage.

Governmental pressure

There is every likelihood that the clear governmental pressure to insulate buildings and conserve energy will continue. The further revision of the building regulations to impose even higher standards of insulation in new buildings appears certain.

Comfort and monetary benefits

Buildings require to be maintained at a reasonable temperature, not just for comfort in use but to also extend their useful life and prevent rapid, harmful deterioration of the structure. Otherwise, insulation is merely a means of reducing the heating bill and the capital expenditure on it is usually balanced only against the relative monetary benefits. In speculative house building even this may not hold true. A meaningful appraisal of the thermal behaviour of buildings should also compare the possible benefits of an insulated building to the environment against the impact of its production.

Environmental pressure

Following the Montreal Protocol of 1987, the building industry has collectively come under significant pressure over its use of environmentally sensitive materials. It is now appraising the alternatives at institutional and manufacturing levels. Meanwhile, the classical schism between profit and pragmatism ensues.

Physical and hygrothermal implications

The singular approach of assuming that the insulated building is a successful building is fundamentally flawed. The physical and hygrothermal implications should also be considered. This will involve appraising the responsiveness of the envelope to internal and external ambient conditions. A slow response may push the building out of phase with sudden changes.

Hot and cold bridges

Although insulation will reduce the rate of heat loss, it may create temperature differences within the structure which will focus heat loss through cold bridges. Differences in temperature of the envelope will induce differences in relative thermal movement. The joints to accommodate this movement have been seen to fail where sealant has softened or become brittle, and butt joints opened. In certain instances the need to determine and accommodate the working temperature range has not been recognised.

Thermal inertia and decrement

Thermal mass

Materials of high mass take a relatively long time to heat up, and are less responsive to sudden thermal changes. This helps to smooth out the thermal response pattern of the building. The opposite is true for light materials, which can also be good insulators. The restriction of the flow of heat through these materials allows materials on their warm side to be held at stable high temperatures. The respective warm-up times of the high and low thermal inertia materials/structures are obviously very different, and when the modern/new intermittent occupancy patterns are applied to them they behave very differently. Both high and low thermal inertia buildings will smooth out the heating effects of solar gain through the external envelope.

Time lag

The time lag and decrement factor of buildings can be shown diagrammatically. Figure 17.1 illustrates the total heat flow from and into the building over a time scale. As external temperatures increase during the day, the interior takes time to catch up, indeed it may never catch up in terms of total heat gain.

Construction in the first half of this century was mostly traditional and thermally heavyweight. The construction methods produced a significant thermal inertia in the buildings, which created thermal stability under uniform conditions at the expense of response to rapid changes in the external or internal environment. There was a distinct time lag as the buildings responded slowly, and where the climatic changes occur rapidly and frequently the building easily becomes out of phase with them. To

Fig. 17.1 Representation of time lag and decrement factor.

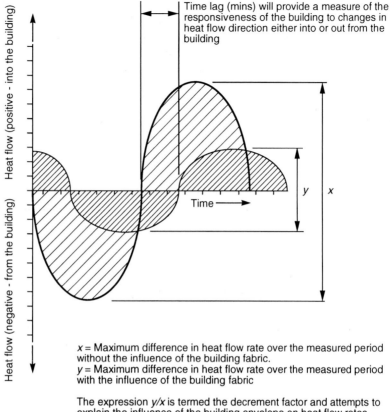

Time lag (mins) will provide a measure of the responsiveness of the building to changes in heat flow direction either into or out from the building

x = Maximum difference in heat flow rate over the measured period without the influence of the building fabric.
y = Maximum difference in heat flow rate over the measured period with the influence of the building fabric

The expression y/x is termed the decrement factor and attempts to explain the influence of the building envelope on heat flow rates into and out of the building. In general lightweight envelopes may have large decrement factors, whereas heavy envelopes have lower factors

some extent the nature of use of buildings at the time redressed this imbalance.

Air change and occupancy

It was normal practice to ventilate buildings extensively, and this was of course assisted by the chimneys associated with the use of the open fire. Air changes were commonly of the order of six per hour. Occupation patterns were generally more constant and energy use or conservation was largely a matter of microeconomics.

Fuel economics and defective buildings

The need to keep fuel economics under control was generally driven by the economics of facility management. Low running costs could be achieved by reducing the heating provision, which in turn reduced the comfort levels in the building. Buildings became colder but their inherent ventilation ensured that they were relatively dry. If the same approach is applied to more modern forms of construction then relative comfort levels are still likely to be reduced, but since the natural ventilation rates are much reduced the incidence of moisture and condensation-related defects increases.

Insulation criteria

Thermal insulation criteria evolved slowly in the first half of the 20th century. During the 1950s recognition of the need for different conditions within dwellings was recognised by the setting of differential thermal insulation criteria between the lounge and the remainder of the house. These early insulation methods merely inserted the material into the envelope to get increased thermal resistance. The effects on the thermal behaviour of the total building were not always effectively thought through, as evidenced by the number of thermally related defects.

The heavyweight trend

The heavyweight construction trend extended into the 1960s with the precast concrete systems. There was some attempt here to modify the thermal properties of the systems at the design stage. This included sandwich panel construction, with thermal insulation as the inner section. This was faced on both sides by structural-grade concrete. Although thermally satisfactory, sharp rises in internal air temperature could initiate surface condensation on the inside concrete surface. Some of the systems were ill-suited to our climate, and the distinct changes in the patterns and nature of use of buildings were already evident. Lightweight structure and low thermal inertia construction, particularly domestic from the era of energy awareness, is characterised by high levels of added insulation. The current and evolving changes in lifestyle have combined to create a number of self-inflicted problems. The thermally lightweight structure with low thermal inertia is likely to respond rapidly to changes in the climate and allow for changes in the internal conditions. There is a reduction in the time lag of the structure, which will provide less of a buffer against changes in the climate.

Trends and phases

The trends of occupation of buildings reflect on the patterns of energy consumption and heating. The effect of energy consumption on internal conditions, reduced ventilation and the increased production of water vapour from domestic activities coincide to increase the potential for condensation. The thermally heavyweight constructions become out of phase with ambient conditions and increase the risk and severity. The thermally lightweight constructions are more responsive to changes. Insulation of these forms will sharpen the thermal differences between inside and outside, increasing the time lag.

Heat gain

Solar radiation

Solar radiation will affect the ageing process of all external materials. This can manifest itself as the embrittlement of plastics and bitumens, or the fatigue associated with thermal movement. Surfaces which are exposed to solar radiation tend to lose their colour as the pigments are irradiated. The general level of solar radiation received at an external surface can be predicted by the use of design charts. Whilst these are used for heat gain calculations, their application for defect prediction appears to be less general.

Shading and the role of projections and surface texture of the building can set up differential solar gains across a facade. This may accentuate any differences in thermal movement between different and similar materials. The depth of window reveals will influence the amount of shading offered to the glazing. The positioning of glazing closer to the outside face of the building may reduce the differential shading pattern whilst increasing the overall solar gain to the building.

High levels of solar gain can increase internal temperatures. This may gain admittance through glazing or through poorly insulated parts of the external envelope. Cavities open to solar gain may act under a localised greenhouse effect and drive up surface temperatures of materials. This can occur where masonry backing walls are positioned behind glazing. The role of the emissivity of the surface in solar gain is critical, with dark matt surfaces being subject to higher surface temperatures and movement. Roofs can receive considerable solar gain. The ageing of unprotected bitumen may be accompanied with substantial movement of the covering, decking and even the structure. The total amount will be dependent on the type of roof construction.

Specific environmental requirements

High solar gain to the building through a poorly responsive external envelope will have comfort implications. This may trigger additional internal defects associated with glare, low humidity and temperature differentials between different parts of the building (draughts and stack effect). The fading of the interior decoration will be more marked. The traditional methods acted as thermal buffers, e.g., thatch and massive masonry. These helped to retain heat into the autumn months, although being defective in other ways.

Special gains

Particular occupancies and processes can produce high humidities. These are frequently outside the normal design range and demand specialist application methods. Materials which behave satisfactorily under these conditions have generally low absorption values or are impervious. Where these processes are associated with high temperatures, the thermal expansion of the finishes must be accommodated by a high humidity-resistant jointing system. Specialist drainage and falls may be required.

Greenhouse effect

The growing awareness of the depletion of the ozone layer has raised several questions concerning the heat gain to buildings. This also affects material selection due to manufacturing methods. The design solutions applied to a new building may be difficult to alter during its life. This may produce an external envelope which becomes progressively defective as the nature and quantity of solar heat gain changes throughout its life.

A micro-greenhouse effect can occur where glass is in front of other high-emissivity surfaces. High temperatures in metal or plastic and glass curtain-walling systems may be boosted by additional temperatures developed at the rear of the walling surface due to the re-emission of absorbed radiation at frequencies to which the surface is not transparent. This produces differential movements which must be accommodated in joint design to retain the weatherproofing integrity of the system. A failure to do so will enable considerable water ingress due to the impermeability of the major surface.

Emissivities

Most building surfaces have a high emissivity. A value of 0.9 is commonly assumed to apply universally. Although for specific surfaces this may be inaccurate, there are sufficient areas of buildings with this value to make general acceptance of this method reasonably accurate. If joints and the general accommodation of movement are constructed adequately, it is not the absolute movement that is a problem. Rather it is the relative movement between components on the surfaces or across a laminated structure.

Slight differences in emissivity can substantially reduce the thermal gain of materials and hence their relative movement, since coefficients of expansion are based on temperature rise. Similar materials with different colours will move by different physical amounts because of different temperature gains, and this can be problematic. Of greater importance is the mismatch between materials of fundamentally different properties, perhaps under different circumstances , e.g., at a corner, or in shade and sunlight. Any movement is three-dimensional also, which can cause delamination to occur.

Temperature-driven moisture movement

Water evaporating from the surface of a wet porous material will continue until a temperature difference no longer exists or the relative humidity of the external air is such that the mass transfer of moisture can no longer occur. Suitable evaporative conditions can occur on the external and internal surfaces of buildings. Where the inner pores of a material are linked, capillary action can cause a movement of moisture through their structure. This can reach the surface and either evaporate into free air or build up water vapour behind impervious coatings and coverings, causing blisters and cracking.

Heat loss

The physics of heat loss from buildings

Any technical appraisal of the thermal performance of buildings brings with it a need for a solid understanding of the mechanisms and physical units, also some appreciation of the constructional forms. These aspects are adequately covered in a range of texts (Smith, Phillips and Sweeney 1987, McMullan 1983). There are three basic mechanisms of heat transfer, 'radiation, convection' and 'conduction'. In buildings, the heat loss through the structure is mainly due to the effects of conduction. This is of course dependent on the relative conductivity of the materials.*

Insulators or conductors

In broad terms the materials can be classified as insulators or conductors. It is one of the inherent paradoxes that the materials needed for constructional stability are often the materials of high thermal conduction. Materials with few free electrons are generally good insulators; those with free electrons or electrons in a metallic bond tend to be good conductors.

Metals and glass are used a great deal in the fabric of buildings, and their presence in the external envelope will make a marked contribution towards the total heat loss of the building. Metal is a good conductor, whereas the relatively thin sections of glass mean that its surface resistance is critical in determining its insulation performance. Double and triple glazing will increase the number of surface resistances in the path of the heat flow as well as providing several dry air cavities. These offer further insulating layers.

Radiation and convection properties

The radiation properties of materials are related to their temperature and the nature of their surface. Dark surfaces absorb and radiate heat well, since they have a high emissivity. Most traditional building materials behave in this way, although the use of low-emissivity foils can be considered an insulating measure. These are commonly incorporated into plasterboards where they may also act as a vapour barrier to prevent moisture entering the fabric of the building.

---

* Thermal conductivity: a measure of a material's ability to transmit heat. Expressed in watts per square metre, degree Kelvin (W/m^2K).

Convection occurs readily in air masses, even in the relatively narrow spaces found in wall and floor cavities. Where secondary glazing contains relatively small gaps, convection currents can add to the heat flow through the window. This may also occur behind curtain-walling systems.

Appearance and performance

All building materials can transfer heat, and the need to balance the amount and its disposition around the building has implications for the appearance and thermal performance of the building. In order to meet the current regulations new buildings usually require the incorporation of specialised insulating materials. The statutory and recommended standards of thermal insulation for buildings are generally based on our ability to achieve the levels practically. As a consequence of the historical increase in prescribed levels of insulation the 'well-insulated' house of the 1950s is the energy wastrel of the 1990s. Ironically, there is some evidence that increasing the insulation value of houses is perceived by the user as an improvement in the standard of comfort of the building, rather than as savings in money and energy.

Total heat loss

The total heat loss of the building envelope is calculated by taking into account the individual conductivity of all of the components in the external envelope. The overall value for one part of the envelope is expressed as a *U*-value.* The lower the *U*-value appears, the better the insulation level and the lower the heat loss. These values are constantly under review and will continue to differ between different building uses; for example, the mandatory requirement to insulate floors in new domestic construction. It should be noted that *U*-values are laboratory-formulated quantities, which limits them to comparative use. They are of course based on steady-state conditions,[†] which may not be the actual conditions occurring in a building.

Night-time heat loss

Steel claddings and coverings of buildings can exhibit temperatures below the ambient air temperature. In marginal conditions this can mean the surface of the covering being frozen when the outside air temperature is close to freezing. This effect can extend to within the envelope since a new

---

* *U*-value: this is the thermal transmittance of an element or component, under steady heat flow conditions. It is measured as the quantity of heat flowing through unit area, in unit time, per unit difference in temperature between the inside and outside environment. It can be calculated as the reciprocal of the total resistance of the element.

† A steady state is assumed to apply when the temperatures on either side of the envelope are not changing and the rate of heat flow through the construction has reached a stable value.

and steep temperature gradient will exist within it. The black night sky is assumed to be a fully absorbent black body at $-40°C$. The shrinkage and embrittlement of a range of materials can occur at low temperatures. The glass transition temperatures of plastics can be within the average UK ambient temperature range.

Heat loss and condensation

Low surface temperatures are more likely in those envelopes containing insulation, particularly where it is placed close to the external face. Insulation will also increase the time lag of the envelope since a standard amount of heat is required to raise the material in the envelope by one degree. High heat loss can be considered a defect at the comfort level since it will mean that the interior of the building will remain cooler for longer. There is an increased risk of surface condensation due to the difference in specific heat capacity (air temperature rises at a greater rate than the solid materials of the building). Absorption of moisture by permeable surfaces will add to degradation and mould growth.

Poorly insulated materials and warm construction detailing produce generally low temperatures in the surface materials. The intermittent use of the building can increase the severity of this effect because there is not a constant heat flow from the internal spaces which could maintain surface temperatures above the dew-point of the air.

Insulation

The practicalities of heat retention

The construction of buildings involves the integration of many different materials and components. Any insulation technique must be compatible with the thermal performance and the other functional aspects of the building. There has been some experimentation with energy-efficient houses. The widespread practicality of a small autonomous house is somewhat limited in the UK due to the rapid and extreme variability of the climate, and the high capital cost. The weather conditions are generally unreliable, and for commercial buildings there appears to be some great difficulty in convincing clients to accept buildings without air conditioning. There have been instances of energy-efficient buildings being fitted with air conditioning merely to make them rentable. Client demands may not always be hygrothermally practical.

Principal techniques

There are three principal techniques for insulating a conventional external wall. These are shown in Fig. 17.2.

These either exploit the cavity or the living space, or are added to the

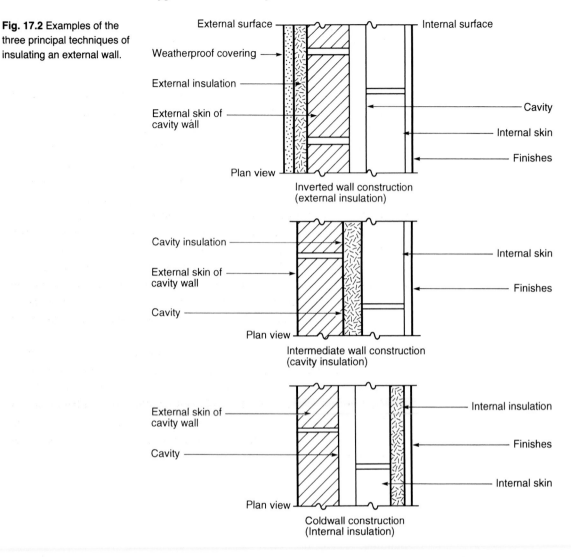

Fig. 17.2 Examples of the three principal techniques of insulating an external wall.

external face. The modern composite construction methods, as used for large-scale buildings, pose problems which require individually engineered solutions.

In new buildings cavity insulation can be installed during the construction stage, and this maintains a cavity albeit of reduced width. The slabs of insulation, termed batts, should be held against the inner (dry) leaf. This is commonly achieved by the use of plastic clips and wall ties. These connect the inner and outer leaves and contribute to the structural stability of the wall. Poor positioning/clipping can allow batts to sag, bridge cavities or drop out completely. The physical thickness of insulation that can be incorporated is obviously dictated by the cavity width. Structural stability limits the width of a conventional cavity to 90 mm; above this the slenderness of the walls and their separation limits structural stability.

Retro-insulating

Additional problems are involved with the older, poorly insulated dwellings. The upgrading of these buildings would perhaps be a more appropriate way to save energy because of their sheer numbers. Where these have cavity walls, one construction solution is to fill them with retrofitted cavity insulation. This was very common in the mid-1970s and the product is enjoying a comeback. The process involved drilling holes in the external walls, or accessing the top of the cavity from the roof space, and filling it with loose materials and foams. There were a range of controversial claims associated with this style of insulation. These included the difficulty of ensuring the complete filling of the cavity, the temporary release of vapours into the house and the consolidation of the material with time. These may affect these early insulation methods although the widespread concern generated by these defects has caused a tightening of standards. However, BRE findings contradict claims that retrofit cavity-fill insulation provides a moisture bridge.

An alternative to these approaches is to insulate the wall internally or externally. This maintains the cavity and allows easy fixing of the insulation. The thickness is not constrained by the cavity dimension. A real limitation is the effect on the openings in the walls and whether the insulation can be detailed around the opening without creating cold bridges.* These systems are of course the only practical methods of upgrading solid wall constructions.

Cold bridging

Hot and cold generally

Temperature differentials are associated with dimensional variations. These will affect joints and their dimensional tolerances. The conductivity of different construction materials varies. When these are assembled into the external envelope they will form hot and cold bridges. These can lead to pattern stains, affecting decoration (Seeley 1987), and encourage the growth of moulds.

Thermal bridging

A specific and frequent agent of condensation occurrence is the cold bridge. This is a location usually brought about by constructional variations such as

---

* A cold bridge can be described as a zone of high conductivity, created by materials of relatively high conduction.

Fig. 17.3 Projecting reinforced concrete beams penetrate the external envelope and are design features inside and outside. Cold bridging is inevitable.

at openings where the conductivity of the fabric is significantly greater than the surrounding structure. The 1960s architectural style of projecting elements was a cold bridge source in its own right. See Figs. 17.3 and 17.4. Standard calculation techniques exist to allow the incidence of cold bridging to be predicted (Burberry 1988).

A common example is the installation of concrete lintels crossing the cavity, a detail still being recommended in the construction textbooks as late as 1989. The U-value of the concrete lintel crossing the cavity and supporting both leaves can be around 2.0 W/m²K. Compared even with an uninsulated brick cavity wall (U-value of 1.3 W/m²K) (Smith, Phillips and

Fig. 17.4 Inset detail showing a continuous beam cold bridging the external envelope.

Sweeney 1987) the difference in U-value is significant. This produces a concentration of heat flow through the lintel and the surface temperature is effectively lowered, making it more susceptible to condensation and mould growth. Examples of cold bridges are shown in Fig. 14.4

Forms and bridges There are a number of other constructional forms which can create cold bridging. An obvious example is the 1930s metal windows, which stream with condensation under the right conditions. The early aluminium frames without thermal breaks could become particularly cold due to the metal's high thermal conductivity. The potential exists for cold bridging in exposed corners of external wall construction, including brick chimneys. Other locations include any bridging of the cavity, such as a wall tie or cavity stopping at junctions between walls. Closing of cavities at reveals can produce two vertical strips of relatively high thermal conductivity. Where bridged cavities exist, the local situation can be worsened by the retrofitting

Fig. 17.5 The concrete sill acts as a cold bridge. This may be a contributory factor to the obvious failure and exposure of the reinforcement.

of cavity-fill insulation which accentuates the relative differences in conductivity. In conjunction with this, the conservation of energy and generally lower internal temperatures increase the risk of condensation, which becomes focused at cold bridges. See Fig. 17.5.

Frames   The timber-frame construction form may use timber studs which punctuate areas of insulation. This may produce pattern staining of interior surfaces where the cold areas exist and create cold bridges. Although the term 'cold' indicates low temperatures, this should be considered as a lower temperature than those of the general surface. A difference of 5°C between one region and another may be sufficient for a cold bridge to occur, although the general temperatures may be above 16°C. The role of vapour barriers in timber-framed construction can be critical.

Relative movement

Thermal movement   The needs of structures to accommodate relative movement is a clear fact from the range of material movements discussed elsewhere. The provision of these at the design stage is an understood concept, although a range of faults can still arise. Seeley, for example notes that where trussed rafters for timber-framed dwellings were built onto the (non-loadbearing) brick cladding, overstressing problems occurred from differential movement within the structure (Seeley 1987).

Relative movement   Basic physical characteristics show that relative movement will occur between the masonry cladding and loadbearing frame of the timber-framed house, although this seems to be a feature that was frequently overlooked or not appreciated. Movement differences can exist across a facade because of material differences and the differences in temperature/heat flow through the materials. Since it is common for many material differences to exist, thermal movement differences will occur.

The scale of consideration can be expanded to include whole buildings. In this instance their relative movement to the ground and adjacent buildings should be considered. This can manifest itself as slippage at DPC level, and the cracking patterns associated with short and long returns to masonry walls. See Fig. 17.6.

Joints and   Movement joints can become defective due to insufficient sizing to
accommodation   accommodate the dual effects of moisture and thermal movements. Under high temperatures joints may close completely, putting the jointed and jointing material under compression. Under low temperatures joints may open up to a degree which exposes the main joint surface, allowing water to gain direct entry. The development of mastic sealed joints which should retain their integrity across a range of design movements was in response to the functional performance criteria. Where inadequate allowance has been made when determining low-temperature movements, joints may be unduly opened causing the mastic bond and/or material to break. Under high temperatures the joint may be compressed, causing exudation of the mastic.

Fig. 17.6 Movement joints should be continuous and not stop at the (stepped) DPC.

Generally, joint sizes are small and likewise the movements. Where the dimensional tolerance during erection is exceeded because of cumulative errors or poor supervision, this can rapidly be eroded. This nullifies any advantages of good joint design and the building tends to perform as if there were no joint, although the relative weakness of the structure at the joint can concentrate failure stresses.

Joints and construction climate

One of the principal criticisms of early joints was their limited buildability. Access for mastic/baffle insertion and the practicality of inspection were critical, and can be a major determinant of the durability and effectiveness of the joint. The physical difficulty of inspecting joints on vertical surfaces of high-rise buildings, and the difficulties of application may turn a 20-year mastic into a 5-year mastic.

The joint may need to accommodate changes in ambient air temperatures from −40°C to +40°C in a 24-hour period. The resulting movement may be accommodated by young joints but any embrittlement is likely to increase with the life of the joint. The point at which failure occurs may be difficult to ascertain without inspection, but must also be related to the physical and economic implications of failure.

Other factors

Deterioration

The long-term exposure to solar radiation of external surfaces may cause their ultraviolet degradation. Devitrification and a general embrittlement of materials can also be induced. This was particularly common in the early plastics and flexible coatings. The plastics were found to discolour and become more brittle, causing failure under moderate loads. Paint films can become detached from their background and their elasticity reduced.

Glass transition temperature

Plastics can go through a range of physical changes as their temperature alters. This manifests itself as a hardening and embrittlement of the plastic. It is easily cracked at low temperatures. At high temperatures it can become soft, ductile and malleable. The relative performance of the material will therefore also change, and these changes will occur on a continuing cyclic basis. The temperature which acts as the boundary between these soft and hard performance characteristics is termed the glass transition temperature. See Fig. 17.7.

Mastics being placed into tension as the surrounding material contracts may crack at low temperatures which approach their glass transition temperature.

Composite effects

Problems have occurred in composite construction when materials of different thermal coefficients of linear expansion are combined. This may cause disruption or dislocation at either extremes of temperature. Some years ago there was a well-publicised example of curtain glazing falling out of its framing in hot weather, where the greater movement of the aluminium frame used up all the bearing tolerance and allowed the glass to pop out under wind buffeting.

Bitumen skirtings and upstands can be pulled out, especially those over

Fig. 17.7 The general physical property changes associated with the glass transition temperature zone of materials.

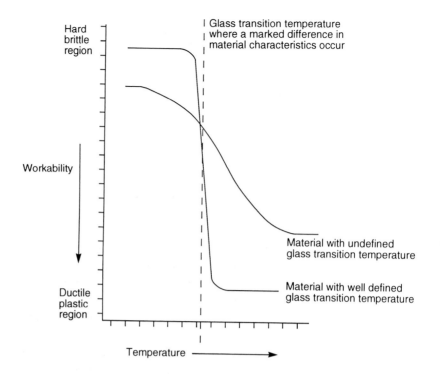

300 mm high. They can be pulled out in hot weather because of the slumping of the material, and pulled out in cold weather by contraction. In the early evening this may be significant because the upstand may lose less heat than the main roof. The degree of physical movement when the large roof area shrinks can distort or pull the flexible upstand free of the chase or fixing.

Shadows and projections

Shadows can exacerbate this problem and give a moving dimension to it. This can affect glazing in deep reveals. Roof areas can be influenced by projections and the shading from surrounding buildings. Projections through external surfaces may act to localise stresses and focus them into areas where jointing is critical. Projections may require upstands for weatherproofing which may move with the projecting structure in a different way to that of the main roof, causing failure at the junction.

References

Adams H 1912 *Building Construction* Cassells London

Addleson L 1978 Building failures technical study 6: Condensation *Architects' Journal* **167** (23): 1023

Addleson L 1982 *Building Failures – A Guide to Diagnosis, Remedy and Prevention* The Architectural Press

Aldous T (ed) 1979 Trees and buildings: Complement or conflict? *RIBA Publications*

Alexander S J, Lawson, R M 1981 *Design for Movement in Buildings* Technical Note 107, CIRIA

Allen W 1989 *Failures, Research, and Feed-back* Proc. Conf. Building Defects: Avoidance, Investigation and Litigation. Royal Garden Hotel, Kensington, London, 10 February 1989

Allen W 1990 Private correspondence 1 June 1990

AMA 1983a *Defects in Housing* Part 1: *Non-Traditional Dwellings of the 1940s and 1950s* Association of Metropolitan Authorities, July 1983

AMA 1983b *Defects in Housing* Part 1: *Non-Traditional Dwellings of the 1940s and 1950s. Technical Appendices* Association of Metropolitan Authorities, September 1983

AMA 1984 *Defects in Housing* Part 2: *Industrialised and System Built Dwellings of the 1960s and 1970s* Association of Metropolitan Authorities, March 1984

AMA 1985 *Defects in Housing* Part 3: *Repair and Modernisation of Traditional Built Dwellings* Association of Metropolitan Authorities, March 1985

Architects' Journal 1984 New technology paints for exterior timber *Architects' Journal* 25 July, **180** (30): 51–4

Architects' Journal 1985 *Failures of industrial roofs* Falconer P, Falconer R *Architects' Journal* **181** (24): 73–4

Architects' Journal 1986a Construction risks and remedies: Thermal insulation *Architects' Journal* **183** (25): 61–71

Architects' Journal 1986b Construction risks and remedies: Condensation *Architects' Journal* **83** (16): 69–81

Architects' Journal 1986c Construction risks and remedies: Thermal insulation *Architects' Journal* **183** (24): 51–64

Architects' Journal 1986d Paints and stains *Architects' Journal* **183** (26): 3–38

Architects' Journal 1987a Products in practice *Architects' Journal Supplement*, 28

January 1987, **185** (4): 15–20

Architects' Journal 1987b Construction risks and remedies: Movement *Architects' Journal* **185** (3): 57–68

Architects' Journal 1989 Thermal insulation Part 1: The risks *Architects' Journal* **183** (24): 51–64

Ashurst J, Dimes F G 1977 *Stone in Building: Its Use and Potential Today* Architectural Press

Atkinson A B 1985 Cavity wall questions. Letter to the *Chartered Surveyor Weekly* **10** (11): 776

Battrick G N 1989 *Prototype Testing* Proc. Conf. Cladding: Current Practice in Specification, Design and Construction. Regent Crest Hotel, London, 7 November 1989

BBC Television 1984 'The Great British Housing Disaster' 4 September 1984

BBC Television 1989 'Built in Britain: Made Abroad' *Panorama* 26 June 1989

Beards C F 1983 *Structural Vibration Analysis* Ellis Horwood, London.

Bell F G 1975 Introduction. In Bell F G (ed) *Site Investigations in Areas of Mining Subsidence* Newnes-Butterworths, London: 11, 18.

Bennett W 1987 *Radon in Buildings* Building Control May/June 1987: 15–16

Benson J 1984 Soft building and trees *Architects' Journal* 179 **(13)**: 55–8

Benson J, *et al.* 1980 *The Housing Rehabilitation Handbook* Architectural Press.

Bickerdike, Allen, Rich and Partners, O'Brien T 1971/72 House built in a wood. In *Design Failures in Buildings* 1st series. George Godwin, Building Failure Sheet 7

BRE 1964a *Design and Appearance–1* Digest 45, Building Research Establishment, HMSO

BRE 1964b *Design and Appearance–2* Digest 45, Building Research Establishment, HMSO

BRE 1971a *Sulphate Attack on Brickwork* Digest 89, Building Research Establishment, HMSO

BRE 1971b *An Index of Exposure to Driving Rain* Digest 127, Building Research Establishment, HMSO

BRE 1973 *Principles of Modern Building* Volume 2 3rd edn Building Research Establishment, HMSO

BRE 1975a *Principles of Modern Building.* Volume 1 3rd edn Building Research Establishment, HMSO

BRE 1976 *Sound Insulation of Lightweight Dwellings* Digest 187, Building Research Establishment, HMSO

BRE 1977a *Cracking in Buildings* Digest 75, Building Research Establishment, HMSO

BRE 1977b *Painting Walls* Part 1: *Choice of Paint* Digest 197, Building Research Establishment, HMSO

BRE 1978a *Dust Explosions* Information Sheet IS 11/78, Building Research Establishment, HMSO

BRE 1978b *Wall Cladding Defects and Their Diagnosis* Digest 217, Building Research Establishment, HMSO

BRE 1979a *Estimation of Thermal and Moisture Movements and Stresses* Part 1 Digest 227, Building Research Establishment, HMSO

BRE 1979b *Estimation of Thermal and Moisture Movements and Stresses* Part 3 Digest 229, Building Research Establishment, HMSO

BRE 1979c *Thermal, Visual and Acoustic Requirements in Buildings* Digest 226, Building Research Establishment, HMSO

BRE 1981a *Assessment of Damage in Low-Rise Buildings, with Particular Reference to Progressive Foundation Movement* Digest 251, Building Research Establishments, HMSO

BRE 1981b *Failure Patterns and Implications* Digest 251, Building Research Establishment, HMSO

BRE 1981c *Concrete in Sulphate-Bearing Soils and Groundwaters* Digest 250, Building Research Establishment, HMSO

BRE 1981d *Maintenance Cost of Flat Roofs* Current Paper CP 4/81, Building Research Establishment, HMSO

BRE 1981e *Rising Damp in Walls: Diagnosis and Treatment* Digest 245, Building Research Establishment, HMSO

BRE 1982a *The Durability of Steel in Concrete* Part 1. *Mechanism of Protection and Corrosion* Digest 263, Building Research Establishment, HMSO

BRE 1982b *The Durability of Steel in Concrete* Part 2. *Diagnosis and Assessment of Corrosion-Cracked Concrete* Digest 264, Building Research Establishment, HMSO

BRE 1982c *Sound Insulation of Party Floors* Digest 266, Building Research Establishment, HMSO

BRE 1982d *Common Defects in Low-Rise Traditional Housing* Digest 268, Building Research Establishment, HMSO.

BRE 1982e *Inspection and Maintenance of Flat and Low Pitched Roofs* Information Paper IP 15/82, Building Research Establishment, HMSO

BRE 1983a *External Walls: Rendering* Defect Action Sheet DAS 37, Building Research Establishment, HMSO

BRE 1983b *Aids to Identification: Boot Houses* Building Research Establishment, HMSO

BRE 1984 *Dust Explosions* Digest 288, Building Research Establishment, HMSO.

BRE 1985a *The Influence of Trees on House Foundations in Clay Soils* Digest 298, Building Research Establishment, HMSO

BRE 1985b *Surface Condensation and Mould Growth in Traditionally-Built Dwellings* Digest 297, Building Research Establishment, HMSO

BRE 1985c *External Walls: Bricklaying and Rendering When Weather May be Bad* Defect Action Sheet DAS 64, HMSO

BRE 1985d *Corrosion of Metals by Wood* Digest 301, Building Research Establishment, HMSO.

BRE 1985e *Mould and its Control* Information Paper IP 11/85, Building Research Establishment, HMSO

BRE 1987a *Concrete* Part 1: *Materials* Digest 325, Building Research Establishment, HMSO

BRE 1988a *Alkali Aggregate Reactions in Concrete* Digest 330, Building Research Establishment, HMSO

BRE 1988b *Loads on Roofs from Snow Drifting Against Vertical Obstructions and in Valleys* Digest 332, Building Research Establishment, HMSO

BRE 1989a *The Assessment of Wind Loads* Digest 346, Building research Establishment, HMSO

BRE 1989b *Site Investigation for Low-rise Building: The Walk-over Survey* Digest 348, Building Research Establishment, HMSO

The Brick Developement Association 1979 Some Common Detailing Faults Bulletin 122 (2nd Series): 4/3

British Board of Agrément 1983 *Method of Assessment for Nuclear Shelters and Ancillary Equipment* BBA MOAT No.20, December 1983, British Board of Agrément

British Geological Survey Representative 1990 *Discussions* 4th May 1990

British Standard 1970 *Specification for Materials for Damp Proof Courses (Amended)* BS 743: 1970, British Standards Institution, London

British Standard 1974a *Specification for Clay Bricks* BS 3921: 1974, British Standards Institution, London

British Standard 1976a *Code of Practice for Stone Masonry* BS 5390: 1976, British Standards Institution, London

British Standard 1976b *Windows and Structural Gasket Glazing Systems* BS 4315: Part 1: 1976, British Standards Institution, London

British Standard 1983a *Specification for Aggregates from Natural Sources for Concrete* BS 882: 1983, British Standards Institution, London

British Standard 1989 *Environmental Cleanliness in Enclosed Spaces* BS 5295: 1989, British Standards Institution, London

BRS 1964a *Design and Appearance – 1* Digest 45, Building Research Station, HMSO

BRS 1965a *Soils and Foundations – 1* Digest 63, Building Research Station, HMSO. Second series (reprinted 1973)

BRS 1965b *Protection Against Corrosion of Reinforcing Steel in Concrete* Digest 59, Building Research Station, HMSO

BRS 1969a *Zinc-Coated Reinforcement for Concrete* Digest 109, Building Research Station, HMSO

Buller P S J 1988 *The October Gale 1987: Damage to Buildings and Structures in the South East of England* Report, Building Research Establishment, HMSO

Burberry P 1979 Condensation and how to avoid it *Architects' Journal* **170** (40): 723–39

Burberry P 1988 Cold bridges: construction, calculation *Architects' Journal* **187** (4): 59–65

Burns J 1989 *Discussions* Proc. Conf. Cladding: Current Practice in Specification, Design and Construction. Regent Crest Hotel, London, 7 November 1989

Butlin R N 1989a Acid deposition and stone *Structural Survey* **7** (3): 316–21

Butlin R N 1989b *The Effects of Acid Deposition on Buildings and Building Materials in the United Kingdom* Building Research Establishment, HMSO

Byrd T 1981 Ronan Point: The final verdict, *New Civil Engineer*, Thomas Telford, 16 July: 15, 16.

Campbell J 1989 *Cladding Specification* Proc. Conf. Cladding: Current Practice in Specification, Design and Construction. Regent Crest Hotel, London, 7 November 1989

Cheney J E 1988 25 years' heave of a building constructed on clay, after tree removal. *Ground Engineering* : 13–27

Chandler I 1988 *Low Rise PRC System* The Birmingham TERN Project Figures

Chevin D 1990 *The Grim Seeper* Building 15 June 1990: 62–5

CIRIA 1989 *The Eighth Report of the Standing Committee on Structural Safety* Construction Industry Research and Information Association

Coggins C R 1980 *Decay of Timber in Buildings* Rentokil Limited

Colomb P, Jones G 1980 *The Housing Rehabilitation Handbook* The Architectural Press, Technical Study 4: 32–9

Colomb P, Jones G 1980 *The Housing Rehabilitation Handbook* The Architectural Press, Technical Study 11: Structures: 205–11

Cook G K 1976 *The effects of Buildings on Wind* undergraduate thesis, University of Aston

Cornwell P B 1979 *Pest Control in Buildings* 2nd edn Rentokil Limited.

Cutler D F, Richardson I B K 1981 *Tree Roots and Buildings* Construction Press

Davey A, Heath B, Hodges D, Milne R, Palmer M 1981 *The Care and Conservation of Georgian Houses* The Architectural Press and Edinburgh New Town Conservation Committee

Davies C 1985 Hammersmith's houses of horror *Architects' Journal* **181** (6): 56–61

Davison L R, Davison C J 1986 Building defects associated with ground conditions: A review. *Chartered Institute of Building Technical Information Service*, R1

Denton M 1989 Recognising problems in the ground. *Housebuilder* **48** (3): 89,92,96

Desch H E, Desch S 1970 *Structural Surveying* Griffin

Diamant R M E 1965 *Industrialised Building 2 50 International Methods* Second Series. Iliffe Books Ltd: 7

Diamant R M E 1968 *Industrialised Building 3 70 International Methods* Third Series. Iliffe Books Ltd: 7

Diamant R M E 1990 Private communication

Dore E W (1989) Cladding safety – Investigation and feedback *The Structural Engineer* **67** (21): 388

Driscoll R 1983 The influence of vegetation on the swelling and shrinking of clay soils in Britain *Geotechnique* **33** (2): 93–106

Duell J, Lawson F 1983 *Damp Proof Course Detailing* The Architectural Press

Edwards M J 1986 *Weatherproof Joints in Large Panel Systems: 1 Identification and Typical Defects* Information Paper IP 8/86, Building Research Establishment, HMSO

Eldridge, H J 1976 *Common Defects in Buildings* HMSO

Everett A 1975 *Materials* Mitchell's Building Series, Batsford

Fernyhough R 1989a *Litigating Defects: Making a Successful Case* Proc. Conf. Building Defects: Avoidance, Investigation and Litigation. Royal Garden Hotel, Kensington, London, 10 February 1989

Fernyhough R 1989b *Discussions* Proc. Conf. Building Defects: Avoidance, Investigation and Litigation. Royal Garden Hotel, Kensington, London, 10 February 1989

Fidler J 1988 Stain reaction *Traditional Homes* November 1988: 127–31.

Foster J S 1973 *Structure and Fabric* Part 1 Mitchell's Building Construction Series, Batsford, p104.

Freeman I L 1975 *Building Failure Patterns and their Implications* Current Paper CP 30/75, Building Research Establishment, HMSO

Gibson A P, Lothian M T 1982 Surveying for timber decay *Structural Survey* **1** (3): 262–8

Gilder P J 1989 *Lessons From Cladding* Proc. Conf. Cladding: Current Practice in Specification, Design and Construction. Regent Crest Hotel, London, 7 November 1989

Grant B 1989 Methane menace *New Builder* 21 September 1989: 27.

Graystone J 1984 What's in a paint *Architects' Journal* **179** (25): 75–7

Gratwick R T 1968 *Dampness in Buildings* (2 vols) Crosby Lockwood

Green S 1989 *Choice of System – American Current Practice* Proc. Conf. Cladding: Current Practice in Specification, Design and Construction. Regent Crest Hotel, London, 7 November 1989

Groak S, Krimgold F 1988 The practitioner-researcher in the building industry *Building Research and Practice* **16** (5): 52–9

Harding J R, Smith R A 1986 *Brickwork Durability* BDA Design Note 7, Brick Development Association, England

Hatchwell L M 1985 Letter to the *Chartered Surveyor Weekly* **10** (12): 856

Hickin N E 1981 *The Woodworm Problem* Rentokil Limited

Hill J 1988 Corrosion in car-parking structures *Structural Survey* **6** (4): 361–6

Hill W F 1982 Tree root damage to low rise buildings. *Structural Survey* **1** (3): 254, 256–61

Hillel M B 1984 NHBC steps into gap *Chartered Surveyor Weekly* 19 (11): 767

Hinks A J, Cook G K 1988a Modern building methods require careful supervision *Building Today* **196** (5785): 28–30

Hinks A J, Cook G K 1988b Feedback from site means better designs *Housebuilding Today* December 1988: 12–14

Hinks A J, Cook G K 1988c Weepholes' pros and cons *Building Today* **196** (5777): 24–5

Hinks A J, Cook G K 1988d Detailing at the waters edge *Building Today* **196** (5781): 24–5

Hinks A J, Cook G K 1988e Victorian basements *Building Today* **196** (5772): 44–6

Hinks A J, Cook G K 1989a Identify and combat defects in stone walls *Building Today* **197** (5788): 20–3

Hinks A J, Cook G K 1989b What causes failures in domestic flat roofs *Building Today* **197** (5797): 20–2.

Hinks A J, Cook G K 1989c Metal roofs need regular inspection for damage *Building Today* **197** (5806): 28–31

Hinks A J, Cook G K 1989d Defects in paintwork *Building today* **198** (5824): 20–6

Hodgkinson A 1983 *AJ Handbook of Building Structure* 2nd edn The Architectural Press Figure 3:79

Hollis M 1982a *Residential Surveys of Victorian Terraced Brick Houses* Structural Survey 1 (3): 283–9

Hollis M, Gibson C 1986 *Surveying Buildings* 2nd edn Surveyors Publications

Honeybourne D B 1971 *Changes in The Appearance of Concrete on Exposure* Digest 126, Building Research Station, HMSO

Hudson R M 1988 The effect of environmental conditions on the performance of steel in buildings *Structural Survey* **6** (3): 215–23

Hugentobler P 1989 *Design and Development* Proc. Conf. Cladding: Current Practice in Specification, Design and Construction. Regent Crest Hotel, London, 7 November 1989

Humphreys B 1988 Defects undermine the industry *Building Today* **196** (5783): 39

Hunton D A T, Martin O 1989 *Curtain Wall Engineering* (presented by Angus Wilson) Proc. Conf. Cladding: Current Practice in Specification, Design and Construction. Regent Crest Hotel, London, 7 November 1989

Hutchinson B D, Barton J, Ellis N 1975 *Maintenance and Repair of Buildings* Butterworth

Johnson A 1984 *How to Restore and Improve Your Victorian House* David and Charles

King H, Everett A 1971 *Components and Finishes* Mitchell's Building Construction, Batsford

Lee R G 1982 *House Longhorn Beetle Survey* Information Paper 12/82, Building Research Establishment, HMSO

Legrand C 1989 *Carrying out a Building Survey* Presented at Institute of Maintenance and Building Management Course: Building Defects – Cures and Value for Money, Dorset Institute of Higher Education, Bournemouth, 12 December 1989

Lenczner D 1981 *Movements in Buildings* 2nd edn Pergamon Press

McKee N 1973 *The Construction Industry Handbook* 2nd edn Medical and Technical Publishing: 152

McMullan R 1983 *Environmental Science in Building* Macmillan Press Ltd

Mainstone R J 1976 *The Response of Buildings to Accidental Explosions* Current Paper CP 24/76, Building Research Establishment, HMSO

Mainstone R J 1978 *Accidental Explosions and Impacts* Current Paper CP 58/78, Building Research Establishment, HMSO

Marchant E W 1989 *Fire Engineering and Smoke Control* Paper presented at Institute of Building Control Fire Engineering Seminar 14 November 1989, Birmingham Centennial Centre

Marsh P 1977 *Air and Rain Penetration of Buildings* The Construction Press

May P 1986 Taking the lid off paint *Building* **33** (7458): 46–7

Melville I A, Gordon I A 1979 *The Repair and Maintenance of Houses* Estates Gazette Figure 31

Melville I A, Gordon I A, Boswood A 1985 *Structural Surveys of Dwelling Houses* 2nd edn Estates Gazette

Mika S L J, Desch S C 1988 *Structural Surveying* 2nd edn Macmillan Education

Mitchell A M 1903 *Building Construction: Advanced Course* Batsford

Mitchell G R 1976 *Snow Loads on Roofs – An Interim Report on a Survey* BRE Current Paper 33/76, Building Research Establishment, HMSO

Moore J F A 1983 *The Incidence of Accidental Loadings in Buildings 1971–1981* Information Paper IP 8/83, Building Research Establishment, HMSO

Munday D 1989 Modern methods of repairing structural timbers *Building Today* **198** (5813): 32–4

Murdoch R 1987 Avoiding corrosion under metal roofs *Building Technology and Management* **25** (5):4–6

Muschenheim A, Burns J 1989 *Specification I – The Specification of Cladding at Skidmore, Owings and Merrill* Proc. Conf. Cladding: Current Practice in Specification, Design and Construction. Regent Crest Hotel, London, 7 November 1989

National Coal Board 1975 *Subsidence Engineers Handbook* National Coal Board, London

Neville A M 1973 *Properties of Concrete* 2nd edn Pitman Publishing

Parker T W 1966 Maintenance of buildings 4. The technology of maintenance. *Building* **211** (6434): 171–6

Penn C 1954 *Houses of Today* Batsford

Penwarden A D and Wise A F E 1975 *Wind Environment Around Buildings* BRE Report, Building Research Establishment, HMSO

Pitts G Yeomans D 1988 Condensation control: Rooms in the roof *Architects' Journal* **187** (3): 57–9

Porges G 1977 *Applied Acoustics* Edward Arnold

Preston J 1989 The surface restoration of buildings – An investment in the present as well as in the future *Structural Survey* **7** (4): 450–60

Price S 1984 Housing rehabilitation 11: Structural alterations *Architects' Journal* **179** (13): 63–70

Pritchard D C 1978 *Lighting* 2nd edn Longman

Ragsdale L A, Raynham E A 1972 *Building Materials Technology* Edward Arnold

Pryke J 1980 Structural stability. In Benson J, *et al.* 1980 *The Housing Rehabilitation Handbook* Architectural Press

Rainger P 1983 *Movement Control in the Fabric of Buildings* Mitchell's Series, Batsford Academic and Institutional Table 2.1

Ransom W H 1981 *Building Failures: Diagnosis and Avoidance* E & F N Spon

Reece R A 1979 Trees and insurance *Arbor Journal* **3**: 492–9

Reid D A G 1973 *Construction Principles. 1: Function* Principles of Modern Building Series. George Godwin

Reith I H 1989 *Preventing Defects, by Design and Construction* Proc. Conf. Building Defects: Avoidance, Investigation and Litigation. Royal Garden Hotel, Kensington, London, 10 February 1989

Richardson C 1985a Structural surveys: 1. Technique and report writing *Architects' Journal* **181** (26): 57–65

Richardson C 1985b Structural surveys: 2. Data sheets – general problems *Architects' Journal* **182** (27): 63–71

Richardson C 1985c Structural surveys: 4. Data sheets – common problems 1850–1939. *Architects' Journal* **182** (29): 63–9

Richardson C 1988 Distorted walls: Survey, assessment, repair *Architects' Journal* **187** (2): 51–6

Rostron R M *Light Cladding of Buildings* Butterworth Architecture

Sealey P C 1985 Letter to the *Chartered Surveyor Weekly* **10** (12): 856

Seeley I H 1987 *Building Maintenance* 2nd edn Macmillan Building and Surveying

Shadbolt C H 1975 Mining subsidence. In Bell F G (ed) *Site Investigations in Areas of Mining Subsidence* Newnes-Butterworths

Smith B J, Phillips G M, Sweeney M 1987 *Environmental Science* Longman Scientific and Technical

Staveley H S, Glover P V 1983 *Surveying Buildings* Butterworths

Taylor G D 1983a *Materials of Construction* Construction Press

Taylor G 1983b Frost resistance of porous building materials *Building Technology and Management*, Part 1: **21** (4): 30–1; Part 2: **21** (5): 27–8; Part 3: **21**(6): 8–9

Thorn A 1990 Devils daughters *New Builder*, **19**,1 February 1990: 24–5

Thorogood R P 1979 *Resistance to Air Flow Through External Walls* Information Paper IP 79/14, Building Research Establishment, HMSO

Tietz S B 1989 Cladding: current practice in specification and design *The Structural Engineer* **67** (21): 389

Underwood G 1988 Security in buildings: Vandalism *Architects' Journal* **171** (24): 1171–4

Urbanowicz C 1987 Weaponry in structural surveys – 2. Common building defects and their diagnosis *Structural Survey* **5** (2): 109–20

de Vekey R C 1979 *Corrosion of Steel Wall Ties: Recognition, Assessment and Appropriate Action* BRE Information Paper IP 28/79, Building Research Establishment, HMSO

Verney P 1979 *The Earthquake Handbook* Paddington Press

Walker E. Morgan S 1975 *Construction Science 1* Hutchinson Educational

Webb S 1984 Not the ideal home show *Architects' Journal* **179** (13): 38

West H W H 1970 Clay products. In Simpson J W, Horrobin P J (eds) 1970 *The Weathering and Performance of Building Materials* Medical and Technical Publishing: 105–33

White R F 1989 *The Benefits of Prefabrication* Proc. Conf. Cladding: Current Practice in Specification, Design and Construction. Regent Crest Hotel, London, 7 November 1989

Whittaker B N and Reddish D J 1989 *Subsidence Occurrence Prediction and Control* Elsevier

Whittick A 1957 *The Small House* 2nd edn University Press

Wigglesworth G 1976 Building defects: Who is guilty? *Architects' Journal* **164** (32): 252–3

Wilcockson J 1980 Better liaison can cut down failures *Construction News* (5647) 17 January 1980: 22

Williams G T 1982 Basic facts about concrete II – Faults *Structural Survey* **1** (2): 170–5

Williams P H 1986 *Cavity Wall Tie Corrosion* Surveyors Publications

Wilson A 1989 presenting *Curtain Wall Engineering* (Hunton D A T, Martin O) 1989 Proc. Conf. Cladding: Current Practice in Specification, Design and Construction. Regent Crest Hotel, London, 7 November 1989

Whitley Moran T 1978 Strengthening earthquake-damaged structures. In *The Assessment and Mitigation of Earthquake Risk* UNESCO, chapter 11: 228

Further reading

Allen W 1983 Building defects – What went wrong? *Architects' Journal* **178** (50) 63–6

Atkinson G 1987 New Materials – Innovations are not a main cause of building failure *Structural Survey* **6** (2): 113–17

Bickerdike, Allen, Rich and Partners, O'Brien T 1971/72 Condensation control in flats *Design Failures in Buildings* 1st series. George Godwin, Building Failure Sheet 7

Bishop D 1989 *Build: The NEDO Report on Latent Defects Insurance* Proc. Conf. Building Defects: Avoidance Investigation and Litigation. Royal Garden Hotel, Kensington, London, 10 February 1989

Bowyer J 1977 *Guide to Domestic Building Surveys* Architectural Press

BRE 1971 *Repair and Renovation of Flood Damaged Buildings* Digest 152, Building Research Establishment, HMSO

BRE 1978 *Sound Insulation Performance between Dwellings built in the Early 1970s* Current Paper 20/78, Building Research Establishment, HMSO

BRE 1979 *Sound Insulation of Floating Party Floors with a Solid Concrete Structural Base* Information Paper IP 9/79, Building Research Establishment, HMSO

BRE 1981 *Field Measurements of the Sound Insulation of Timber-Joist Party Floors* Information Paper IP 5/81, Building Research Establishment, HMSO

BRE 1981 *Failure Patterns and Implications* Digest 251, Building Research Establishment, HMSO

BRE 1982 *Cavity Trays in External Cavity Walls: Preventing Water Penetration* Defect Action Sheet DAS 12, Building Research Establishment HMSO

BRE 1983 *British Standard BS 5821: 1980. Ratings of the Sound Insulation of Floating Party Floors* Information Paper IP 10/83, Building Research Establishment, HMSO

British Standard 1976 *External Rendered Finishes* BS 5262: 1976, British Standards Institution, London

British Standard 1976 *Windows and Structural Gasket Glazing Systems* BS 4315: Part 1: 1976, British Standards Institution, London

BRE 1976 *Sound Insulation: Basic Principles* Digest 143, Building Research Station, HMSO

Building 1987 Architects question warm deck *Building Supplement*, 29 May 1987.

Chandler I 1989 *Building Technology 2: Performance* Mitchell/Chartered Institute of Building Construction and Technology Series

Di Pasquale R A 1987 Building failures: Redundant systems *Progressive Architecture* **68** (8): 60–1

Duell J 1986 Element design guide. External walls 1 Masonry: Part 1 *Architects' Journal* **184** (27): 45–55

Elder A J, Vandenburg M (eds) 1984 *A J Handbook of Building Enclosure* The Architectural Press

Gloyn B 1989 *Insuring Building Defects: New Possibilities* Proc. Conf. Building Defects: Avoidance, Investigation and Litigation. Royal Garden Hotel, Kensington, London, 10 February 1989

Insall D 1972 *The Care of Old Buildings Today* Architectural Press

Keyworth B 1987 Flat Roof Construction *Structural Survey* 6 (2) 119–23

Liddament M 1986 *Building Airtightness and Ventilation – An Overview of International Practice* Building Services Research and Information Association

Maidstone J R, Nicholson H G, Alexander S J 1978 *Structural Damage in Buildings Caused by Gaseous Explosions and Other Accidental Loadings 1971–1977* Report, Building Research Establishment, HMSO

Mills R L 1983 Structural remedial work for listed buildings and ancient monuments *Repair and Renewal of Buildings* Thomas Telford: 13–22

NBA 1983 *Common Building Defects* The National Building Agency, Construction Press

NEDO 1985 *Latent Defects in Buildings: An Analysis of Insurance Possibilities* Report for Building Economic Development Committee (EDC) Insurance Feasibility Steering Committee, National Economic Development Office

Porteous W A 1985 Perceived characteristics of building failure – A survey of the recent literature *Architectural Science Review* **28** (2): 30–40

Read S J 1986 *Controlling Death Watch Beetle* Information Paper 19/86, Building Research Establishment, HMSO

Reith I H 1989 *Preventing Defects, by Design and Construction* Proc. Conf. Building Defects: Avoidance, Investigation and Litigation. Royal Garden Hotel, Kensington, London, 10 February 1989

Rickards M 1987 If you can't see it, it probably isn't rising damp *Structural Survey* **1** (3): 233–6

Roe M 1989 *Latest Developments on Liability for Defects* Proc. Conf. Building Defects: Avoidance Investigation and Litigation. Royal Garden Hotel, Kensington, London, 10 February 1989

Schild E, Oswald R, Rogier D. Schweikert H 1978 *Structural Failure in Residential Building* Volume 1: *Flat Roofs, Roof Terraces, and Balconies* Granada Publishing

Schild E, Oswald R, Rogier D, Schweikert H 1979 *Strutural Failure in Residential Building* Volume 2: *External Walls and Openings* Granada Publishing

Schild E, Oswald R, Rogier D. Schweikert H 1980 *Structural Failure in Residential Buildings* Volume 3: *Basements and Adjoining Land Drainage* Granada Publishing

Schild E, Oswald R, Rogier D, Schweikert H 1981 *Structural Failure in Residential Buildings* Volume 4: *Internal Walls, Ceilings and Floors* Granada Publishing

Townsend J, Bennett D 1985 Patent glazing *Architects' Journal* **182** (33): 50–1

Wigglesworth G 1976 Building defects: Who is guilty? *Architects' Journal* **164** (32) 252–3

Index